CONT' ---

PREFACE TO VOLUME 2

Chapter 1 PROTECTION

1.1. Introduction
1.2. Distance (Impedance) Protection
    1.2.1. Fault impedance as a func
    1.2.2. Compensation of distance ~~~~~~
        Phase-fault compensation
        Earth-fault compensation   9
    1.2.3. Polarised mho, off-set mho and reactance relays   11
    1.2.4. Acceleration of stage 2 time   19
    1.2.5. Switched distance systems   21
    1.2.6. Accuracy, sensitivity and setting   21
1.3. Phase-Comparison Carrier-Current Protection (high-frequency phase-comparison)   23
1.4. Auto-reclosing Circuit Breakers   30
    1.4.1. Rural power systems   30
    1.4.2. Main transmission lines (Grid)   33
1.5. Current Transformers: Transient Operation   34
    References   40
    Examples   41

Chapter 2 CIRCUIT INTERRUPTION AND SWITCHING
          OVER-VOLTAGES   43

2.1. Introduction   43
2.2. Circuit breaker rating   46
2.3. Arc extinction and reignition   46
2.4. Balanced 3-phase fault at the circuit breaker terminals:—transient recovery voltage   49
2.5. First phase to clear   54
2.6. Multi-frequency transient recovery voltage   56
2.7. Resistance switching   60
2.8. Rate of rise of restriking (or transient recovery) voltage   61
2.9. Chopping magnetising current   63
2.10. Interrupting the charging current of a long unloaded line with a breaker near the supply source   64
2.11. Interrupting the charging current of a long unloaded line with a breaker at some distance from the supply source   67
2.12. Switching surges on closing or reclosing an unloaded line   72
2.13. Short-line or kilometric fault   74
2.14. Present position in circuit breaking   80
2.15. The conditions for equivalence between synthetic and direct tests   81
    References   84
    Examples   85

CW01500673

v

Chapter 3   STABILITY OF POWER SYSTEMS

3.1. Introduction   88
3.2. Power transfer: transfer reactance   89
    3.2.1. Synchronous generator   89
    3.2.2. General case of the 2-port, 4-terminal network   95
3.3. Equal-area criterion for transient stability   101
    3.3.1. Change in load   102
    3.3.2. Change in transfer reactance due to switching   105
    3.3.3. Change in transfer reactance due to a fault   105
3.4. Mechanics of angular motion   107
    3.4.1. Momentum (M) and inertia constant (H)   108
    3.4.2. Reduction of a power system to one machine connected to an
           infinite busbar   110
    3.4.3. Proof of the equal-area criterion   111
3.5. Swing curve (load angle/time curve)   112
3.6. Small oscillations   117
3.7. Steady-state stability   119
3.8. Methods of improving power system stability   120
    3.8.1. Resonant links   121
    3.8.2. Excitation and automatic voltage regulators   125
    3.8.3. Turbine governors   137
    References   139
    Examples   140

Chapter 4   TRAVELLING WAVES IN TRANSMISSION LINES

4.1. Introduction   144
4.2. Single conductor line with nearby earth plane   144
    4.2.1. Loss-free line   147
    4.2.2. Two conductors with nearby earth plane   152
4.3. Matrix methods applied to two conductors with nearby earth plane   153
4.4. Effects of various terminations   158
    4.4.1. General termination   159
    4.4.2. Resistance termination; junctions of lines and cables   163
    4.4.3. Reactive terminations   166
    References   168
    Examples   168

Chapter 5   INSULATION CO-ORDINATION

5.1. Introduction   172
5.2. Over-voltages   173
    5.2.1. System-frequency over-voltages   173
    5.2.2. Transient over-voltages   174
    5.2.3. Over-voltages due to lightning   175
5.3. Impulse test levels of equipment   177
5.4. Switching-surge strength of insulation in E.H.V. systems   181
5.5. Insulation levels for overhead lines   184
5.6. Protection of equipment in high-voltage outdoor sub-stations   185
5.7. Insulation co-ordination in h.v.d.c. systems   190
    References   191

£ 8·75

PERGAMON INTERNATIONAL LIBRARY
of Science, Technology, Engineering and Social Studies
The 1000-volume original paperback library in aid of education,
industrial training and the enjoyment of leisure
Publisher: Robert Maxwell, M.C.

ELECTRICAL
POWER SYSTEMS
VOLUME 2
Second Edition (SI/Metric Units)

Other titles of interest in the
PERGAMON INTERNATIONAL LIBRARY

ABRAHAMS & PRIDHAM
Semiconductor Circuits: Theory Design and Experiment
Semiconductor Circuits: Worked Examples

BADEN FULLER
Engineering Field Theory
Worked Examples in Engineering Field Theory

BROOKES
Basic Electric Circuits, 2nd Edition

CHEN
Theory and Design of Broadband Matching Networks

COEKIN
High Speed Pulse Techniques

DUMMER
Electronic Inventions and Discoveries, 2nd Revised

FISHER & GATLAND
Electronics – From Theory into Practice, 2nd Edition

GATLAND
Electronic Engineering Application of Two Port Networks

HAMMOND
Applied Electromagnetism
Electromagnetism for Engineers, 2nd Edition (in SI/Metric Units)

HANCOCK
Matrix Analysis of Electrical Machinery, 2nd Edition

HINDMARSH
Electrical Machines and Their Applications, 3rd Edition
Worked Examples in Electrical Machines and Drives

MURPHY
Thyristor Control of AC Motors, 2nd Edition

RODDY
Introduction to Microelectronics, 2nd Edition

YORKE
Electric Circuit Theory

ELECTRICAL POWER SYSTEMS

In two volumes

A. E. GUILE
D.Sc.(ENG.), PH.D., B.Sc.(ENG.), C.ENG., F.I.E.E.
Professor of Electrical Engineering,
University of Leeds

the late W. PATERSON
B.Sc., B.Sc.(ENG.), C.ENG., F.I.E.E.
one-time Principal Lecturer in Electrical Engineering,
Leeds Polytechnic

VOLUME 2
Second Edition (SI/Metric Units)

PERGAMON PRESS
OXFORD · NEW YORK · TORONTO · SYDNEY · PARIS · FRANKFURT

U.K.	Pergamon Press Ltd., Headington Hill Hall, Oxford OX3 0BW, England
U.S.A.	Pergamon Press Inc., Maxwell House, Fairview Park, Elmsford, New York 10523, U.S.A.
CANADA	Pergamon Press Canada Ltd., Suite 104, 150 Consumers Rd., Willowdale, Ontario M2J 1P9, Canada
AUSTRALIA	Pergamon Press (Aust.) Pty. Ltd., P.O. Box 544, Potts Point, N.S.W. 2011, Australia
FRANCE	Pergamon Press SARL, 24 rue des Ecoles, 75240 Paris, Cedex 05, France
FEDERAL REPUBLIC OF GERMANY	Pergamon Press GmbH, 6242 Kronberg-Taunus, Hammerweg 6, Federal Republic of Germany

---

First edition 1972
Second edition 1977
Reprinted 1978, 1980, 1982

Library of Congress Cataloging in Publication Data

Guile, Alan Elliott.
Electrical power systems.
(Pergamon international library of science, technology, engineering and social studies)
Includes bibliographical references and indexes.
1. Electrical power systems. 2. Electric power transmission. I. Paterson, William, B.Sc., joint author. II. Title.
TK1001.G84   1977   621.319   77-1789

ISBN 0-08-021730-3 (Hardcover)
ISBN 0-08-021731-1 (Flexicover)

Printed in Great Britain by A. Wheaton & Co. Ltd., Exeter

Chapter 6   RECTIFICATION, INVERSION AND HIGH-VOLTAGE
            D.C. SYSTEMS

6.1. Introduction   192
6.2. Rectification   192
      6.2.1. Double-star connection   192
      6.2.2. 3-phase bridge connection   194
      6.2.3. Grid or gate control and commutation   196
      6.2.4. Voltage and current ratios   197
6.3. Inversion   203
6.4. High-voltage d.c. systems   208
      6.4.1. Operation   209
      6.4.2. Applications   211
      References   214
      Examples   215

Chapter 7   3-PHASE NETWORK MATRICES AND CO-ORDINATE
            SYSTEM TRANSFORMATIONS

7.1. Introduction   217
7.2. Transformations with invariance of power   219
      7.2.1. Application to mesh currents   221
      7.2.2. Application to nodal voltages   221
7.3. Mesh current connection matrices applied to a 3-phase network   222
      7.3.1. Single-phase representation of a balanced 3-phase network   224
7.4. General transformation   226
7.5. Equal transformation matrices for currents and voltages and ortho-
      gonal transformations   227
7.6. Symmetrical component transformation matrix   228
      7.6.1. General unbalanced 3-phase circuit   230
      7.6.2. Completely balanced 3-phase circuit up to the point of fault 231
      7.6.3. Fault which is asymmetrical with respect to the reference phase 233
      7.6.4. System with synchronous machines on both sides of the fault   235
7.7. Transformation to rotor fixed axes (direct and quadrature axes)   238
7.8. α, β and 0 components (Clarke's components)   240
      References   241

Chapter 8   POWER SYSTEM ANALYSIS

8.1. Introduction   242
8.2. Mesh current method and connection matrices   243
      8.2.1. Basic equations   243
      8.2.2. Interconnection of networks into which a system has been sub-
             divided   247
8.3. Nodal voltage method and connection matrices   250
      8.3.1. Basic equations   250
      8.3.2. Node elimination by means of sub-matrices   255
8.4. Application of nodal voltage method to the solution of power system
      load flow problems   259
8.5. Direct methods involving inversion of the nodal admittance matrix   262
      8.5.1. Gaussian elimination   264
      8.5.2. Partitioning or factorised inverse matrix   265

8.6. Modification of the inverse of the nodal admittance matrix   267
    8.6.1. Application to the addition of a new load admittance at one
        node   270
    8.6.2. Application to the addition of a new impedance between two
        nodes   271
    8.6.3. Application to transformer tap changing   272
8.7. Iterative methods   276
    8.7.1. Gauss–Seidel method   276
    8.7.2. Newton–Raphson method   279
    8.7.3. Convergence and acceleration factors   282
    8.7.4. Hybrid methods   283
8.8. Tearing   284
    8.8.1. Simple application of tearing   285
    References   297
    Examples   297

Chapter 9   ASPECTS OF SYSTEM INTEGRATION AND
            DEVELOPMENT

9.1. Introduction   303
9.2. Transient Stability   304
9.3. Comparison of system conditions at high and low loads   308
9.4. Computer applications in power systems   313
9.5. Load prediction   318
9.6. Problems associated with E.H.V. A.C. transmission   319
9.7. Superconducting cables   321
9.8. Fluctuating loads   322
    References   323

Appendix A.1

MATRIX ALGEBRA

A.1.1. Introduction   327
A.1.2. Addition or subtraction of matrices   328
A.1.3. Multiplication of a matrix by a real or complex number   328
A.1.4. Multiplication of two or more matrices   328
A.1.5. Transpose of a matrix   330
    A.1.5.1. Transpose of the product of two matrices   330
    A.1.5.2. Transpose of partitioned matrices   330
A.1.6. Diagonal matrix   331
A.1.7. Unit matrix   332
A.1.8. Inverse (or reciprocal) of a matrix   332
    A.1.8.1. Inverse of the product of a number of matrices   332
A.1.9. Matrix inversion   333
    A.1.9.1. Orthogonal matrix   334
A.1.10. Eigenvalues of a matrix   334
    References   335

Appendix A.2

ECONOMICS OF ELECTRICAL POWER GENERATION

A.2.1. Introduction   336
A.2.2. Forward estimates of future load growth   336
A.2.3. Economic principles   339
A.2.4. Economic operation of generating plant   341
A.2.5. Choice of fuel   343
        References   346

INDEX   347

CONTENTS OF VOLUME 1

CONTENTS OF VOLUME 2   ix

PREFACE TO FIRST EDITION   xv

PREFACE TO SECOND EDITION   xvi

Chapter 1   TRANSMISSION

1.1. Introduction   1
1.2. Short Lines   4
    1.2.1. Analytical methods   4
    1.2.2. Graphical methods; performance charts; operation of a
           power system   9
    1.2.3. Power formulae: steady state power limit   17
    1.2.4. Voltage control of a radial line   20
    1.2.5. Control of parallel lines and rings: control of the grid system 22
1.3. Long Lines   24
    1.3.1. Nominal Π and T circuits   24
    1.3.2. Uniform long line   27
    1.3.3. Measurement of parameters   36
    1.3.4. Equivalent Π and T circuits   37
    1.3.5. Propagation coefficient γ: characteristic impedance $\mathbf{Z}_c$ 37
    1.3.6. Travelling waves   39
    1.3.7. Distortion-free line   41
    1.3.8. Loss-free line   41
1.4. General 4-terminal (2-port) Network: ABCD Constants   43
    1.4.1. Voltage drop and inherent voltage regulation   44
    1.4.2. Symmetrical, passive network   44
    1.4.3. Measurement of parameters   45
    1.4.4. Application of matrix methods   47
    1.4.5. Performance chart for the 4-terminal network   51
    1.4.6. Power formulae: power limits   53
1.5. Reactive Power Compensation   55
    1.5.1. Static compensation   57
    1.5.2. Synchronous compensators   60
1.6. Natural Load: Surge Impedance Load   60
    Examples   64

Chapter 2   TRANSMISSION LINE INDUCTANCE

2.1. Introduction   72
2.2. Two-wire Line   73
    2.2.1. Flux linkages within the conductor producing the flux   74
    2.2.2. Flux linkages outside the conductor producing the flux   77
2.3. General Multi-conductor System   78
2.4. Effective Conductor Self-inductance   83
2.5. Positive- or Negative-sequence Inductance of a Transposed
     Single-circuit 3-phase Line   85

2.6. Zero-sequence Inductance of a Transposed Single-circuit 3-phase
     Line   86
2.7. Stranded and Steel-cored Line Conductors   89
2.8. Bundle or Multiple Conductors   90
2.9· Double-circuit 3-phase Lines   92
     Examples   96

Chapter 3   TRANSMISSION LINE CAPACITANCE

3.1. Introduction   99
3.2. Two-wire Line   103
3.3. General Multi-conductor System   105
3.4. Effect of Earth   108
3.5. General Single-circuit 3-phase Line   110
3.6. Positive- or Negative-sequence Capacitance to Neutral of a
     Transposed Single-circuit 3-phase Line   113
3.7. Zero-sequence Capacitance to Neutral of a Transposed Single-
     circuit 3-phase Line   116
     3.7.1. Without earth wire   116
     3.7.2. With earth wire   117
3.8. Corona   118
3.9. Bundle or Multiple Conductors   123
     Examples   128

Chapter 4   OVERHEAD LINES

4.1. Insulators for Overhead Lines   131
     4.1.1. Voltage distribution over an insulator string   134
4.2. Conductor Materials for Overhead Lines   137
     4.2.1. Virtual modulus of elasticity for S.C.A.   141
     4.2.2. Virtual coefficient of linear expansion for S.C.A.   142
4.3. Sag and Stress Calculations   143
     4.3.1. Parabola   143
     4.3.2. Spans of unequal length: equivalent span   146
     4.3.3. Supports at different levels   147
     4.3.4. Catenary   148
     Examples   151

Chapter 5   UNDERGROUND CABLES

5.1. Introduction   154
5.2. Capacitance of Single-core Cable   155
5.3. Capacitance of 3-core Belted Cable   158
5.4. 3-core Cables   159
     5.4.1. Solid-type cables   159
     5.4.2. Oil-filled cables   160
     5.4.3. Gas-pressure cables   161
5.5. Conductor Inductive Reactance   162
5.6. Effective Conductor Resistance   162
5.7. Breakdown   165
     5.7.1. Thermal instability   165
     5.7.2. Tracking following void ionisation   167

5.8. Thermal Considerations and Current Rating   169
5.9. D.C. Cables   172
     Examples   175

Chapter 6   TRANSFORMERS

6.1. Equivalent Circuit of a Two-winding Transformer   179
6.2. 3-phase Two-winding Transformer Types   181
6.3. Phase Shift   182
6.4. Parallel Operation of 3-phase Transformers   186
     6.4.1. Equal ratios   186
     6.4.2. Unequal ratios   193
6.5. Harmonics   195
6.6. Magnetising Inrush Current   198
6.7. Tap Changing   200
6.8. Unbalanced Loading   202
6.9. Three-winding Transformer Equivalent Circuit   203
6.10. Zero-sequence Impedance   206
     Examples   212

Chapter 7   SYNCHRONOUS MACHINES IN POWER SYSTEMS

7.1. Introduction   218
7.2. Steady-state Operation of a Single Unsaturated Cylindrical-rotor
     Alternator   219
7.3. Synchronous Impedance and Short-circuit Ratio   226
7.4. Two-axis Theory of a Salient-pole Alternator   228
7.5. Transient Conditions   229
7.6. Unbalanced Loading or Faults: Sequence Reactances   244
7.7. Operation of Synchronous Machines Connected to Infinite Bushbars 246
7.8. Stability Margin and Operational Limits for a Cylindrical-rotor
     Alternator   252
7.9. Operation at Leading Power-factor with Fast Continuously-acting
     Voltage Regulators   256
7.10. Governors and Frequency Control   258
     Examples   267

Chapter 8   POWER SYSTEM FAULT CALCULATIONS

8.1. Introduction   273
8.2. Symmetrical 3-phase Short-circuit   274
     8.2.1. Basic calculations   274
     8.2.2. Current-limiting reactors   277
     8.2.3. Fault in-feeds: source impedance   280
8.3. Symmetrical Components   283
     8.3.1. General principles   283
     8.3.2. Conditions for no mutual coupling between the three
            sequence networks   287
     8.3.3. The three sequence networks   289
     8.3.4. Z_1, Z_2 and Z_0 for an interconnected power system 290
     8.3.5. Phase shift through a delta/star transformer   292

8.4. Asymmetrical Faults 294
    8.4.1. Earth fault on phase *a* 294
    8.4.2. Line-to-line (phase-fault) across lines *b* and *c* 298
    8.4.3. Line-to-line-to-earth short-circuit (double-earth-fault) on
        phases *b* and *c* 300
    8.4.4. General case of double-earth-fault on phases *b* and *c* 301
    8.4.5. General unbalanced three-phase-earth fault 303
    8.4.6. Series fault on phase *a* 305
    8.4.7. Simultaneous shunt and series faults on phase *a* 306
    8.4.8. Fault volt-amperes 307
    8.4.9. Comparison of fault levels 308
    8.4.10. Asymmetrical fault whilst system is on load 309
8.5. Symmetrical Component Filter Networks 313
    8.5.1. Sequence current 314
    8.5.2. Sequence voltage 315
    Examples 316

Chapter 9   PROTECTION

9.1. Introduction 329
9.2. Current Transformers 330
9.3. Explanation of Protective Terms 336
9.4. Time-graded (non-unit) Protective Systems 338
    9.4.1. Fuses 339
    9.4.2. Inverse definite minimum time lag induction relays 340
    9.4.3. Definite-time relays 346
    9.4.4. Ring feeders: directional relays 347
9.5. Unit Protective Systems 350
    9.5.1. Circulating-current systems 350
        Stabilising resistor 352
        Biased relays 353
        Generator protective schemes 355
        Transformer protective schemes 358
    9.5.2. Balanced voltage protective systems 364
        Summation transformers 365
        Pilots 366
        Feeder protective systems 367
    Examples 370

APPENDICES

1. The Per-unit System 377
2. Star-delta Transformation 384
3. h-operator 385
4. Volt-amperes, Power and Reactive Power 387

INDEX 393

PREFACE TO VOLUME 2

This book is the second of two volumes written for students of Electrical Power Systems at Universities, Polytechnics and Colleges of Technology, who are preparing for such qualifications as Honours and Ordinary degrees, Council of Engineering Institutions Examination Part II, H.N.D. and H.N.C. This volume extends the subject to include material commonly found in a Final Year Honours Degree course. Courses vary so much, however, that some students on M.Sc. courses may find parts of the book of help, while students on courses below Honours Degree level should be able to benefit from a suitable choice of sections. As it is impossible to cover the entire field of power system engineering, the authors have made what they regard as a reasonable choice of base material. It is hoped that students will do a substantial amount of reading in these two books by themselves, thereby leaving the lecturer and students time for tutorial discussion and to extend the scope of the lectures to suit their own interests.

In this volume it has been possible to emphasise more than in the first volume, the fact that power system problems involve the interaction of many items of equipment, so that a systems approach to the whole subject is desirable. Much greater use is made of matrix methods than in Volume 1, and more is said about the use of digital computers and the systematic presentation of a power system in a form suitable for modern computer methods. The last chapter surveys some of the most significant system aspects now undergoing development. Many references have been given to enable knowledge of these developments to be extended further, and to illustrate the great challenge and opportunities which exist for engineers choosing to work in this field.

The authors wish to thank the following members of staff of the North Eastern Region of the Central Electricity Generating Board: Mr. F. H. Birch for his very great help with Chapter 9, and Messrs. C. C. Baxendale, W. Fairney and K. G. Warner for their help with

Chapters 1, 3 and Appendix A2 respectively. Siimlarly they thank Mr. N. S. Ellis of A. Reyrolle and Co. Ltd. for reading Chapter 2. As with Volume 1, they thank the Institutions from whose examination papers they have reproduced questions.

Chapter 1

PROTECTION

1.1 Introduction

The subject of protection was started in Volume 1, Chapter 9, which finished by discussing the protection of feeders (transmission/distribution) using pilots up to about 30 km in route length, and balanced-voltage (opposed-voltage) systems of protection. This present chapter now continues the discussion of the protection of overhead lines which are longer than the above limit. The basic idea is to eliminate the pilot. Two such systems will be discussed:

(a) Distance protection, and

(b) Carrier protection.

Carrier protection does not actually eliminate the pilot: it uses the power line itself as the pilot.

One of the simplest systems of feeder protection achieves discrimination by time-grading (see Volume 1, section 9.4), but has the disadvantage of the longest fault clearance times for faults nearest the source where the fault levels are highest. The damage which could be done during such long fault clearance times is quite unacceptable for important high-voltage lines so this protective system is not used, not even as back-up protection. Balanced-voltage unit protection (see Volume 1, section 9.5.2) gives instantaneous (about 0·1 s) tripping times for internal faults (within the zone defined by the positions of the two sets of C.T.s), and stability (non-tripping) for external (through) faults, but the system requires a continuous pilot round which current can flow. For the longer lines, this pilot is usually a rented Post Office telephone cable which is expensive (being charged per mile, per annum) and subject to interference. The practical limit is about 30 km and this is the pilot route length which is often longer than the line route length.

For practical details of the protective systems to be discussed, the student should consult the reference Electricity Council, Power System Protection (3 volumes) and the manufacturers' pamphlets.

1.2 Distance (Impedance) protection

Fig. 1.1(a) shows a simple radial line. Assume initially, for simplicity, that there is generation at the busbar A only and no fault infeed at the other busbars. The figure shows a distance system, installed in the

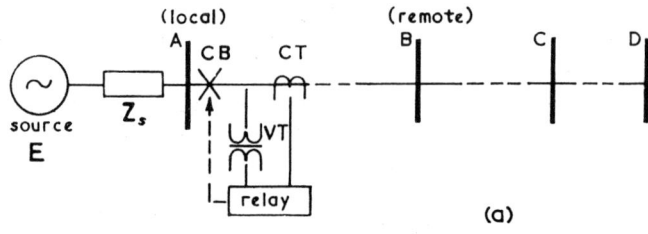

(a)

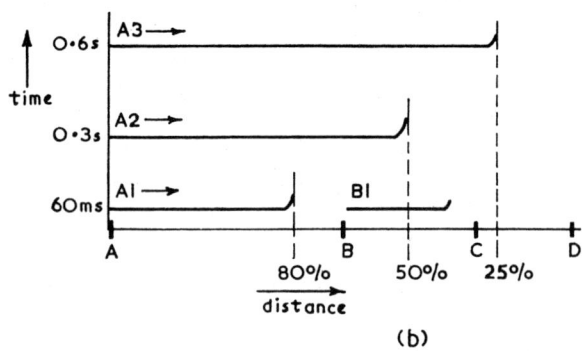

(b)

Fig. 1.1. Distance protection, (a) Schematic. (b) Time/Distance graph.

line AB at end A (the local end), whose purpose is to protect the line AB (the protected zone) against fault infeeds at A: end B is the remote end. Assume a 3-phase short-circuit on the line AB, and let Z_f be the impedance of the line from the relaying point at A to the point of fault (the primary impedance). The voltage on the V.T. (voltage transformer) primary side is

$$\mathbf{V} = \mathbf{E}\mathbf{Z}_f/(\mathbf{Z}_s+\mathbf{Z}_f) \qquad [1.1]$$

The fault current is

$$\mathbf{I} = \mathbf{E}/(\mathbf{Z}_s+\mathbf{Z}_f) \qquad [1.2]$$

The relay compares (and is often called a comparator) the secondary values of V and I, in a circuit shown in Fig. 1.2, so as to measure their ratio which is an impedance Z_m where

$$Z_m = \frac{V/\text{V.T. ratio}}{I/\text{C.T. ratio}}$$

$$= Z_f \frac{\text{C.T. ratio}}{\text{V.T. ratio}} \qquad [1.3]$$

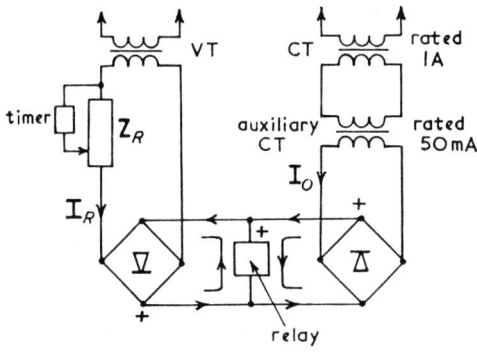

Fig. 1.2. Distance relay circuit (moving-coil relay).

Z_m, the measured impedance, or apparent impedance at the relay terminals, is called the secondary impedance. The auxiliary C.T. is treated as part of the relay, so in [1.3], C.T. means the line C.T. The reach of a relay is the distance from the relaying point to the point of fault and can be given as a distance, or as a primary, or secondary impedance, and generally refers to the relay at the threshold of operation, i.e. at its setting. Because of the circuit rectifiers, the secondary impedance is a scalar. The comparator compares like quantities, namely current, so the V.T. output is converted to a current, called the restraint current I_R, due to $\mathbf{Z}_R$, the restraint impedance. The C.T. provides the operating current I_O. The rectified currents form a circulating-current system (see Volume 1, section 9.5.1) and the relay current is their numerical difference. The relay is a permanent-magnet, moving-coil, loud-speaker type, similar to that used in Fig. 9.20, Volume 1. A current, greater than the setting current, entering at the positive terminal will cause the relay to close its contacts and initiate the tripping of the circuit-breaker at A (and only at A). Thus the C.T. output tends to operate the relay while the V.T. output tends to restrain it. For a uniform line, Z_f, and hence Z_m, will be proportional to the length of line from the relaying point to the point of fault: thus the distance relay locates the position of the fault. The setting current is that relay current which creates just sufficient force to overcome the actions of inertia, friction and resetting spring. During an internal fault, the relay current should be several times the setting current to ensure fast operation. The contact travel distance is small, so operation is instantaneous (60 ms). The coil is wound on an aluminium former to provide anti-vibration damping against harmonic currents or external forces.

It would appear that the relay could discriminate between faults internal and external to AB, by adjusting the circuitry so that the operating and restraining currents were equal for a fault at B. If the point of fault moved into AB, $\mathbf{Z}_f$ would decrease, so would $\mathbf{V}$, but $\mathbf{I}_f$ would increase and the relay would operate. Similarly, a fault in BC would restrain the relay. But, due to tolerances in the circuit components, the measuring accuracy cannot be perfect so it is usual to set the relay for a secondary impedance corresponding to 80% of the impedance of AB. The relay is then referred to as stage 1 of the A relay (marked A1 in Fig. 1.1(b)).

The problem of protecting the remaining 20% of AB is solved by changing the setting of the relay to reach 50% into the zone BC, by opening the timer contacts shown in Fig. 1.2 and so increasing the restraint impedance $\mathbf{Z}_R$ (in practice other methods are also used). This reduces the restraint current, and to regain relay balance, the operating current and hence the fault current must be reduced by increasing $\mathbf{Z}_f$. The relay is now referred to as the A relay, stage 2, and is marked A2. But the A2 relay must discriminate with the B1 relay so the change from A1 to A2 is time delayed by 0·3 s in order that faults in the first 50% of BC are cleared by B1 in 60 ms plus the total time of the circuit breaker at B (0·15 s), plus a reasonable safety margin. Discriminative time margins were discussed in Volume 1, p. 356. Thus a distance system has the disadvantage of being a non-unit, time-graded system with slow stage 2 fault clearances, but it does give instantaneous clearance at *both* ends of the zone over 60% of the zone and is much superior to an inverse-time (I.D.M.T.L.) system, since the times are the same for every zone and are not cumulative towards the source.

Since, for economic reasons, the same relay is usually used for both stages, a failure of the first stage also involves failure of the second stage. To give back-up protection, a separate distance system is added, with a reach 25% into CD, called stage 3, which is time delayed a further 0·3 s. This 3-stage system is called a 3-step distance system. If a fault occurs within the reach of A3 it starts a timing relay. If the fault is within the first 80% of AB, A1 should operate instantaneously and trip the circuit breaker at A; but if not, after 0·3 s, the timer will change A1 into A2 by opening the normally-closed contacts in Fig. 1.2 and this should clear faults in the end 20% (again at the circuit breaker at A). If after 0·6 s the fault has not been cleared, the timer trips the circuit breaker at A. The student should complete this discussion for the remainder of the A3 reach, assuming that 3-step distance systems are installed at B and at C. The figures of 80%, 50% and 25%, quoted above, can be varied and the student should determine the range of variation, subject to the restriction that discrimination must not be lost, i.e. the time characteristics must not overlap. The quoted time margin of 0·3 s can also be varied (to match the circuit-breaker total-time) but is not likely to be less than 0·25 s.

Distance systems would be installed at both ends of all feeders (each feeder would then be a protected zone), all reaching into their own zone: those reaching to the right can be shown with positive times, and those to the left with 'negative' times, in Fig. 1.1(b). Distance systems are of the non-unit type since they trip only their own local circuit breaker. The system discussed above is non-directional as it would give the same operation for a given fault current whether flowing from or to the busbars: it is called a plain impedance system.

The discussion above assumed that the only fault infeed was at A.

The student should now consider the effect of a 3-phase fault infeed from generation at B (see Volume 1, section 8.2) by placing the fault within the first 50% of the zone BC, and finding whether the A2 relay over-reaches or under-reaches. A relay is said to over-reach if it indicates a fault distance greater than the true distance (see examples).

1.2.1 FAULT IMPEDANCE AS A FUNCTION OF THE TYPE OF FAULT

So far, only 3-phase faults have been mentioned. Consider now other types of fault. Assume, for simplicity, that the source impedance Z_s is negligible, i.e. that $V = E$. The symbols are those used in Volume 1, Chapter 8, and the phase sequence is *abc*. Let Z_1, Z_2, Z_0 be the positive-, negative-, and zero-sequence impedances of the line (see Volume 1, Chapter 2) from the relaying point to the point of fault, which occurs at a fixed point on the line for all the faults. The formulae will be developed in terms of primary impedance, i.e. the apparent impedance as measured on the primary side of the V.T. and C.T. The object of the exercise is to show that the primary impedance is a function of the type of fault and then to devise methods whereby the relay need only be set in terms of the 3-phase fault condition, i.e. in terms of the positive-sequence impedance.

(*a*) 3-phase fault

$$I_a = E_{ae}/Z_1$$

primary impedance $= E_{ae}/I_a = Z_1$

(*b*) Earth fault on phase *a*

$$I_{a1} = I_{a2} = I_0 = E_{ae}/(Z_1 + Z_2 + Z_0)$$

$$I_a = I_{a1} + I_{a2} + I_0 = 3E_{ae}/(Z_1 + Z_2 + Z_0)$$

primary impedance $= E_{ae}/I_a$

$$= (Z_1 + Z_2 + Z_0)/3 = (2Z_1 + Z_0)/3 = Z_E \qquad [1.4]$$

since $Z_1 = Z_2$ for a line: Z_E is called the earth-loop impedance. If $k = Z_0/Z_1$, then

$$Z_E = \left(\frac{2+k}{3}\right)Z_1 \qquad [1.5]$$

The above analysis is only valid if all the zero-sequence current passes through the C.T., i.e. if the source is earthed at only one point: it is not valid for multiply-earthed power systems (see later). Thus if (and only if) $Z_0 = Z_1$, then the primary impedance for an earth fault is the same as that for a 3-phase fault: otherwise it is a function of k. Normally X_0 is about 2 to 4 times X_1 and it should be noted that Z_0 includes the resistance of the earth-return via the earth and/or the aerial earth wires. This problem is solved using separate relays for phase faults and for earth faults, and three of each for the three phases.

(c) Phase-to-phase fault across phases b and c

$$I_{a1} = -I_{a2} = E_{ae}/(Z_1 + Z_2)$$

$$I_0 = 0$$

$$I_b = -I_c = h^2 I_{a1} + h I_{a2} = (-j\sqrt{3}) I_{a1}$$

primary impedance $= E_{ae}/I_b$

$$= (2/(-j\sqrt{3}))Z_1 \simeq 115\% \text{ of } Z_1 \text{ numerically}$$

This discrepancy of 15%, plus a tolerance of about 5% for the accuracy of measurement *was* the reason for the choice of about 80% for the stage 1 reach of an uncompensated (see section 1.2.2) distance system applied to a single-circuit feeder fed from one end only.

(d) Double-earth fault on phases b and c
The student should analyse this problem, noting that there is a choice of current, i.e. I_b or I_c.

1.2.2 COMPENSATION OF DISTANCE RELAYS

(i) *Phase-fault compensation*

The 15% discrepancy mentioned above can be removed from the phase-fault relays by using the phasor difference of two line currents and this can be done by connecting the C.T.s in delta. Also, the voltage used will be the line-to-line voltage (not the line-to-earth voltage).

(a) 3-phase fault

$$\text{primary impedance} = \mathbf{E}_{ab}/(\mathbf{I}_a - \mathbf{I}_b)$$
$$= ((\sqrt{3}\underline{/30^\circ})\mathbf{E}_{ae})/((\sqrt{3}\underline{/30^\circ})\mathbf{I}_a)$$
$$= \mathbf{Z}_1$$

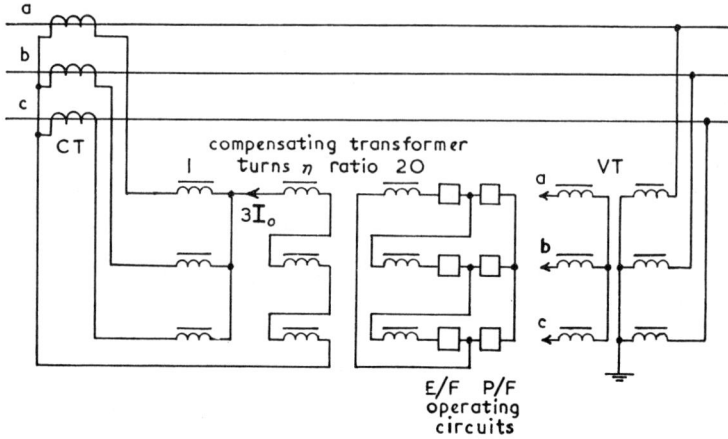

Fig. 1.3. Residual compensation. Earth-fault (E/F) relays take phase voltage. Phase-fault (P/F) relays take line voltage.

(b) Phase-to-phase fault on phases b and c

$$\text{primary impedance} = \mathbf{E}_{bc}/(\mathbf{I}_b - \mathbf{I}_c)$$
$$= \frac{(-\mathrm{j}\sqrt{3})\mathbf{E}_{ae}}{(-\mathrm{j}2\sqrt{3})(\mathbf{E}_{ae}/2\mathbf{Z}_1)}$$
$$= \mathbf{Z}_i$$

(c) Phase-to-phase-to-earth fault on phases b and c

Again, this analysis is left to the student to show that the primary impedance has a magnitude of Z_1.

Thus a compensated, phase-fault distance system (see Fig. 1.3) uses auxiliary C.T.s, connected in delta on their secondary sides to supply the relay operating circuit, which is star connected: the restraint circuit is delta connected so as to take the line voltage from the secondary side of the V.T. The V.T. secondary is star connected, rated at 110 V across the lines and 63·5 V from line to neutral. The primary is also star connected and solidly earthed at its neutral point. The V.T. must be a bank of three single-phase transformers, or their magnetic equivalent, but not a 3-limb, core type (see Volume 1, section 6.10).

(ii) Earth-fault compensation

Distance systems are mainly used with power systems operating at 132 kV and above, where every star point is solidly earthed (multiple, solid earthing). For such systems the zero-sequence current to earth at the point of fault distributes its return over more than one neutral (see Volume 1, Chapter 8, page 336, example 32). Thus the amount of zero-sequence current at each relaying point depends on the earthing arrangements. This complication is overcome by using earth-fault compensation with the earth-fault relays: two methods will be discussed.

Residual compensation

For an earth fault on phase a, the voltage at the relaying point is

$$\mathbf{V}_{ae} = \mathbf{I}_{a1}\mathbf{Z}_1 + \mathbf{I}_{a2}\mathbf{Z}_2 + \mathbf{I}_0\mathbf{Z}_0$$

The residual compensation circuit uses auxiliary C.T.s to add, to the fault current $\mathbf{I}_a$, a fraction n of the residual current,

$$\mathbf{I}_a + \mathbf{I}_b + \mathbf{I}_c = 3\mathbf{I}_0$$

so chosen that the relay measures $\mathbf{Z}_1$. Thus

$$\frac{\mathbf{V}_{ae}}{\mathbf{I}_a + n(3\mathbf{I}_0)} = \frac{(\mathbf{I}_{a1} + \mathbf{I}_{a2})\mathbf{Z}_1 + \mathbf{I}_0\mathbf{Z}_0}{\mathbf{I}_{a1} + \mathbf{I}_{a2} + \mathbf{I}_0 + 3n\mathbf{I}_0} = \mathbf{Z}_1$$

Hence

$$n = (k-1)/3 \qquad [1.6]$$

(where $k = Z_0/Z_1$),

and this result is independent of $\mathbf{I}_0$. From [1.5], substituting for k,

$$n = (Z_E/Z_1) - 1 \qquad\qquad [1.7]$$

Fig. 1.3 shows a circuit giving both phase-fault and residual earth-fault compensation. Clearly n must be a scalar. The student should examine the above analyses, remembering that phasor operations must be used except that, because of the rectifier circuits, the final results must be stated as a scalar relationship.

Sound-phase compensation

When an earth fault occurs on a multiply-earthed system, fault current flows in the sound (healthy) phases. Earth-fault compensation can be obtained by adding to the fault current in the faulted phase, a fraction n of each of the two sound-phase currents, so that the measured primary impedance is the earth-loop impedance $\mathbf{Z}_E$, defined in [1.4]. The circuit is shown in Fig. 1.4: it replaces the compensating transformer shown in Fig. 1.3.

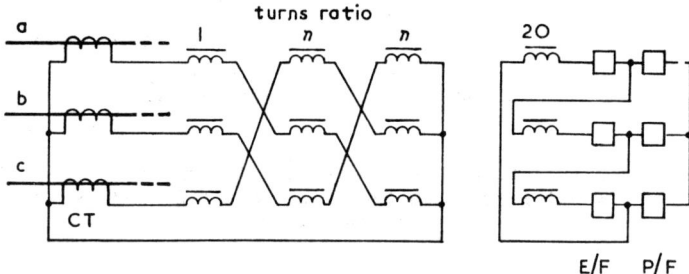

FIG. 1.4. Sound-phase compensation.

It is left as an exercise for the student to show that

$$n = \left|\frac{\mathbf{Z}_0 - \mathbf{Z}_1}{\mathbf{Z}_0 + 2\mathbf{Z}_1}\right| = \left|\frac{\mathbf{k} - 1}{\mathbf{k} + 2}\right|. \qquad\qquad [1.8]$$

It might seem that a fully-compensated distance system could be set for a stage 1 reach of 95% of the feeder length, but in practice a setting of only about 70% is used for double-circuit lines with generation at all busbars to allow for relay over-reach caused by

(a) the mutual inductance coupling on double-circuit lines, (b) the effect of transients on the relay, V.T., and C.T. (see section 1.5), and (c) the throttling effect due to infeeds from teed circuits and the effect of different magnitudes of fault infeeds at both feeder ends (see examples).

1.2.3 POLARISED MHO, OFF-SET MHO AND REACTANCE RELAYS

The characteristic curve of a relay defines the threshold of operation and is usually plotted as the locus of the point (R, X) where $Z = R + jX$ is the setting of the relay: see Fig. 1.5. Care must be taken

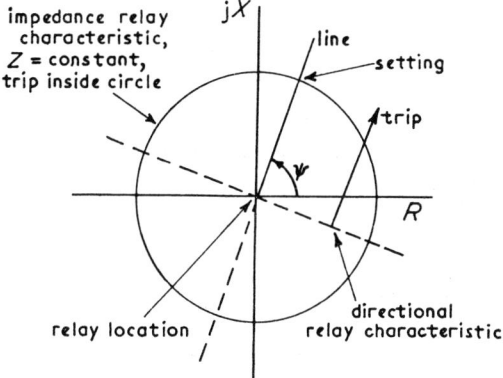

Fig. 1.5. Plain impedance relay; tripping area inside circle with directional control to restrict the tripping area to the upper semi-circle. .

to make clear which side of the locus is trip and which is stabilise. The impedance of the protected line is plotted on the same axes and is a straight line inclined at the line-impedance angle ψ to the R-axis where

$$\tan \psi = \frac{\text{line reactance}}{\text{line resistance}}.$$

The plain impedance relay already discussed in section 1.2 is non-directional and measures only the magnitude of the fault-loop impedance (regardless of phase angle). Thus its characteristic (Fig. 1.5) is a circle, centre at the origin, and radius equal to its setting. The relay trips for any fault-loop impedance (regarded as a point) inside the circle. The relay could be made directional by adding a wattmetrical relay (see Volume 1, section 9.4.4), but such systems have been replaced by the relays now to be discussed.

Polarised-mho relay

A mho relay is a distance measuring relay which is inherently directional. The circuit is given in Fig. 1.6(a), and is derived from

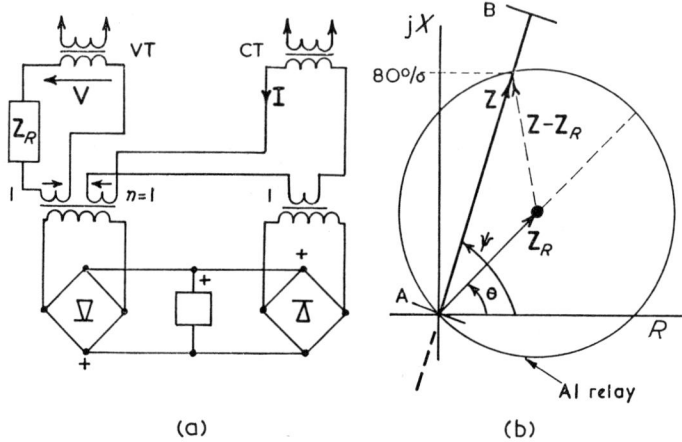

Fig. 1.6. Mho relay. (*a*) circuit. (*b*) characteristic. (Polarising circuit not shown).

Fig. 1.2 by adding mixing transformers so that currents which are suitable proportions of the transmission line voltage and current can be added as phasors. A mho (or admittance) relay is obtained by subtracting the current output of the C.T. from that of the V.T. to obtain the total restraint current. The relay will operate when

$$|\mathbf{I}| \geqq \left|\frac{\mathbf{V}}{\mathbf{Z}_R} - \mathbf{I}\right|$$

$$|\mathbf{Z}_R| \geqq \left|\frac{\mathbf{V}}{\mathbf{I}} - \mathbf{Z}_R\right|$$

$$|\mathbf{Z}_R| \geqq |\mathbf{Z} - \mathbf{Z}_R|, \qquad\qquad [1.9]$$

where $\mathbf{Z} = \mathbf{V}/\mathbf{I}$ is the fault impedance measured in secondary ohms. (The student is warned to be very careful when dealing with inequalities, especially those involving phasors.) The relay characteristic is a circle, passing through the origin, whose centre is given by $\mathbf{Z}_R = Z_R\underline{/\theta}$ relative to the R-axis: θ is called the characteristic angle (or maximum torque angle) of the relay. The setting of the relay (usually in secondary ohms) is given by the diameter $(2Z_R)$ of the circle. Thus

$$\text{relay reach} = \text{relay setting} \times \cos{(\psi - \theta)} \qquad [1.10]$$

The relay trips for any fault impedance inside the circle.

If [1.9] is taken as an equation, it can be converted to the equation of a circle in polar co-ordinates. If the fault is behind the relay, the polarity of the C.T. current is reversed, relative to the V.T. voltage. Thus the restraint action is increased while the operating action is unchanged: the relay is inherently directional.

When $\mathbf{V} = 0$, i.e. the fault is close to the relaying point (close-up fault), the operation of the relay becomes uncertain (see [1.9]). This problem is solved by adding, to the restraint side, a voltage called the polarising voltage which is obtained from a pair of sound (healthy) phases. The resulting relay is then called a polarised-mho relay: but the word 'polarised' is frequently omitted and in practice a mho relay means a polarised-mho relay. The circuitry and analysis of the polarised-mho relay are involved: these can be found in the references, e.g. Electricity Council, Volume 1, page 385. The mho relay is so called because its characteristic is a straight line when plotted to (G, B) axes where $\mathbf{Y} = G - jB$: the geometrical inversion of a circle is a straight line. (Since the introduction of S.I. units, the mho has been replaced by the siemen as the unit of admittance.)

If the fault at the V.T. is a 3-phase fault, then there is no polarising voltage available. A solution, called memory action, is to insert a series capacitor in the polarising circuit so as to tune that circuit to 50 Hz with a short-circuit on the V.T. primary terminals. The oscillatory circuit maintains a polarising current for a few cycles after fault initiation, and so enables the relay to measure correctly. Even this fails if the V.T. is on the line side of the circuit breaker when it is closed on to a 3-phase fault.

Off-set mho relay

This is a mho-type relay, as in Fig. 1.6, except that only a fraction $n < 1$ of the C.T. output current is injected into the restraint circuit. Thus the relay operates when

$$|\mathbf{I}| \geqq \left|\frac{\mathbf{V}}{\mathbf{Z}_R} - n\mathbf{I}\right|$$

$$|\mathbf{Z}_R| \geqq |\mathbf{Z} - n\mathbf{Z}_R| \qquad [1.11]$$

The relay characteristic is shown in Fig. 1.7: the tripping area lies inside the circle. The off-set is given by $(1-n)\mathbf{Z}_R$. This relay

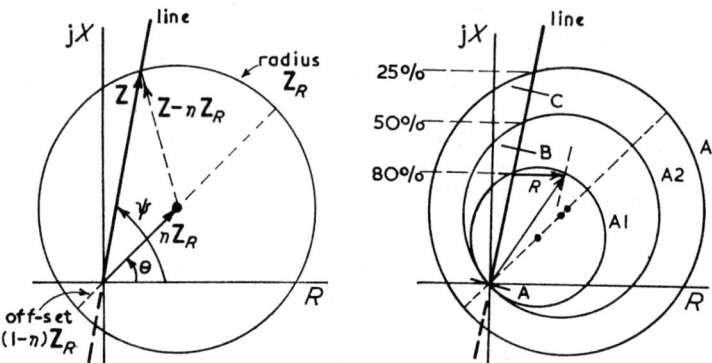

Fig. 1.7. Off-set mho.   Fig. 1.8. Typical phase-fault scheme showing arc resistance (R).

operates when $\mathbf{V} = 0$, and will also operate for faults a short distance behind itself, i.e. it gives some back-up protection for the busbars. A typical value of the off-set is 10% of the protected zone length.

Typical distance system for phase and earth faults

This is shown in Fig. 1.8. Stage 1 is a (polarised) mho distance measuring relay set to about 80% of the length of the protected line AB. It is converted, after about 0·3 s, into the stage 2 relay when its setting is about $1\frac{1}{2}$ zones ahead. The stage 3 relay, an entirely separate 3-pole relay, is an off-set mho relay, set to reach about $2\frac{1}{4}$ zones ahead, which, for a fault within its reach, starts two timer-relays. The first timer-relay converts stage 1 to stage 2 measurement about 0·3 s while the second trips the circuit breaker at A about 0·6 s after fault initiation (if the fault has not been cleared already). Flag indicators on the relays indicate the type of fault and its zone location.

Worked example 1.1

A 132-kV line is in two sections, AB 96·5 km and BC 64 km long, and $Z_1 = 0·435\underline{/70°}\,\Omega$/phase-km. The minimum source MVA (3-phase fault level) at A is 1500 MVA (neglect resistance), the V.T. ratio is 132 000/110 V, the C.T. ratio is 600/1 A, and the relay characteristic angle is 60°. Calculate the setting of the phase-compensated, polarised-mho relay at A which will give a stage 1 reach of 80% of AB. Assume a 3-phase fault.

SOLUTION   (pry. = primary; secy. = secondary)

$$\text{Source maximum reactance} = \frac{132/\sqrt{3}}{1500/(\sqrt{3} \times 132)}$$

$$= 132^2/1\,500 = 11·62 \text{ pry. } \Omega/\text{ph.}$$

Reach of stage $1 = 0·8 \times 0·435 \times 96·5\underline{/70°}$

$$= 33·6\underline{/70°} = 11·48 + j31·6 \text{ pry. } \Omega/\text{ph.}$$

Total fault impedance $= 11·48 + j43·22$

$$= 44·6\underline{/71·18°} \text{ pry. } \Omega/\text{ph.}$$

Minimum fault current $= (132/\sqrt{3})/44·6 = 1·71$ pry. kA

or $1710/600 = 2·852$ secy. A.

Line voltage at relaying point

$$= \sqrt{3} \times 1·71 \times 33·6 = 99·2 \text{ pry. kV (line)}$$

or $99·2/(132/110) = 82·7$ secy. V (line)

Reach of *compensated* relay $= 82·7/(\sqrt{3} \times 2·852)$

$$= 16·8 \text{ secy. } \Omega/\text{ph.}$$

This result could have been obtained by substitution in [*1.3*]:

$$\text{Reach} = 33 \cdot 6 \times \frac{600/1}{132\,000/110} = 16 \cdot 8 \text{ secy. } \Omega/\text{ph.}$$

Relay stage 1 setting $= 16 \cdot 8/\cos(70 - 60)° = 17 \cdot 0\,\Omega$
(Typical settings available are from 3 to 20 Ω.)

Arc resistance

If a line flashes over from phase to phase, (or phase to tower (earth)), the arc resistance can be appreciable, especially for the higher-voltage lines with their larger spacings, and also for faults not cleared in stage 1. This arc resistance is added to the impedance of the line, up to the point of fault, to give the impedance measured by the relay. Fig. 1.8 shows that this effect can be minimised by choosing a relay with a characteristic angle (θ) less than the impedance angle of the line (ψ). (The student could redraw the figure with $\theta = \psi$.) The reference, Electricity Council, Volume 2, page 44, gives the following values for θ:

System voltage, kV:	400,	275,	132,	66,	33,	11
$\theta°$:	75,	75,	60,	60,	45,	45

Warrington gives the following formula for estimating the approximate arc resistance:

$$R = 16\,300(1 \cdot 75S + vt)/I^{1 \cdot 4} \text{ ohm}$$

where

$S =$ conductor spacing, m.

$v =$ wind velocity, km/hour.

$t =$ time, s.

Reactance relay

The earth-fault loop impedance is a function of the resistances of the fault arc and of the earth-return path, both of which can be variable and indeterminate. This affects the impedance measured by the relay and the apparent location of the fault. A method of overcoming this problem is to measure the fault-loop reactance by using a reactance relay.

In the general form of this relay, called the ohm relay, the restraint circuit is supplied from the V.T. output while the operating circuit is supplied by the phasor difference between the C.T. output and the V.T. output. Thus the ohm relay operates when

$$\left| \mathbf{I} - \frac{\mathbf{V}}{\mathbf{Z}_R} \right| \geq \left| \frac{\mathbf{V}}{\mathbf{Z}_R} \right|$$

$$|\mathbf{Z}_R - \mathbf{Z}| \geq |\mathbf{Z}| \qquad [1.12]$$

The relay characteristic is shown in Fig. 1.9: it is the perpendicular bisector of the restraint impedance $\mathbf{Z}_R = Z_R\underline{/\theta}$ (inductive). The slope of the characteristic can be chosen by a choice of θ and the relay

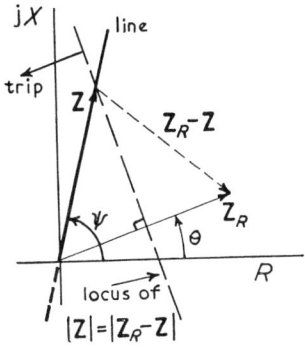

Fig. 1.9. Ohm relay.

trips when the measured impedance lies to the 'left' of the characteristic. Ohm relays are occasionally used as blocking relays which prevent a circuit breaker from tripping by opening normally-closed contacts in the circuit breaker trip circuit.

In a reactance relay, the restraint impedance is a pure inductance ($\theta = 90°$), but this ideal is not physically realisable, so phase-shifting circuits are required. The characteristic is a horizontal straight line and is shown in Fig. 1.10. The relay trips when the measured impedance lies below the characteristic, including the whole of the third and fourth quadrants. Thus a reactance relay is non-directional.

A reactance relay has some merit as an earth-fault distance measuring relay when the protected line is short, since any variable arc resistance can be then an appreciable fraction of the loop impedance.

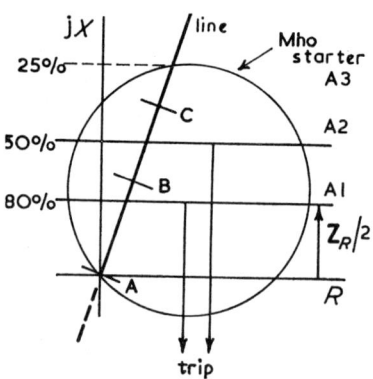

Fig. 1.10. Typical earth-fault scheme.

Earth-fault distance scheme

A possible scheme is shown in Fig. 1.10. Stages 1 and 2 comprise a reactance relay. These relays, being non-directional, are liable to trip for faults behind the relaying point, and possibly even for load current flowing from the protected line AB into the busbars at A. Therefore a starting mho relay (A3) is necessary. It is not just a question of putting the two sets of relay contacts in series: the starting relay contacts are normally-closed contacts short-circuiting the output of the auxiliary C.T. to the reactance relay. Thus the reactance measuring relay cannot start to measure until the starting relay has checked that the fault is in front of the relay.

In the U.K. all grid lines are fitted with aerial earth wires which tend to minimise and stabilise the resistance of the earth-return path. Also, reactance relays are subject to inaccuracies when an earth-fault arc has appreciable resistance and there is fault infeed from both ends of the line. The arc impedance, as measured by the reactance relays, appears to have a reactive component so that the relay at one end of the line over-reaches while that at the other end under-reaches (see example 5). For these reasons reactance relays are not used and mho relays are used, even for short lines, since the latter have a better accuracy.

1.2.4 ACCELERATION OF STAGE 2 TIME

A major disadvantage of a distance system is the slow stage 2 clearance of a fault in the remote-end 30% of the protected zone AB. For important, very-high-voltage lines, such long clearance times cannot be tolerated because of the damage caused by such a fault and the possible loss of power system stability (see chapter 3). However such a fault is within the reach of the stage 1 relay at B (protecting the other end of the line AB) which trips the circuit breaker at B and also initiates the transmission of a signal to end A using a telephone pilot, or high-frequency carrier equipment (see section 1.3) or a radio link. Three contrasting systems will be described briefly. For each, the basic components are the 3-step distance system already described and, in the event of the absence or failure of the signal channel, each gives the normal time-discriminating margins between the three stages.

Accelerated distance system

In this system the signal trips the circuit breaker at A if, and only if, the stage 3 relay at A has already operated (or stage 2 if that is a separate relay). Thus the protection at A should operate in the stage 1 time of B (60 ms) plus the signal relay and transmission times of about 70 ms. The signal channel need not have high integrity since any interference in the channel cannot cause tripping unless there is a fault on the system at the same time.

Intertripped distance system

Two circuit breakers are said to be intertripped if the tripping of the first initiates the tripping of the second. In the intertripped distance system, the signal from B trips the circuit breaker at A without reference first to the protection at A. The signal channel must have high integrity and is usually more expensive than an accelerated distance channel due to the precautions which must be taken to prevent spurious tripping due to channel interference (see Electricity Council, Volume 2, pp. 204–218). In the case of a carrier channel the interference is due mainly to electromagnetic radiation ('noise') from the arcs in adjacent circuit breakers, isolators and spark gaps when these operate. A disadvantage of carrier intertripping is that the signal has to be transmitted through the fault when the fault is near the end of the line: it can be shown that the signal attenuation is then a maximum. A high speed intertripping time is about 25 ms.

Interlocked (blocked) distance system

This system could be regarded as the dual of the previous system since now the signal from B prevents (blocks) the circuit breaker at A from tripping. The stage 2 relay at A is set to about $1\frac{1}{2}$ zones (i.e. AB plus about half of BC) but now can instantaneously trip the circuit breaker at A, providing a blocking signal has not been received. Thus for a fault anywhere in the zone AB, the circuit breakers at A and B are tripped instantaneously and the system is virtually a unit system. To prevent the circuit breaker at A from tripping for a fault in the first half of the zone BC, a starter (fault-detecting) relay at B in the zone BC initiates a blocking signal which opens normally-closed contacts in the trip circuit of the circuit breaker at A (and elsewhere). This starter, which also starts the timing relays, is usually a plain impedance relay (Fig. 1.5) with a setting of about 225%: it need not be very accurate or immune to over-reach but is very fast (1 ms). An advantage of the blocking system is that for faults external to AB, the carrier signal is transmitted over the healthy line AB. Interference can, at worst, only delay tripping for an internal fault by times not exceeding the normal stage times, since the stage timers bypass the blocking relay contacts. The system also comprises a normal instantaneous stage 1 relay set for about 70% of AB (for a double-circuit, ring system) whose operation is independent of any blocking signal and hence of interference. The blocking signal can be used for the shorter lines and where transformers are connected solidly to the line (i.e. without any controlling circuit breakers). In this last case a transformer fault causes the circuit breakers at A and B to trip instantaneously, a motorised isolator at the transformer then opens the infeed and delayed auto-reclosing circuits (see section 1.4.2) reclose the circuit breakers at A and B. For further details of a blocking system the student is referred to the G.E.C.–E.E. pamphlet on 'Interbloc' carrier-distance relaying.

1.2.5 SWITCHED-DISTANCE SYSTEMS

A complete distance system at both ends of a line could each comprise an earth-fault, and a phase-fault relay for each of the three phases and all triplicated for the three stages, a total of 18 relays per end (plus two timing relays, one for stage 2 and one for stage 3, plus auxiliary relays and d.c. circuitry). This expense could only be justified for an important high-voltage line. A common arrangement, used at 132 kV and above, combines stages 1 and 2, thus reducing the distance relays to 12 per end. To make a distance system economic at 66, 33 or even 11 kV, switched-distance systems are available. In these, only one mho measuring relay is used, and attracted-armature, over-current starting relays in the three phases determine the nature of the fault and apply the appropriate voltage and current to the mho relay, with only a small increase in the total relay time. Some complications are the possibility of the fault changing type before being cleared, and the requirement that the peak load current must be less than the minimum fault current. If the minimum plant fault level is too low, overcurrent starters cannot be used and impedance-starting elements are required and these are more expensive.

1.2.6 ACCURACY, SENSITIVITY AND SETTING

Errors occur in the distance-measuring relays, C.T., V.T., line impedance data and route length. The stage 1 setting is critical and is chosen so that, under the worst possible conditions, it will not reach a fault on the busbars at the remote end of the protected line.

For any relay to operate within a specified accuracy, the manu-facturer specifies a minimum value of energising voltage (e.g. 3% of 110 V (line) or 63·5 V/phase). It is convenient to use a reciprocal function, namely nominal voltage/relay voltage, and since the minimum value of the ratio is 1, the characteristic impedance ratio (C.I.R.) of the relay could be defined as

$$\text{C.I.R.} = \frac{\text{nominal voltage}}{\text{relay voltage}} - 1 \qquad [1.13]$$

(e.g. C.I.R. $= (1/0{\cdot}03) - 1 = 32{\cdot}3$).

From [1.1], for any value of $\mathbf{E}$,

$$\frac{\text{source e.m.f.}}{\text{relay voltage}} - 1 = \frac{\mathbf{Z}_s + \mathbf{Z}_f}{\mathbf{Z}_f} - 1 = \frac{\mathbf{Z}_s}{\mathbf{Z}_f}$$

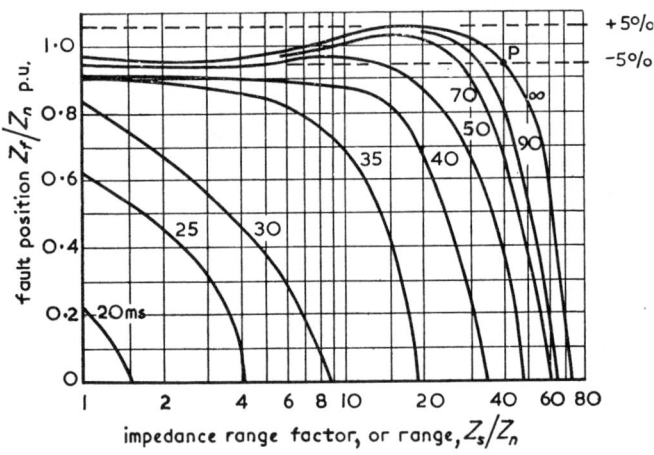

Fig. 1.11. Accuracy curves for a mho phase-fault relay relating relay operating time (ms) to source impedance and fault impedance. (From Reyrolle pamphlet 1297).

The numerical value of this ratio is defined as the system impedance ratio (S.I.R.), i.e.

$$\text{S.I.R.} = Z_s/Z_f \qquad [1.14]$$

Thus for any relay to be within its specified accuracy limits,

$$\text{S.I.R.} \leqq \text{C.I.R. of relay} \qquad [1.15]$$

B.S. 3950 defines C.I.R. from [1.15], i.e. as the maximum permissible system impedance ratio, rather than as a characteristic of the relay.

Fig. 1.11 shows Z_f to a base of Z_s for a family of relay operating times, the outermost curve being for the relay just on the threshold of operation. Both Z_f and Z_s are expressed in per-unit form using, as a base impedance, Z_n the nominal or marked setting of the relay. Point P shows that if the relay is to be accurate to within 5% of its setting, the source impedance must not exceed 40 times the relay setting (both in secondary ohms) and even then the time of operation is uncertain. To allow for effective time grading, an operating time of 70 ms would be more reasonable, giving a ratio of just under 30 for the same 5% tolerance.

1.3 Phase-comparison carrier-current protection (high-frequency phase-comparison)

This protective system, known as carrier protection for brevity, is an alternative to the distance system for the protection of long, important, high-voltage lines. For both systems, the basic principle is simple but the practical application is complicated and expensive. Carrier protection is a unit system of protection, whose protected zone is defined by the location of the two sets of C.T.s, whose operating time is instantaneous and which uses the power line itself as a channel to transmit information, in the form of a modulated high-frequency (h.f.) signal, regarding the conditions at the two ends of the line. A carrier system requires expensive solid-state electronic equipment to generate the h.f. (50—700 kHz) signal, and expensive high-voltage capacitors to inject the signal into the power line: see Figs. 1.12 and 1.13.

The basic principle of carrier protection is that for load and external fault conditions the currents $\mathbf{I}_A$ and $\mathbf{I}_B$ entering and leaving the zone AB are nearly equal and in phase, with respect to the system voltage (subject to later qualifications), while for an internal fault the two currents are approximately in antiphase (or one current is zero). These currents initiate the transmission of h.f. signals from each end to the other. At both ends of the line, the modulations of the local and remote h.f. signals are compared and, if they indicate an external fault, the protection stabilises; otherwise it trips the local circuit breaker. Since the power line, when carrying the h.f. signal, acts as an aerial radiating signal energy, so that the signal is attenuated during transmission and by a variable amount dependent on the weather, the signal modulation received at one end cannot give

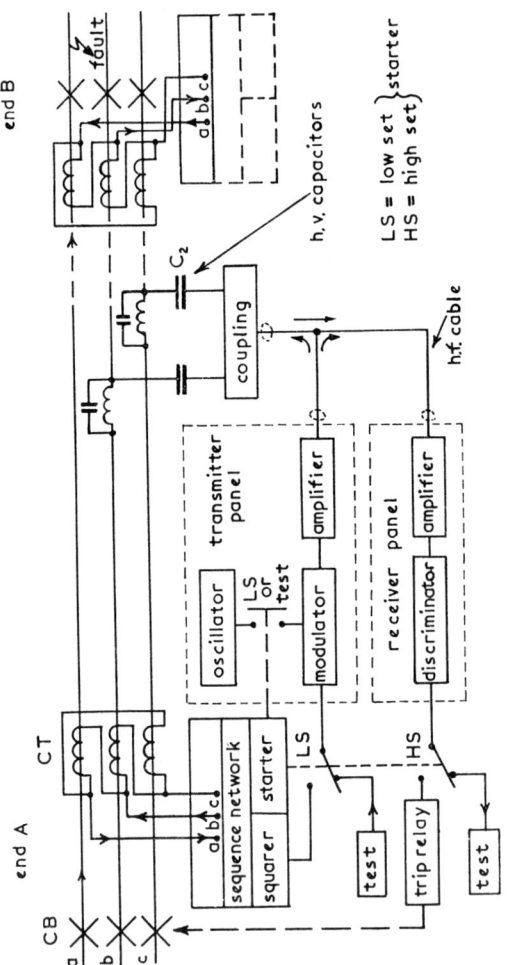

Fig. 1.12. Phase-comparison carrier-current protective system.

information regarding the amplitude of the current at the other end. Thus the signal transmits information regarding the phase only of the current. In this context, it is convenient to measure phase as a function of time by using the instantaneous zero-points of the current/time wave.

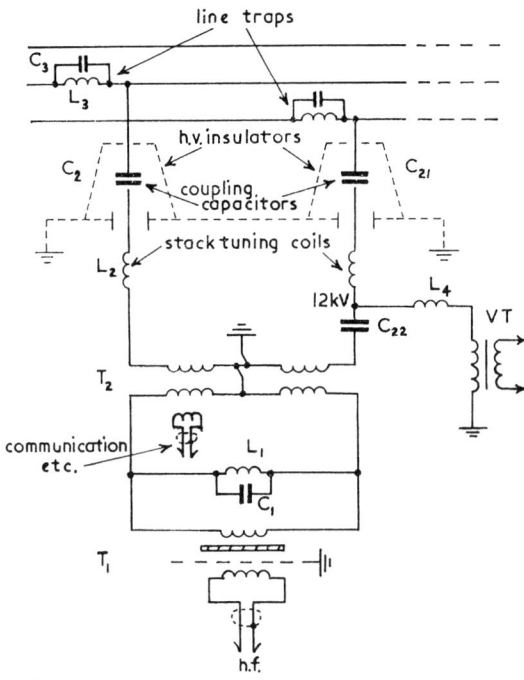

Fig. 1.13. Details of coupling equipment.

Fig. 1.12 shows that the two sets of C.T.s are identically connected with respect to their own busbars. Thus during an external fault, although the C.T. primary currents are nearly in phase, their second-ary equivalents are nearly in antiphase when fed into the sequence networks. In an ideal carrier protection system, these two secondary currents would each initiate the transmission of an h.f. signal, while each current was instantaneously positive, lasting half a cycle and in adjacent half cycles (both at power system frequency) so that at each receiver there was a continual h.f. input signal. The protection is designed to trip due to an absence of h.f. signal exceeding the time equivalent of about 40° (at power frequency). During an internal fault the transmission from the two ends takes place during the same half cycle so tripping action occurs during the half cycle of zero signal.

Consider now some practical complications.

Sequence network. In the U.K., to reduce interference with the many users of the h.f. wavebands, the Post Office regulations do not permit the power line(s) to be permanently energised at full signal power input (about 10 W) so the signal is transmitted only during fault conditions, and this requires a starting (fault-detecting) relay. A problem is that, for the Grid network, the minimum fault current (in summer) can be less than the maximum load current (in winter) so the protection cannot be started by starting relays dependent on current amplitude alone. The starting relays are energised by the output of the sequence network.

The sequence network (see Volume 1, section 8.5) gives a single-phase output which should be linearly dependent on the positive- and negative-sequence components of the (3-phase) line currents (the delta-connected C.T.s suppress the zero-sequence component): reference Electricity Council, Volume 2, p. 179, quotes $5I_2 - I_1$ as being the least sensitive to the type of fault. The positive-sequence component is necessary to detect a 3-phase fault, but consideration must also be given to the network output due to maximum load current and to line charging current. Use is made of circuits which in effect measure the step change of current with a setting of at least 25% of full load current. Even then the circuit is liable to give an output if a heavy load is suddenly switched on. Since load is equivalent to an external fault, the protection will stabilise and the carrier will only be on for the short interval during which the load is changing. Some circuits can also be made load dependent (biased): e.g. a 3-phase fault setting of 40% on no-load but 170% of rated current at rated load current. The network can if required give a 3-phase fault setting less than the maximum load and yet should cause no carrier transmission for a steady load greater than the fault setting.

To reduce the network output due to the d.c. component of the fault current and to line surges, etc., the sequence network output is taken through a bandpass filter (about 25 to 100 Hz).

Starting (fault-detecting) relay. For maximum fault current, *single* starter relays at A and B should both operate with certainty and the carrier protection will discriminate between internal and external faults. For minimum fault currents, the possibility arises that, because of slight differences in the relay operating values, only one starter will operate. The protection will treat this condition as an internal fault and will trip, although the fault could be an external one. This problem is overcome by using two starter relays at each end, one low-set (LS) and the other high-set (HS) and set about 25% above the low-set, i.e. above the error tolerance of the low-set relays. The low-set relay switches the carrier on to the line while the high-set relay controls the trip circuit. Consider an increasing line current. If the low-set relay at A operates first, the carrier at A is transmitted to both receivers, and both phase detectors operate but neither circuit breaker trips because neither high-set relay has operated. If the line current increases so that both low-set relays operate, the protection will discriminate between internal and external faults but still cannot trip to the former. If, with increased line current, the high-set relay at A operates, then the circuit breaker at A will trip if (and only if) the fault is internal. The effective setting of the protection is the higher of the high-set relays. An internal fault fed from one end only, and exceeding the setting, will trip its local C.B.: it cannot trip the remote C.B. It is essential to prevent tripping to an external fault if, for any reason, only one starter operates. Starters add about 20 ms to the total fault clearance time.

Squarer. This pulse-shaping circuit converts the sequence network output into a square wave output. Ideally the input and output zeros should be time-coincident and the output should have unity mark/space ratio over 360° (at power frequency). Fig. 1.14(b) shows the action of a simple limiter circuit: the duration of the output pulse is very dependent on the amplitude of the input current relative to the setting of the limiter.

Oscillator. The high-stability oscillator is permanently energised but is only switched in to the modulator if the low-set starter relay has operated (or if the test signal is being transmitted). The frequency is chosen to give minimum interference with adjacent carrier systems: the lower frequencies are allocated to the longer lines to reduce signal attenuation. The lowest carrier frequency is limited due to the coupling capacitor C_2 becoming too large and costly, and the highest is limited due to signal attenuation along the line. Attenuation increases with frequency, line length and ice thickness on the conductors.

Modulator. In effect this is an off-on switch (100% modulation), the carrier being transmitted when the output of the squarer is instantaneously positive: see Fig. 1.14(c).

Transmitter amplifier. This gives an output in the range of 10 to 20 W and is usually, for economic reasons, a standard wide bandpass type, covering the full range of carrier frequencies, with a narrow bandpass input filter tuned to the oscillator frequency.

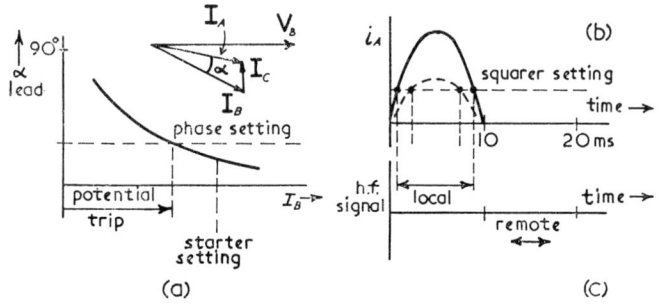

Fig. 1.14. Phase-comparison, showing:
    (*a*) the effect of charging current on phase angle
    (*b*) the effect of limiter setting on pulse duration
    (*c*) the pulse durations for an external fault.

Line coupling equipment (see Fig. 1.13). The capacitor C_2 (2000 pF) is a series-connected stack rated for the system line to neutral voltage and enclosed within a high-voltage insulator. It has a high reactance at power frequency and a low reactance at carrier frequency and this last is reduced to zero by series-tuning L_2, C_2 to the carrier frequency. The line trap (wave trap), L_3(100 μH), C_3, is normally fitted on top of the h.v. insulator and tuned to the carrier frequency so that the carrier signal is directed into the protected zone. L_3 is rated to carry the load current continuously and mechanically designed to withstand the maximum fault current. It is possible to use only one phase, with an earth-return, but standard practice in the U.K. is to use two phases. This reduces the radiation from the line and hence the signal attenuation but nearly doubles the cost of the coupling equipment. L_1, C_1 is tuned to the carrier frequency, and is virtually a short circuit at power frequency, thus minimising any infeed from the power system to the h.f. equipment. For the same reason the transformer is fitted with barrier insulation and an earthed screen. The transformers give impedance matching of the coupling components for minimum signal reflection. The coupling equipment, as a complete unit, is equivalent to a wide bandpass filter (which can cover the carrier frequency for protection and communication) feeding into the characteristic impedance of the line.

If (as is usual for economic reasons) the coupling equipment is to be used for communications and data transmission, then use is made of a tertiary winding on the transformer T_2 (hybrid transformer). Fig. 1.13 also shows, on the right, how C_2 can be divided into C_{21} and C_{22} so as to act as a voltage divider, which, together with the wound V.T., comprises a capacitor voltage transformer (C.V.T.). The use of the coupling equipment for protection, communication and voltage measurement reduces the cost to be charged to each.

Receiver amplifier. This is a standard, wideband amplifier covering the range of carrier frequencies, with a narrow bandpass input filter tuned to the carrier-protection frequency in use: it must reject communication signals. For a transmitter amplifier output of 10 W and a total transmission attenuation of 30 dB, the receiver input would be 10 mW. An input limiter reduces the local and remote signals to the same amplitude.

The receiver should be insensitive to interference (radio noise) from the arcs in load-breaking isolators, circuit breakers and faults, and from noise from the lines themselves, since such interference might delay tripping for an internal fault.

Phase-detector (demodulator). The demodulator gives a square-wave voltage output, for an h.f. input, and feeds an R, C integrator followed by a level detector (voltage measuring relay) across the capacitor, the circuitry being so arranged that, if an h.f. signal is fed into the receiver the capacitor discharges, but otherwise it charges. In a practical system during an external fault (Fig. 1.14(c)), the h.f. signal must be on for nearly the whole of a complete cycle. During an internal fault the local and remote signals are fed to the receiver in the same half cycle, so that the capacitor charges more than it discharges and the capacitor voltage rises (integrates) in a sawtooth waveshape until, after 1 or 2 cycles, it exceeds the setting of the level detector.

The total line-charging current, I_C, causes a phase shift between I_A and I_B (see Fig. 1.14(a)), the shift being greater for the smaller load or through-fault currents: a typical allowance might be 10°. The protection would be liable to trip for these small currents were it not inhibited by the starter setting. The signal transmission time for a 160 km line (about the upper limit for carrier protection) is about 0·5 ms or about 10° of phase shift. Allowing for about 5° phase shift in the equipment and a margin of safety, a typical setting would be $\pm$ (30° to 40°): the protection would operate for a phase shift in excess of the setting, providing the magnitude of the fault current was sufficient to operate the high-set starter relay.

The student should consider a possible alternative to Fig. 1.14(c), namely, designing the circuitry so that the h.f. signal would be transmitted except when the output of the squarer was positive.

To check that the carrier protection equipment is functioning correctly, test signals are transmitted automatically every half-hour to the remote end and received back at the local end: this does not check the starter or the trip circuits. Because of the starters, a genuine fault over-rides the test procedure: see Fig. 1.12.

An alternative to carrier transmission along the power line is to use radio links in the v.h.f. or higher bands. Since these are very directional, transmission could be continuous and, since a large number of channels can be accommodated, could be used for protection, control, communication and data transmission. They are restricted to line-of-sight transmission (about 60 km) unless expensive repeaters are used.

1.4 Auto-reclosing circuit breakers

1.4.1 RURAL POWER SYSTEMS

A typical rural distribution system comprises a radial (single infeed) 11-kV overhead line with several tee-off or spur lines, and covers a large area, much of it inaccessible to road transport. The lines are mounted on insulating wood-poles (B.S. 1320) so that earth-faults are unlikely unless the conductor breaks and falls to the ground. In earlier protective systems, fuses were fitted at the entry to the main line and to the spur lines, forming a time-graded protective system (see Volume 1, section 9.4). The practical problem was that nearly 90% of all the faults were transient faults due to lightning flashovers and to debris blown across the phases, i.e. the fault did not exist after the ionised air had dispersed: but replacing the fuse(s) was an expensive, time-consuming job, especially after a severe lightning storm.

Automatically-reclosing circuit breakers (auto-reclosers) have now been developed to increase security of supply and reduce outages: circuit breakers are dealt with in Chapter 2. The auto-recloser is a small, 3-phase, oil circuit breaker mounted below the line on the wood pole. It can be programmed mechanically to give up to four trips, either instantaneous or time-delayed, or both, to suit the power system. The necessary energy for the opening- and closing-springs is obtained from the power line itself by an electromagnet shunt-connected through auxiliary contacts across the incoming supply. When the auto-recloser is open (tripped) the electromagnet is energised and its armature partially charges the closing spring to a point where the armature opens the auxiliary contacts in the electro-magnet circuit. The armature resets and the cycle is repeated until the spring is fully charged, at which point it closes the auto-recloser, and this action in turn opens the auxiliary contacts.

Phase-fault (over-current) protection is given by three series trip-coils which carry the line currents, any one of which provides the energy to trip the auto-recloser (3 phases) when the line current exceeds the setting (about twice rated current). Time delay in the tripping action is obtained by using an oil dash-pot, the size of the orifice through which the oil is ejected controlling the delay time. Instantaneous tripping is obtained by a large bypass valve. The necessary mechanical adjustments are pre-set to give any sequence of instantaneous or delayed trips up to a total of four. Both types of trip vary inversely with fault current. The instantaneous-trip time decreases from about 0·3 s to 0·06 s and the maximum delayed-time trip from 40 s to 0·7 s for currents of twice and twenty times rated current respectively, but the delayed-trip times can be reduced by increasing the orifice size, in order to match the fuse characteristic.

The following is a typical sequence for an auto-recloser, namely, two instantaneous trips followed by two time-delayed trips. On the occurrence of a fault, the auto-recloser trips instantaneously to reduce the potential damage due to the fault current. It then stays open—the dead time or open time—long enough for the fault arc to deionise. The dead time can be varied depending on the type of load, e.g. motors, but is typically 1 s. The auto-recloser then recloses. If the fault were a transient one, the supply should now be restored to all consumers, but otherwise the auto-recloser again trips instantaneously, waits a further 1 s, then recloses. The majority of transient faults are cleared at the first trip, while the remainder are cleared at the second trip. If, however, the fault is a permanent one, the third trip is time-delayed long enough to melt the slow-melting spur fuse carrying the fault current, and if successful the auto-recloser remains closed and supply is restored to all consumers except those beyond the fuse. The second delayed trip is available to clear stubborn, permanent faults at the fuse. If the fault has not yet been cleared, the auto-recloser completes its sequence with a second delayed trip and locks-out (remains open), in which case the mechanism must be manually reset. If a fault is cleared before lock-out, the mechanism resets automatically and the full sequence is again available, unless another fault occurs before it is fully reset.

Earth-fault protection is given by three residually-connected (i.e. paralleled) C.T.s in the auto-recloser bushing insulators (see Volume 1, p. 358). The setting is about 20 A, and all the trips are instantaneous until the auto-recloser locks-out. (A dangerous earth-fault condition occurs when a conductor breaks and falls on to dry ground, hedges, etc. The fault current could be less than 20 A, and is an almost unprotectable condition.)

Automatic sectionalisers, which are load-breaking, fault-making, 3-phase oil switches, are now available to replace the spur fuses. The sectionalisers are manually closed and spring opened. The pulses of fault current are counted or integrated by an oil dash-pot whose plunger is operated by an electromagnet energised by the line current. Tripping is set mechanically to occur after 1, 2 or 3 pulses, but is time controlled so that the sectionaliser opens only during the dead time of the auto-recloser. Earth-fault protection is obtained by using residually-connected C.T.s, feeding a pulse-counting relay. Suppose that the auto-recloser is set for three instantaneous trips followed by one time-delayed trip, and that the sectionaliser is set to trip after three pulses of fault current. Transient faults should be cleared, at latest, during the second dead time. The sectionaliser, since it has not completed its sequence of three pulse counts, will reset its counter within a few minutes. For a permanent fault, the sectionaliser will complete its sequence of three pulse counts, trip during the third dead time of the auto-recloser, and lock-out. Two sectionalisers can discriminate with each other by setting the remoter one to two pulse counts and the one nearer the source to three pulse counts. Sectionalisers (or fuses) and auto-reclosers must discriminate with each other and with the time-graded protection (Volume 1, section 9.4.2) on the 11-kV side of the 33/11-kV substation feeding the rural lines: the pulsing of the current complicates the calculation.

1.4.2 MAIN TRANSMISSION LINES (GRID)

The principle of the auto-reclosing circuit breaker can be applied to the Grid lines, and the earlier type used is referred to as a high-speed auto-recloser since its dead time was less than one second. In most cases these have now been replaced by slow-speed or delayed auto-reclosers due to the problems created by the high-speed type.

The fault level on the Grid is so high that considerable damage can be done even by one pulse of fault current, so only single-shot reclosing was (and is) used. If a loaded Grid line is opened, the sending-end area has a surplus of power and accelerates, while the receiving-end area decelerates (see Chapter 3). Thus the phase angle between sending- and receiving-end voltages increases, and if the circuit breaker were closed a current surge would result. The maximum time a Grid line can be left open (circuit reconnection time) is about 0·5 s, while the minimum time for the fault arc to deionise (dead time) is about 0·3 s. Auto-reclosers can only be used with high-speed, unit protection (including accelerated-distance) since the circuit breakers at both ends of the Grid line must open instantaneously. A typical protection operating-time is 3 cycles (60 ms) and a circuit-breaker total-time is 6 cycles (0·12 s). The circuit breakers— bulk-oil or air-blast—were specially designed for a rapid break-make-break operation: e.g. the closing sequence was started soon after the start, and before the finish, of the tripping sequence. The use of the high-speed auto-recloser has been largely discontinued because, if it reclosed on to a permanent fault, the system swing during two pulses of fault current could cause loss of system stability.

A delayed-, or slow-speed, auto-recloser is one which, due to interlocks, is not allowed to reclose until checks have been made that it is safe to do so; e.g.

1. Dead-line-charge end
    (a) neither the busbar nor the line V.T. protections have operated (the fault was on the line)
    (b) busbar V.T. is alive but the line V.T. is dead (remote C.B. is open)
    (c) continue these checks for several seconds and, if no change takes place, close the C.B.

At the remote end of the line, the checks would be as follows:

2. Check-synchronise closed end
    (a) as above
    (b) busbar and line V.T.s both alive
    (c) check-synchronising relay indicates that the voltages on both sides of the C.B. are within prescribed limits of phase difference (35° for 3 seconds)
    (d) as for (c) above
    (e) if line V.T. is not energised within a pre-set time then the circuit is deadline charged from this end.

Automatic switching

When a fault occurs on a transformer teed on to a feeder fitted with delayed auto-reclose, the auto-reclose equipment is stopped (using the intertripping channel) until the transformer motorised isolator is automatically opened. Then reclosure takes place as above. If the isolator fails to open, the intertripping signal resets the delayed auto-reclose equipment and locks out the circuit breakers.

1.5 Current transformers: transient operation

With the advent of static (electronic) protection relays and very fast circuit breakers (e.g. the vacuum type), the total times of the protection and the circuit breaker are being reduced to about 2 cycles each, giving a total fault-clearance time of about 4 cycles or 80 ms. Thus the behaviour of a C.T. during the first few cycles of a fault, when it carries the transient component of the fault current, is becoming of increasing interest. (Similarly the transient behaviour of a V.T. is important when connected to a very fast distance relay which has to determine the fault impedance during the first few cycles of a fault.)

If a short-circuit occurs on a power system (considered as an R, L series circuit) the current waveform is given in Volume 1, Fig. 7.9: it consists (generally) of a decaying a.c. current superimposed on a decaying d.c. current, the decay in the former being due to the increase in alternator reactance as it passes through the subtransient, transient and synchronous reactance stages, while the decay in the latter is due to the resistance of the power system. An alternator is the only component of a power system whose reactance varies with time, so this effect is significant only for faults close to the alternator, and will not be allowed for in the present context, except that the transient value will be used.

The equations for the fault current on pp. 238–9 can be re-written as

$$i_p = (E_{pm}/Z_p)[\sin{(\omega t + \alpha - \phi_p)} - \exp{(-R_p t/L_p)} \sin{(\alpha - \phi_p)}] \quad [1.16]$$

where subscript p indicates the primary side of the C.T. This equation can be written down by inspection: the first term is the steady-state current while the second is the transient current—a single-exponential decay function, since the circuit has only a single energy-storage element— such that $i_p = 0$ when $t = 0$. α is a parameter dependent upon the instant on the voltage wave at which the fault occurs. The transient current will have a maximum value when $\alpha - \phi_p = -\pi/2$ rad and [1.16] reduces to

$$i_p = I_{pm}[-\cos{\omega t} + \exp{(-R_p t/L_p)}] \quad [1.17]$$

where $I_{pm} = E_{pm}/Z_p$. Normally, as the system rated voltage rises, $\phi_p \rightarrow \pi/2$ rad (or 90°) and $\alpha \rightarrow 0$. In the limit where $\phi_p = \pi/2$ rad, a fault at zero voltage gives rise to the maximum fault current asymmetry (current doubling).

The equivalent circuit (model) of a current transformer is given in Volume 1, Fig. 9.1(a). By inserting typical values, it can be shown that the insertion of the C.T. has no significant effect on the analysis above: e.g. take the worst case of minimum Z_p and maximum Z_s, where subscript s indicates the secondary side of the C.T. Typical fault levels can be obtained from B.S. 116 on Circuit Breakers, e.g.

250 MVA at 11 kV gives $Z_p \simeq 0.5\,\Omega$. A typical 11-kV C.T. is 400/5 A with a rated burden (load) of 15 VA *at rated current*, so $Z_s = 0.6\,\Omega$. The secondary equivalent of Z_p is $Z_p' = 0.5(400/5)^2 = 3200\,\Omega \gg 0.6\,\Omega$. A prime or dash (') indicates a value transferred to the other side of the C.T. The student should repeat this calculation using a 400/1 A 15 VA C.T.

The complete transient analysis of a C.T. is difficult so, for the present context, some simplifying assumptions will be made. The burden (relay) will be assumed to be a pure resistance: it can be shown that this is one of the worst cases that can occur. Protective systems normally use bar-primary, ring-core C.T.s with a uniform secondary winding covering most of the perimeter, i.e. low-leakage reactance type (see Volume 1, pp. 348–9): so the secondary leakage reactance will be neglected. The C.T. burden will be assumed to be a pure resistance, and combined with the resistance of the leads between the C.T. and its burden (relay), and then combined with the iron-loss resistance of the C.T. exciting circuit to give a total resistance R_s, carrying a secondary current i_s. The exciting shunt inductance is L_e and takes a current i_e. Thus the secondary circuit now comprises a shunt R_s, L_e circuit, whose input is the primary fault current referred to the secondary side, i.e.

$$i_p' = i_p(N_p/N_s): \text{ also let } I_{pm}' = I_{pm}(N_p/N_s).$$

The circuit elements will be assumed to be constants so that linear analysis can be used: due to the saturation of the C.T. core this assumption is far from being valid (especially if R_s includes the iron-loss resistance) but it will give a first approximation to the true solution.

In order to estimate the transient flux in the core, the magnetising current will be neglected (i.e. $L_e \to \infty$), as it can be shown that this is one of the worst cases; also [*1.17*] will be used, i.e. the case of maximum transient primary current.

The secondary voltage is

$$v_s = R_s i_s = N_s(\mathrm{d}\phi/\mathrm{d}t)$$

and the maximum core flux is

$$\Phi_m = (1/N_s) \int v_s \mathrm{d}t$$

Thus for the steady-state component of i_s, and integrating over a quarter-cycle,

$$\Phi_m = -(1/N_s) \int_{t=0}^{\pi/2\omega} R_s I'_{pm} \cos \omega t \, dt$$

$$= -(R_s I'_{pm}/N_s \omega) \qquad\qquad [1.18a]$$

For the transient component of i_s, and integrating from $t = 0$ to $t \to \infty$,

$$\Phi_m = (1/N_s) R_s I'_{pm} \int_{t=0}^{\infty} \exp(-R_p t/L_p) \, dt$$

$$= (R_s I'_{pm}/N_s \omega)(\omega L_p/R_p) \qquad\qquad [1.18b]$$

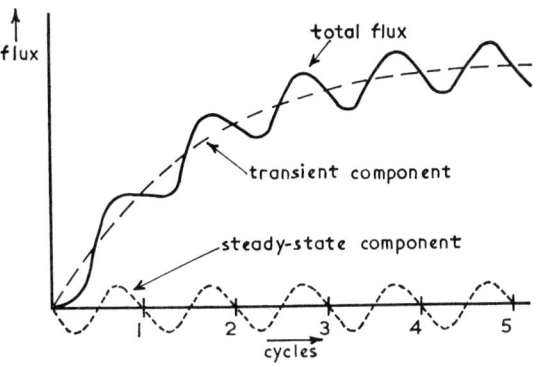

Fig. 1.15. C.T. core fluxes during transient period.

Thus the component of core flux due to the transient component of the fault current is equal to that due to the steady-state component multiplied by the factor $\omega L_p/R_p$ where, taking the worst case of a fault near a large power station, $\omega L_p/R_p$ could be as high as 30, corresponding to a primary-circuit time-constant of $L_p/R_p \simeq 0\cdot1$ s, or about 5 cycles. Assuming the two fluxes can be added numerically (the worst case) the total core flux is equal to the steady-state component multiplied by the factor

$$1 + (2\pi f L_p/R_p) = 1 + 2\pi T_p$$

where T_p is the primary-circuit time-constant in cycles. If saturation of the C.T. core is to be avoided, in systems having high values of T_p, the steady-state flux must be almost negligible, and the transient flux will mainly determine the section area of the iron core. The flux waveforms are sketched in Fig. 1.15.

Any flux will require an exciting current i_e, so a finite exciting inductance L_e will now be assumed; the three currents in the C.T. secondary side are related by

$$i'_p = i_e + i_s.$$

$$L_e(di_e/dt) = R_s i_s = R_s(i'_p - i_e)$$

$$di_e/dt + (R_s/L_e)i_e = (R_s/L_e)i'_p$$

This equation can be integrated, after substitution from [1.17]—the case of maximum transient primary current: the solution is

$$i_e = I'_{pm}R_s\left[\frac{-\cos(\omega t - \phi_s) + \{\exp(-R_s t/L_e)\}\cos\phi_s}{(R_s^2 + \omega^2 L_e^2)^{\frac{1}{2}}}\right.$$

$$\left. + \frac{L_p\{\exp(-R_s t/L_e) - \exp(-R_p t/L_p)\}}{R_p L_e - R_s L_p}\right] \qquad [1.19a]$$

$$= I'_{pm}\left[\left\{-\frac{R_s}{(R_s^2 + \omega^2 L_e^2)^{\frac{1}{2}}}\right\}\{\cos(\omega t - \phi_s) - (\exp(-t/T_s))\cos\phi_s\}\right.$$

$$\left. + \left\{\frac{T_p}{T_s - T_p}\right\}\{\exp(-t/T_s) - \exp(-t/T_p)\}\right] \qquad [1.19b]$$

where T is a time-constant.

The Q-factor $= X_p/R_p$, for a power system during a fault varies from about 30 for a fault near a large power station to about 10 on a Grid transmission line. Taking $Q = 20$ as an average value, $T_p = L_p/R_p = 0.064$ s, (say 0.06 s). If an open-circuit test on the secondary of a 1 A, 15 VA C.T. gave the results $V = 100$ V, $I = 150$ mA, p.f. $= 0.71$ lagging then $L_e = 3$ H and $\omega L_e = 942\ \Omega$. If the loop resistance of the leads is taken as $2\ \Omega$, then $R_s = 16.7\ \Omega$, $\phi_s = 89°$, $T_s = 0.18$ s and $T_p/(T_s - T_p) = 0.5$. Thus, in [1.19], $\cos\phi_s$ can be put equal to zero, and the steady-state component of i_e is

$$\text{steady-state } i_e \simeq -\frac{I'_{pm}R_s \sin\omega t}{(R_s^2 + \omega^2 L_e^2)^{\frac{1}{2}}}$$

$$\simeq -\frac{I'_{pm}R_s \sin\omega t}{\omega L_e} \qquad [1.20]$$

the peak value of which is about $0.0177\ I'_{pm}$.

The transient component of the exciting current reaches a maximum value when

$$t = \frac{T_s T_p}{T_s - T_p} \ln \frac{T_s}{T_p} \qquad [1.21]$$

and the maximum value is

$$I'_{pm} \exp\left(-t/T_s\right) \qquad [1.22]$$

where t is given by [1.21]. Fig. 1.16 shows plots of the transient components of the exciting current. The student should consider the case of $T_s = T_p$.

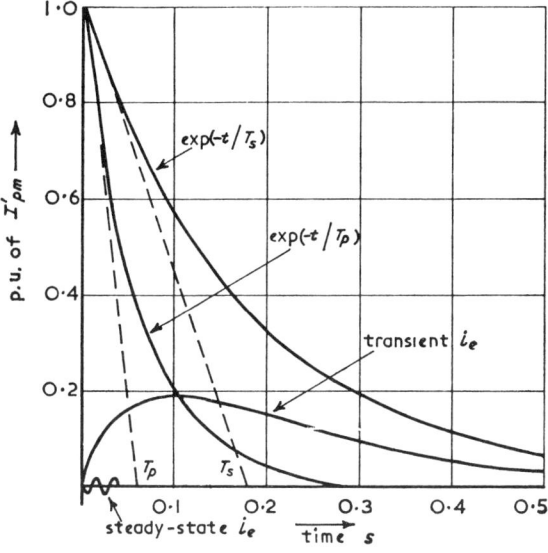

Fig. 1.16. Transient currents in a C.T. secondary.

The secondary current is given by $i_s = i'_p - i_e$ and can be obtained by subtracting [1.19] from [1.17], or more directly by eliminating i_e from the differential equation at the beginning of the above analysis. The flux waveshape can be obtained by integrating $i_s R_s$, and should be compared with the waveshape of the exciting current i_e.

For further information on the transient analysis of circuits containing C.T.s, the student is referred to the references: e.g. Wright, A., *Current Transformers*. A comparison of the above theoretical analysis and practical tests can be summarised as follows. A constant shunt resistance correctly simulates the eddy-current iron loss but not so for the hysteresis loss. This resistance parallels the C.T. burden to reduce the effective value of R_s and hence also the transient flux. Iron losses tend to reduce the secondary time constant. Any inductance in the burden, the leads and the C.T. secondary is negligible compared with the shunt exciting inductance L_e, and so has little effect on the secondary time constant and transient flux.

The effect on the secondary current of saturation of the iron core is fully discussed by the reference above.

REFERENCES

ANIL KUMAR, N. M. 1970. Composite instantaneous comparators: basis for design and their transient performance. *Proc. I.E.E.* **117**, 147–156.

ELECTRICITY COUNCIL (Ed.) 1969. *Power System Protection*. 3 vols. Macdonald.

GRAY, C. B. & MUSTAPHI, K. K. 1969. Transient response of networks employed in protective relaying. *Proc. I.E.E.* **116**, 1349–1355.

HAMILTON, F. L. & PATRICKSON, J. B. 1967. Application of transistor techniques to relays and protection for power systems. *Proc. I.E.E.* **114**, 213–227.

HOEL, H. *et al.* 1966. Composite polar characteristics of multizone systems of phase-comparison distance protection. *Proc. I.E.E.* **113**, 1631–1642.

HUMPAGE, W. D. & KANDIL, M. S. 1968. Discriminative performance of distance protection under fault operating conditions. *Proc. I.E.E.* **115**, 141–152.

HUMPAGE, W. D. & KANDIL, M. S. 1970. Measuring accuracy of distance protection with particular reference to earth-fault conditions on 400 kV looped-circuit interconnections. *Proc. I.E.E.* **117**, 431–438.

HUMPAGE, W. D. & KANDIL, M. S. 1970. Distance-protection performance under conditions of single-circuit working in double-circuit transmission lines. *Proc. I.E.E.* **117**, 766–770.

HUMPAGE, W. D. & LEWIS, D. W. 1967. Distance protection of teed feeders. *Proc. I.E.E.* **114**, 1483–1498.

INSTITUTION OF ELECTRICAL ENGINEERS. 1968. Some present-day protection problems. *Colloquium digest*. No. 1968/19.

JACKSON, L. *et al.* 1968. Distance protection: optimum dynamic design of static relay comparators. *Proc. I.E.E.* **115**, 280–287.

KHINCHA, H. P. *et al.* 1970. Developments in amplitude-comparator techniques for distance relays. *Proc. I.E.E.* **117**, 1118–1124.

KHINCHA, H. P. *et al.* 1970. New possibilities in amplitude- and phase-comparison techniques for distance relays. *Proc. I.E.E.* **117**, 2133–2141.

MATHEWS, P. & NELLIST, B. D. 1963. Transients in distance protection. *Proc. I.E.E.* **110**, 407–418.

WARRINGTON, A. R. C. Vol. 1. 1968; Vol. 2. 1969. *Protection Relays*. Chapman & Hall.

WELLMAN, F. E. 1968. *The Protective Gear Handbook*. Pitman.

WHEELER, S. A. 1970. Influence of mutual coupling between parallel circuits on the setting of distance protection. *Proc. I.E.E.* **117**, 439–445.

WRIGHT, A. 1968. *Current Transformers*. Chapman & Hall.

Examples

1. A power station at A contains six identical generator-transformer sets. The generators are rated at 11 kV, 120 MW, 150 MVA, and $X_1' = X_2 = 18\%$ (on rating). The transformers are rated at 150 MVA, 11/132 kV, delta/solidly-earthed star-connected, with all star points earthed and $X = 12\%$ (on rating). The 132-kV line AB is 96 km long and $X_1 = 0.44\Omega$/phase-km. Assume the system is running on no-load at rated voltage and neglect system resistance. The plain impedance relays at A require a minimum of 3% of nominal voltage for satisfactory operation. For 3-phase faults, calculate the minimum distance of the fault from the relay for satisfactory operation of the relay for (*a*) 1 set, (*b*) 6 sets running.

(L.P.) (1·54, 0·413 km)

2. All six sets in the power station at A are running (see example 1). The stage 1 distance relay at A is now a (polarised) mho relay with a characteristic angle of 90° and is correctly set to reach 80% of the line AB. A small power station, having a source impedance of j158·6 Ω/phase at 132 kV, is now connected in at the midpoint of the line AB. Determine the location of a 3-phase fault which will just operate the relay and hence calculate the under-reach of the relay. Assume the source e.m.f.s are all equal to rated voltage and in phase.

(L.P.) (5·42%)

3. The stage 2 relay at A (see example 2) is correctly set to reach 50% into the line BC, which is 64 km long, when there is no infeed at B. If now a power station, identical to that at A, is connected in at B, and all 12 sets are running, calculate the under-reach of the relay at A. There is no infeed at the midpoint of AB.

(L.P.) (22·3%)

4. In a compensated phase-fault mho protection system each relay is supplied with the difference between two phase voltages and the corresponding difference between the two phase currents, i.e. the *a* relay measures $(\mathbf{V}_a - \mathbf{V}_b)/(\mathbf{I}_a - \mathbf{I}_b)$. The source impedance is negligible. For a phase-to-phase fault across phases *b*–*c* show that the impedances measured by the *a*, *b* and *c* relays are respectively $2\mathbf{Z}_1\underline{/-60°}$, $\mathbf{Z}_1$

and $2\mathbf{Z}_1\underline{/60^\circ}$ where $\mathbf{Z}_1 = R_1 + jX_1$ is the positive sequence impedance of the line from the relay to the fault. If the resistance of the line can now be neglected show (i) that if the relays each have a setting of X_1 and a characteristic angle of 90°, only the b relay will operate and (ii) that if the relay characteristic angles are each 30° and the settings such that the b relay is on the threshold of operation, then the a relay is liable also to operate.

(L.P.)

5. A line AB has a total impedance Z_L. The source impedances feeding in at A and B are both Z_G and the source e.m.f.s. can be assumed equal and in phase. To protect the line AB a distance relay is installed at A. If an earth-fault of resistance R occurs on the busbar at B, show that the impedance measured by the relay $(\mathbf{V}_{an}/\mathbf{I}_a)$ is given by

$$\mathbf{Z}_L + R(2 + (\mathbf{Z}_L/\mathbf{Z}_G))$$

Assume that both source star points are solidly earthed and that all plant has the same impedance to all three sequence currents.

If the relay is a reactance relay set to 80% of the total line reactance, and $\mathbf{Z}_L = 15\underline{/60^\circ}\ \Omega$ and $\mathbf{Z}_G = 6\underline{/80^\circ}\ \Omega$, show that the relay will trip for $R \geqq 3 \cdot 04\ \Omega$. Comment on this result.

(L.P.)

Chapter 2

CIRCUIT INTERRUPTION AND SWITCHING OVER-VOLTAGES

2.1 Introduction

Apart from normal load switching, circuit breakers are required in power systems to give rapid fault clearance, in order to avoid over-current damage to equipment and loss of system stability. The fault tripping signals to circuit breakers are derived from protective systems (see Chapter 9 of Volume 1 and Chapter 1 of this volume) via a final tripping relay connected to the breaker trip coil.

When any type of circuit breaker is manufactured it is necessary to check in a short-circuit testing station that (among other features) the breaker clears satisfactorily various fault currents. Fig. 2.1 (taken from B.S. 116) shows an oscillogram of an oil circuit breaker connected to a 3-phase supply, with a 3-phase short-circuit suddenly applied at its terminals. Following this instant, the closed contacts of the three phases of the breaker carry currents with equal alternating components but differing d.c. components as shown. Arc voltages are recorded as soon as the contacts separate, and when the arcs are extinguished they contain transient high-frequency components which cannot be properly observed in an oscillogram of long duration such as that of Fig. 2.1. These transient voltages are discussed extensively throughout this chapter, and the additional voltage rise which occurs in phase B of Fig. 2.1, due to the current in that phase being interrupted before that in the other phases, is discussed in section 2.5.

Among operating times which are significant in circuit breaking are:—

$$\text{clearing time} = \text{relay and protective system} + \text{breaker interrupting time;}$$

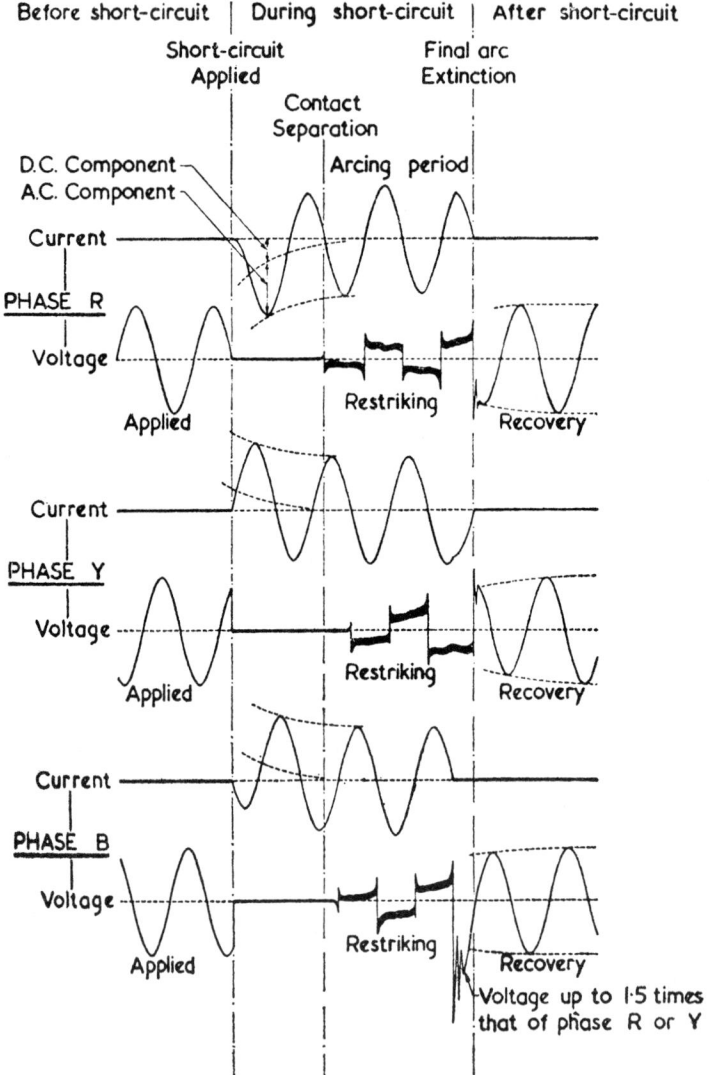

Fig. 2.1. Typical oscillogram of three-phase short-circuit.
(From B.S. 116).

relay and protective system time = time from fault inception to the closing of the breaker trip coil circuit;

breaker interrupting time = time for the current in the trip coil to rise to the value at which the moving contact is released + contact opening time + arcing time.

For electro-mechanical relays, the relay time can vary from about 20 ms to as much as about 2 s in some parts of existing systems. Static electronic relays, which use semiconductor devices such as transistors and thyristors in place of electromechanical elements, are now being used in power systems, and can give very fast operation.

The arcing time is now generally between one and two half-cycles, depending upon the instant in the current half-cycle at which the contacts part, but some oil circuit breakers still in service may arc for several half-cycles at some values of fault MVA. Contact opening may occupy between about one and three cycles. It is therefore possible, with fast relays and modern circuit breakers, to achieve fault clearance in as little as about three cycles of the 50-Hz current, but the time varies considerably from system to system and in some cases in different parts of one system.

If the fault were to cause damage to equipment such that the supply could not be at once restored it would be said to be persistent. The majority of faults in power systems, however, such as lightning flashover of overhead line insulators, or line conductors 'dancing' close together under extreme wind and ice conditions, set up an arc in air and are transient. After current interruption in such faults, the arc path is rapidly deionised, and the circuit can be automatically reclosed after a small time interval, so that the interruption of supply is so short that it does not affect the majority of loads (e.g. lights dip only slightly and motors continue to run). The reclosing time, which is the interval from the energisation of the trip coil to the subsequent reclosing of the circuit breaker contacts, could be as little as twelve cycles in high-voltage systems, but is now generally much longer (see section 1.4). The breaker can be set so that if the fault has not been cleared by the time it recloses, then it may remain open, or it may again reclose a second or more times before finally locking open if the fault is persistent. The uses of auto-reclosing circuit breakers are dealt with in section 1.4.

2.2 Circuit breaker rating

A circuit breaker will be able to make and break currents between 10 and 100 times the normal current, which is the load current that it can carry continuously; e.g. breaker capacities now reach 35 000 MVA while generator ratings are just advancing past 500 MVA. The breaking capacity in MVA = $\sqrt{3} \times$ rated line voltage (kV) $\times$ maximum rated breaking current (kA, r.m.s.). In British ratings this interrupting current is the r.m.s. value of the alternating component only at the instant of contact separation, while in American ratings it is the r.m.s. value of a.c. and d.c. components at that instant (see Fig. 2.1). Thus for a given breaker the American rating would be higher than the British rating. For a 3-phase breaker which is interrupting current in an inductive circuit, there must be a d.c. component of current in two and probably all three of the phase contacts, and then its value at interruption will depend upon the instant in the cycle at which the fault occurs, the circuit time constant and the breaker speed. Proving tests on breakers take account of this, e.g. B.S. 116 requires oil circuit breakers to undergo tests where the d.c. component of current at the moment of contact separation is not less than 50% of the a.c. component in one of the phases. Similarly when a circuit breaker closes there will be a d.c. transient current and the peak current in one phase may reach $2.55 \times$ the r.m.s. value of symmetrical current if the power factor is 0.15 or less. The factor of 2.55 is 10% below the theoretical peak current of $2\sqrt{2}$ given by a d.c. component equal to the peak of the alternating component, because the d.c. component has fallen from its initial value due to losses by the time the alternating component reaches its peak value. Some discussion of the transient d.c. component of current and the changing a.c. component of current flowing when a fault occurs in a system appears in Chapter 7 of Volume 1.

2.3 Arc extinction and reignition

As the moving contact parts from the fixed contact, an arc is formed between them. In certain kinds of circuit interruption devices such as fuses or d.c. circuit breakers (which are not dealt with here), the current is forced suddenly to zero (or chopped). In an a.c. circuit breaker, current chopping, which could cause severe over-voltages in the system, is avoided as far as possible. The breaker is designed to prevent reignition of the arc at the first or earliest possible subsequent power-frequency current zero, after the contacts have parted and initiated the arc.

An arc has anode and cathode fall regions extremely close to the contacts, and electrons emitted from a non-refractory cathode (by a mechanism as yet imperfectly understood), gain sufficient energy in the very high electric field (which may be of order 10^7 V/cm) in the cathode fall, to ionise neutral or excited vapour and gas atoms at the end of that region. These fall regions and the neighbouring transition regions are connected by a column which has a relatively low electric field in it which is sufficient to make good the losses of charged particles and heat from it. Arc column voltage gradient is increased from about 10 V/cm for an unconfined arc, to more than 100 V/cm for an arc subjected to severe convection and conduction, as occurs due to the supersonic flow of gas in the nozzle of a gas-blast breaker. Conditions are somewhat similar in an oil circuit breaker because the high arc temperatures rapidly vaporise the oil, and arc extinction takes place in gases such as methane, acetylene and hydrogen, and, except for very small ratings, these breakers have an arcing chamber, so shaped as to produce a high-pressure blast of gas around the arc. As a result of the high rate of convection at the arc periphery, the arc becomes much narrower, so that though the core temperature is much increased (up to about 50 000 K), the region of high electrical conductivity is reduced in size, and the column voltage gradient is increased by about an order of magnitude. Satisfactory interruption is achieved if, at a natural (i.e. power-frequency) current zero, electrons and positive ions are removed from the gas in the contact gap sufficiently quickly for the circuit voltage which then appears between the contacts to be unable to reignite the arc; i.e. rapid deionisation must take place. This is achieved: (a) by volume recombination of positive ions and electrons in the regions where the temperature has been reduced by convection due to gas blast, and in some cases by conduction where the arc was forced against cooling surfaces, and (b) by diffusion and ejection by high pressure gas movement of ionised particles out of the gap between the contacts.

In breakers, therefore, a gas blast is generally used to reduce the temperatures of much of the region between the contacts, except in the centre of the arc core, to sweep ionised particles out of the gap, and in some cases to lengthen the arc and hold it against cooling surfaces. In some cases magnetic forces are used to drive the arc through the gas.

The critical conditions and processes within a few microseconds around current zero, where reignition may or may not be prevented, are as yet imperfectly understood. The design of circuit breakers, unlike that of most other system equipment, is not amenable to detailed calculation, and may perhaps be said to be as much an art as a science. It is necessary to carry out proving tests on breakers in conditions as near as possible to those encountered in service.

It was thought at one time that arc reignition always occurred by spark breakdown in the gas, and that there was a 'race' between the recovery of electric strength of the gas, and the restriking voltage appearing between the contacts. It is true that reignition occurring more than some hundreds of microseconds after current zero, due to a relatively low natural-frequency restriking voltage transient, is of the spark breakdown type, and that the rate of rise of restriking (or transient recovery) voltage (r.r.r.v.) is an important parameter in assessing the severity of a test or fault in service. Spark breakdown is an electric field effect produced by application of a high voltage to a gap which contains essentially no free electric charge. Since after current zero the temperature is above and the gas density is below the ambient conditions, spark breakdown may occur below the normal withstand voltage for the contact gap. For this form of reignition to occur, however, the gap must first recover to an insulating condition from the highly-conductive arc conditions.

During the early post-arc or post-zero period, the gas in the gap has in fact a finite and varying resistance, and the restriking voltage can produce an oscillatory post-zero current, which may be as high as 20 A in some breakers, and may last for 5 to 100 μs, so that appreciable energy is supplied to the gas. This may cause thermal reignition to occur during the first tens of microseconds after current zero. There is, therefore, an energy balance to be considered for the arc column region. The total energy per unit volume $\varepsilon = \varepsilon_1 + \varepsilon_2$, where ε_1 is the internal energy (thermal, ionisation, excitation, dissociation etc.) and ε_2 is the directed kinetic energy due to the electric field. Each unit volume has an electrical power input p given by the product of post-zero current density and the column region voltage gradient, and a power loss p_L which includes convection (kinetic energy loss), conduction, radiation and enthalpy. The transient energy equation is then $\partial \varepsilon / \partial t = p - p_L$, so that to avoid thermal reignition p must be kept small, i.e. the voltage across the gap should be reduced as much as possible, and p_L must be increased by the means outlined above.

The reignition of an arc, either by thermal or spark breakdown process, therefore, depends greatly upon the voltage between the contacts, and this has two components:—

(1) recovery voltage at supply frequency;

(2) restriking voltage transient of one or more natural circuit frequencies decaying exponentially about the recovery voltage. There is a strong movement to use the expression 'transient recovery voltage' for this component, but since all specifications at present use 'restriking voltage', the latter term is also used in this book.

The restriking voltage transient depends upon the circuit breaker as well as upon the circuit, due to finite inter-contact conductivity and post-zero current, so that the voltage transient and gap deionisation are not independent, and it is a great over-simplification to regard circuit breaking as a race between the rising voltage and the dielectric recovery of the gap, and to regard the rate of rise of restriking voltage as a complete measure of 'circuit severity'. A 'circuit severity' criterion would need to take into account many factors, e.g. the way in which at a circuit breaker location, the r.r.r.v. falls with increasing percentage of rated MVA, and at each location this curve will differ, and this curve should include special fault conditions such as the short-line fault which is discussed later in this chapter.

2.4 Balanced 3-phase fault at the circuit breaker terminals:— transient recovery voltage

If the fault is a balanced 3-phase fault to earth, then only one phase need normally be considered, although in fact interphase reactions cannot always be neglected in switching phenomena. Fig. 2.2 shows the conditions just after current zero in the contacts of the breaker in one phase. The yellow line terminal is marked Y and the earth is E. A fault is shown occurring at or very near the breaker terminals. G consists largely of post-zero conductance in the circuit breaker, together with any circuit shunt conductance.

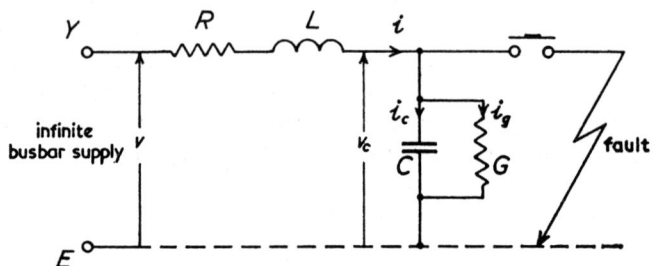

Fig. 2.2. One phase of circuit with a three-phase fault at, or very near, a circuit breaker.

The following assumptions are made:—

(a) The voltage source is equivalent to an infinite bus-bar at YE, maintaining a phase voltage $v = V_m \cos \omega t$ of constant amplitude and frequency.

In practice, though the frequency may be regarded as constant for a fault at a point in a large power system, there will be a fall in the voltage.

(b) R, L, C and G are independent of frequency, current, voltage, time, etc.

This involves, among other assumptions, neglecting the variation of circuit breaker capacitance and conductance after current-zero.

(c) The arc voltage is negligibly small compared with the circuit voltage.

(d) The fault current is given by

$$i_f = \frac{V_m}{Z} \cos (\omega t - \phi)$$

where $Z = \sqrt{(R^2 + \omega^2 L^2)}$ and

$$\phi = \tan^{-1} \frac{\omega L}{R};$$

$\omega = 2\pi f$; $f =$ supply frequency.

Any d.c. transient component of current is thus neglected. Then after i_f has been interrupted

$$v = V_m \cos \omega t = Ri + L\frac{di}{dt} + v_c \qquad [2.1]$$

$$v_c = \frac{1}{C}\int i_c \, dt = \frac{i_g}{G} \qquad [2.2]$$

$$i = i_c + i_g \qquad [2.3]$$

$$i_g = Gv_c \text{ and } i_c = C\frac{dv_c}{dt}$$

Substituting in [2.3] and multiplying by R

$$Ri = RGv_c + RC\frac{dv_c}{dt} \qquad [2.4]$$

and

$$L\frac{di}{dt} = L\left(G\frac{dv_c}{dt} + C\frac{d^2v_c}{dt^2}\right) \qquad [2.5]$$

Substituting [2.4] and [2.5] into [2.1] and dividing both sides by LC

$$\frac{V_m}{LC}\cos \omega t = \frac{d^2v_c}{dt^2} + \left(\frac{R}{L} + \frac{G}{C}\right)\frac{dv_c}{dt} + \frac{1}{LC}(1+RG)v_c \qquad [2.6]$$

When a fault occurs in a power system, the load is by-passed, and the only resistance is that of supply equipment such as generators, transformers, lines, etc., so that $R \ll \omega L$ and can be neglected for some purposes in comparison with ωL.

Putting $R = 0$ in [2.6] gives

$$\frac{V_m}{LC}\cos \omega t = \frac{d^2v_c}{dt^2} + \frac{G}{C}\frac{dv_c}{dt} + \frac{1}{LC}v_c \qquad [2.7]$$

Taking the Laplace transform (see Volume 1 of 'Circuit Theory' by J. O. Scanlan and R. Levy, published in this series) of the right-hand side with $v_c = 0$ at $t = 0$, gives

$$\mathscr{L}(v_c)\left(s^2 + \frac{G}{C}s + \frac{1}{LC}\right)$$

where $\mathscr{L}(v_c)$ is the Laplace transform of v_c, and s is the complex frequency. If the roots of $s^2 + sG/C + 1/LC = 0$ are 'imaginary', i.e.

$G^2/C^2 < 4/LC$, there will be an oscillation in the voltage v_c across the breaker contacts, at a natural frequency of

$$f_n = \frac{1}{2\pi}\sqrt{\left(\frac{1}{LC}-\left(\frac{G}{2C}\right)^2\right)} \text{ Hz} \qquad [2.8]$$

This oscillatory voltage, which is a single-frequency restriking voltage transient, occurs about the recovery voltage which is at supply frequency of $f = \omega/2\pi$ Hz.

If both G and R were to be neglected, then the fault current i_f would lag $\pi/2$ behind the supply voltage v, and when the breaker cleared the fault at a 50-Hz zero, the 50-Hz voltage component across the breaker contacts would rise instantaneously from zero (neglecting arc voltage) to V_m, the peak phase voltage. Since in most power circuits $f_n \gg f$, then for a few cycles at natural frequency the recovery voltage may be assumed to remain constant at V_m.

[2.1] then reduces on differentiation to

$$0 = L\frac{d^2i}{dt^2}+\frac{i}{C}$$

and taking the Laplace transform gives

$$\mathcal{L}(i)\left(s^2+\frac{1}{LC}\right) = 0$$

where $\mathcal{L}(i)$ is the Laplace transform of i.

The roots of $s^2+1/LC = 0$ are

$$s = \pm\,j\sqrt{\left(\frac{1}{LC}\right)}$$

and

$$i = A\exp j\sqrt{\left(\frac{1}{LC}\right)}t + B\exp -j\sqrt{\left(\frac{1}{LC}\right)}t$$

at $t = 0$, $i = 0$ since C is then by-passed by an arc of infinite conductance, therefore $A = -B$ and

$$i = A\left(\exp j\sqrt{\left(\frac{1}{LC}\right)}t - \exp -j\sqrt{\left(\frac{1}{LC}\right)}t\right)$$

$$= 2jA\sin\sqrt{\left(\frac{1}{LC}\right)}t \qquad [2.9]$$

From [2.9]

$$L\frac{\mathrm{d}i}{\mathrm{d}t} = 2jAL\sqrt{\left(\frac{1}{LC}\right)}\cos\sqrt{\left(\frac{1}{LC}\right)}t$$

$$= V_m \text{ at } t = 0 \qquad\qquad \text{(from [2.1])}$$

Substituting the value of A thus determined into [2.9] gives

$$i = V_m\sqrt{\left(\frac{C}{L}\right)}\sin\sqrt{\left(\frac{1}{LC}\right)}t$$

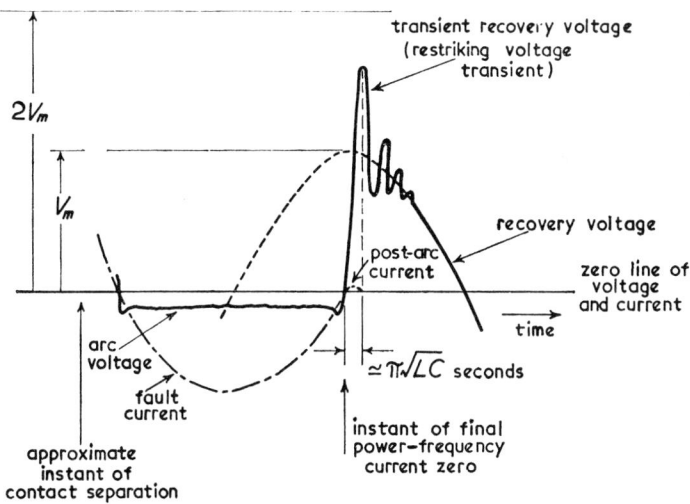

Fig. 2.3. Single-frequency transient recovery voltage.
(Restriking voltage transient).

The voltage across the circuit breaker (and capacitance C) is

$$v_c = \frac{1}{C}\int i\,\mathrm{d}t = -V_m\cos\sqrt{\left(\frac{1}{LC}\right)}t + K$$

and at $t = 0$, $v_c = 0$, neglecting arc voltage drop, so that $K = V_m$
and

$$v_c = V_m\left(1-\cos\sqrt{\left(\frac{1}{LC}\right)}t\right) \qquad\qquad [2.10]$$

[2.10] shows that at $t = \pi\sqrt{(LC)}$ seconds, i.e. after $\frac{1}{2}$ cycle at natural frequency, the voltage across the circuit breaker would reach $2V_m$ if there were no damping. For a fault at or near the breaker there is a single-frequency restriking voltage transient as illustrated in Fig. 2.3. It can be seen that this Figure refers to a circuit in which R and G are not negligibly small, since the oscillation in voltage is fairly rapidly damped out.

Interruption of current in a circuit becomes progressively easier as the $R/\omega L$ ratio of the circuit increases; e.g. if $R \gg \omega L$ then at current zero the voltage would increase from zero as a normal power frequency sine wave without oscillation, and the rate of rise of voltage would then be very small compared with that which occurs when $R \ll \omega L$ as in most fault conditions.

2.5 First phase to clear

Since the three line currents have a phase displacement of $2\pi/3$ radians relative to one another, the contacts in one phase of a 3-phase breaker must interrupt the current at a zero, while the other two sets of phase contacts are still carrying current. If either the star point of the supply is not earthed or if the 3-phase fault is not to earth as in Fig. 2.4(a), then the voltage across the contacts of the first phase to clear rises by a further 50% during the time that the other two phases are still carrying current.

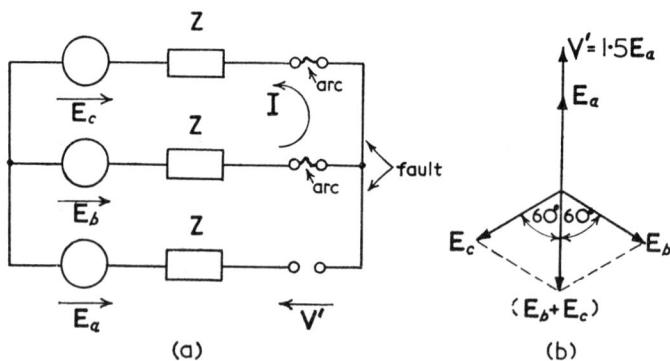

Fig. 2.4. Circuit and phasor diagram for the first phase to clear with no path via earth.

All three phases are assumed to contain equal impedances Z from the fault back to the point of generation, where balanced e.m.f.s E_a, E_b and E_c are generated. (*abc* is used for convenience in place of RYB as denoting the three phases in the order of phase rotation.) The voltage V' across the breaker contacts in phase *a* which is the first one to clear is given by

$$V' = E_a - (E_b - IZ)$$

and

$$I = \frac{E_b - E_c}{2Z}$$

$$V' = E_a - E_b + \left(\frac{E_b}{2Z} - \frac{E_c}{2Z}\right)Z$$

$$= E_a - \frac{(E_b + E_c)}{2}$$

and

$$\frac{(E_b + E_c)}{2} = -\frac{E_a}{2} \qquad \text{(see Fig. 2.4(b))}$$

$$V' = 1 \cdot 5\, E_a \quad \text{where} \quad E_a = V_m/\sqrt{2}$$

Both oscillatory and power-frequency components of the voltage across the first phase to clear, for a 3-phase fault in a circuit without a complete earth path, rise to values 50% greater than in the other two phases until current is interrupted in those phases.

It should be noted than an unbalanced case such as this one should be dealt with by the method of symmetrical components, which will show the conditions involved. It has been shown in Chapter 8 of Volume 1, that when there is an open circuit in one phase (say the *a* phase), the sequence networks may be connected to represent this as shown in Fig. 2.5, where Z_1', Z_2' and Z_0' are the sequence impedances on the supply side of the open circuit and Z_1'', Z_2'' and Z_0'' are on the load or fault side. If the fault (or load) has no connection to earth and/or the supply star point is unearthed as considered above, then Z_0'' or Z_0' respectively is infinity, and $I_0 = 0$. If there is only static equipment on the fault or load side of the open circuit at the circuit breaker, then $Z_1'' = Z_2''$. If the breaker is sufficiently remote from the supply alternator(s) for the difference between the positive- and negative-sequence impedances of the machines to be swamped by that of transformers and lines, then $Z_1' \simeq Z_2'$, so that if $Z_2 = Z_2' + Z_2''$ and $Z_1 = Z_1' + Z_1''$, then $Z_1 \simeq Z_2$. In this case, and only in this case, Fig. 2.5 shows that

$$V_1 = E_a Z_2 / (Z_1 + Z_2) = \frac{E_a}{2}$$

The voltage V' across the first phase to clear is given by

$$V' = V_1 + V_2 + V_0 = 3V_1 \simeq 1 \cdot 5 E_a$$

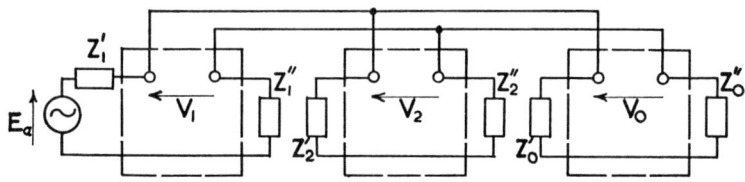

Fig. 2.5 Connection of sequence networks for first phase to clear.

For a few cycles after a fault occurs the positive- and negative-sequence reactances of a synchronous machine will not differ greatly, since the latter is approximately the average of the positive-sequence subtransient direct- and quadrature-axis reactances (see Chapter 7 of Volume 1), so that a voltage rise of about 50% will occur with a fast relay and breaker even for a fault near the alternator.

2.6 Multi-frequency transient recovery voltage

If a 3-phase fault, instead of occurring at or very near the busbar to which the circuit breaker is connected, occurs at a more remote point from the breaker which must isolate it from the rest of the system, then the restriking voltage transient will generally contain more than one natural frequency. The simplest case is where there are two oscillatory circuits, and where R and G are negligible in both circuits, and this is illustrated in Fig. 2.6.

The oscillatory current in the fault-side oscillatory circuit may be written from [2.9] as

$$i_2 = 2jA_2 \sin \sqrt{\left(\frac{1}{L_2 C_2}\right)} t$$

$$= 2jA_2 \sin \omega_2 t \qquad [2.11]$$

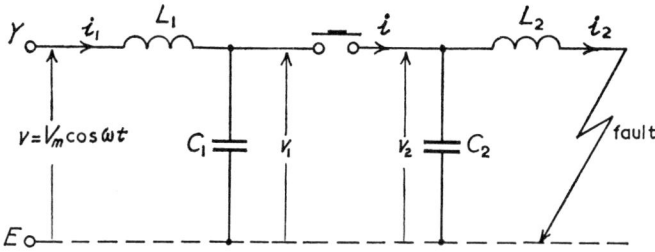

Fig. 2.6. One phase of circuit with a three-phase fault remote from the circuit breaker.

At $t = 0$, the 50-Hz current i has just become zero. Just prior to this, the currents i_1 and i_2 have been approximately equal to i if $1/\omega C_1$ and $1/\omega C_2 \gg \omega L_2$ where $\omega = 2\pi f$ and f is again the supply frequency. Thus under these conditions

$$v_2 \simeq \frac{vL_2}{L_1 + L_2}$$

and at $t = 0$

$$(v_2)_{t=0} = L_2 \left(\frac{di_2}{dt}\right)_{t=0} = \frac{V_m L_2}{L_1 + L_2}$$

From [2.11]

$$2jA_2 \omega_2 = \frac{V_m}{L_1 + L_2}$$

so that

$$i_2 = \frac{V_m}{\omega_2 (L_1 + L_2)} \sin \omega_2 t$$

and

$$v_2 = L_2 \frac{di_2}{dt} = \frac{V_m L_2}{L_1 + L_2} \cos \omega_2 t \qquad [2.12]$$

The oscillatory current in the supply-side circuit is given by

$$i_1 = 2jA_1 \sin \omega_1 t$$

where

$$\omega_1 = \sqrt{\left(\frac{1}{L_1 C_1}\right)}$$

$$L_1 \frac{di_1}{dt} = 2jA_1 \omega_1 L_1 \cos \omega_1 t$$

and with the same approximation as above

$$L_1 \left(\frac{di_1}{dt}\right)_{t=0} = \frac{V_m L_1}{L_1 + L_2}$$

so that

$$2jA_1 = \frac{V_m}{\omega_1 (L_1 + L_2)}$$

and

$$i_1 = \frac{V_m}{\omega_1 (L_1 + L_2)} \sin \omega_1 t$$

$$v_1 = \frac{1}{C_1} \int i_1 \, dt$$

$$= -\frac{V_m}{\omega_1^2 C_1 (L_1 + L_2)} \cos \omega_1 t + K$$

at $t = 0$

$$v_1 = v_2 = \frac{V_m L_2}{(L_1 + L_2)}$$

$$K = \frac{V_m}{(L_1 + L_2)} \left(L_2 + \frac{1}{\omega_1^2 C_1}\right)$$

and substituting for ω_1^2

$$K = V_m$$

$$v_1 = V_m \left(1 - \frac{L_1 \cos \omega_1 t}{(L_1 + L_2)}\right)$$

This is only the natural-frequency part of v_1, i.e. it gives an oscillation about V_m instead of about $V_m \cos \omega t$. Including the power-frequency component and making the approximation that $1/\omega C_1 \gg \omega L_1$, gives

$$v_1 = V_m \left(\cos \omega t - \left(\frac{L_1}{(L_1 + L_2)} \right) \cos \omega_1 t \right)$$

[2.13]

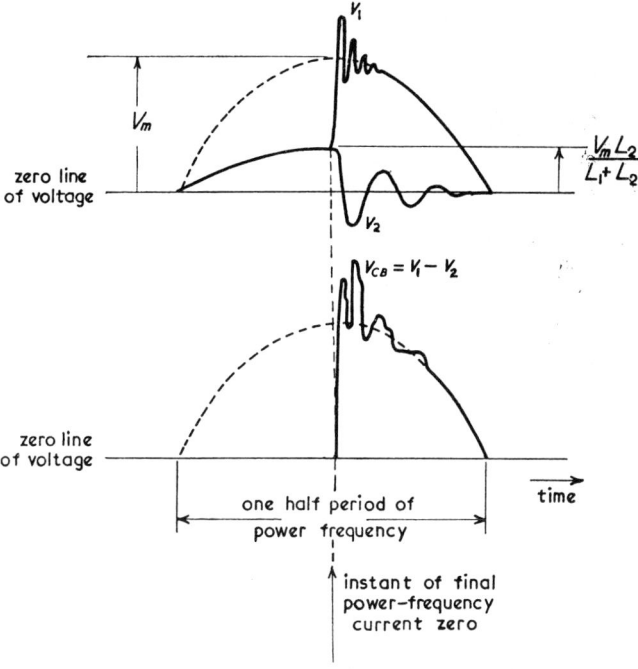

Fig. 2.7. Multi-frequency transient recovery voltage (re-striking voltage transient).

The voltage across the circuit breaker contacts $v_{CB} = v_1 - v_2$ and from [2.12] and [2.13]

$$v_{CB} = V_m \left(\cos \omega t - \left(\frac{L_1}{(L_1 + L_2)} \right) \cos \omega_1 t - \left(\frac{L_2}{(L_1 + L_2)} \right) \cos \omega_2 t \right)$$

[2.14]

Each oscillatory circuit will have some resistance R_1 and R_2

respectively, giving time constants $T_1 = L_1/R_1$ and $T_2 = L_2/R_2$ so that

$$v_{CB} \simeq V_m \left(\cos \omega t - \left(\frac{L_1}{(L_1 + L_2)} \exp(-t/T_1) \right) \cos \omega_1 t \right.$$

$$\left. - \left(\frac{L_2}{(L_1 + L_2)} \exp(-t/T_2) \right) \cos \omega_2 t \right) \qquad [2.15]$$

Fig. 2.7 shows waveforms of v_1, v_2 and $v_{CB} = v_1 - v_2$ appearing just after the final current zero in the breaker. Multi-frequency transients can give a great variety of waveshapes to the restriking voltage beside the one illustrated in Fig. 2.7. Most transients are multi-frequency rather than single-frequency, and the majority of them consist of two frequencies with considerable separation, i.e. they are of the type illustrated here.

2.7 Resistance switching

The rate of rise of restriking voltage can be very considerably reduced by resistance switching. In one arrangement the main gap of an air-blast breaker is fed with fresh un-ionised gas, and at the first current zero after the contacts part, both main and auxiliary gaps have the same voltage across them, but the auxiliary gap receives partially ionised gas. The auxiliary gap is designed so that it breaks

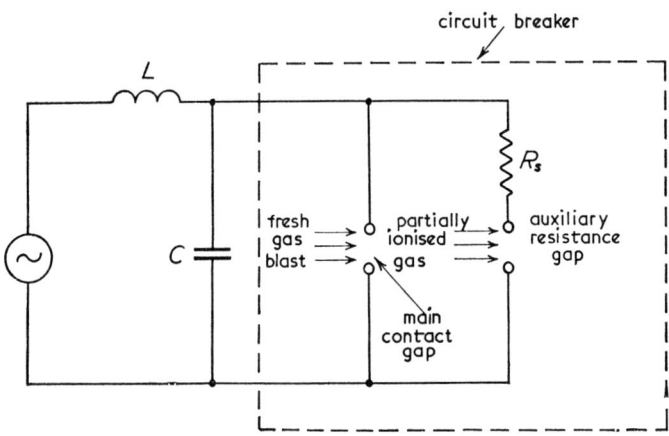

Fig. 2.8. Diagrammatic representation of one method of resistance switching.

down instead of the main gap, thus inserting a resistor R_s into the circuit, across the main gap and in parallel with the circuit capacitance. This is shown diagrammatically in Fig. 2.8. In another arrangement, now in common use, the resistor is switched by a separate contact and nozzle, e.g. 400-kV breakers have a number of resistor gaps in series per phase in addition to a number of main gaps.

It can be seen from [2.8] that if $G^2/C^2 = 4/LC$, the restriking voltage transient is critically damped. Thus if

$$R_s \lesssim \tfrac{1}{2}\sqrt{\left(\frac{L}{C}\right)}$$

the voltage is critically or over-damped, and the r.r.r.v. is very considerably reduced. The current through the resistor and auxiliary arc gap is reduced by R_s, and its power factor is improved, so that interruption of this arc at the first current zero after it forms is relatively easy, as indicated at the end of section 2.4.

2.8 Rate of rise of restriking (or transient recovery) voltage

The rate of rise of restriking or transient recovery voltage, which for a single-frequency transient is taken as the average slope from zero voltage to the first peak of the restriking voltage curve, is to a limited extent a measure of the severity of a test or of duty in service. The higher the r.r.r.v. the less time for the arc products to deionise and the greater the chance of reignition. The r.r.r.v. is more important in practice with gas-blast breakers than with oil breakers, which can handle without difficulty values very much higher than those which they experience in the system. In the case of a multi-frequency transient such as that shown in Fig. 2.7 the r.r.r.v. is less definite, but may be defined empirically depending upon the waveshape (see Trencham, 1953). The concept of r.r.r.v. as a criterion of severity has its limitations, as there is no single curve of dielectric strength for a given breaker: this is modified by the particular circuit parameters.

The r.r.r.v. is affected by

(a) the value of the phase voltage at the final current zero, and this depends upon the power factor of the faulted circuit, and the drop of voltage due to supply-circuit impedance, any loss of speed of generators, and current wave asymmetry.

(b) value of the natural frequency f_n. In test plant it may be as high as 100 kHz, though it is generally rather less. In power systems it is generally between 0·5 and 20 kHz, except for short-line faults where it is of the order of 30 to 100 kHz.

(c) post-arc conductivity. Gap conductance contributes largely to the conductance G shown across C in Fig. 2.2, and the higher the value of G the more the damping of the oscillation and the longer its periodic time.

(d) resistance switching.

For the first phase to clear, there is the following limit which the r.r.r.v. cannot in fact approach closely because R and G are not zero, viz.

$$\text{r.r.r.v.} < \frac{3V_m}{\pi\sqrt{(LC)}} \text{ volts/s}$$

With some oil circuit breakers, the arcing time may be longer for a given r.r.r.v. at a relatively low MVA than when interrupting the rated maximum interrupting capacity. In oil circuit breakers the extinguishing action depends upon the generation of gas by the arc and is therefore sensitive to the value of current. It is important to test at values both below and at the maximum rating. Values of 5, 10, 30, 60 and 100% of rated MVA have commonly been used. It may be noted that other current-interruption devices such as fuses and magnetic blow-out breakers can also experience relatively greater difficulty with low over-currents than with their full rated capacity, though the reasons differ from one device to another. Air-blast breakers are generally able to interrupt the current in one loop or one small loop plus the following half-cycle of arcing over their whole range of MVA, and modern oil circuit breakers can achieve or approach this. (The term loop is used to denote current flow in one direction which has been distorted from a half-cycle by d.c. asymmetry, see Fig. 2.1.)

The over-voltages which can be produced upon opening or closing a switch, may be particularly severe in certain cases. With E.H.V. transmission systems, lightning is less of a problem than it is with the lower voltage levels, because lightning surges rarely reach the impulse withstand voltage of the system equipment, e.g. 400-kV circuit breakers are impulse tested with a 1425-kV 1/50 wave (1 μs wavefront to peak voltage and 50 μs to fall to 50% pf peak voltage). In E.H.V. systems, switching surges thus become relatively more important. Also the breakdown voltage of air gaps and of some equipment insulation (e.g. suspension insulator strings and post insulators) is lower for a surge wavefront of about one to two hundred microseconds, than for shorter or longer wavefronts, and switching surge wavefronts can lie in this critical region. This is discussed in Chapter 5. The main basic causes of severe over-voltages are now discussed, though clearly far more involved cases can occur in practice which are beyond the scope of this book (see Thomas, 1966).

2.9 Chopping magnetising current

The magnetising current of a transformer is very low compared with its full-load current, and is therefore exceedingly small compared with the maximum breaker interrupting current. In an air-blast breaker, the intense deionising action designed to deal with the latter may result in the magnetising current being forced to zero ahead of a normal power frequency zero. This is known as current chopping. Oil circuit breakers are less prone to chopping since the interruption is more 'soft' because the gas pressure and turbulence depend upon the arc current being interrupted.

In the circuit of Fig. 2.6, L_2 and C_2 can represent the transformer magnetising reactance and capacitance, and $L_2 \gg L_1$. There will be cyclic transfer of energy between the electromagnetic and electric fields, if current is suddenly 'chopped' to zero at a value i_c, which may be of the order of 20 A for gas-blast breakers. If the voltage v_2 across C_2 at the instant when the current is chopped has a value v_c, then the total energy is $v_c^2 C_2/2 + i_c^2 L_2/2$. The maximum value of the voltage v_2 will occur when all the energy is stored in the electric field in the capacitance and this energy will be $\hat{v}_2 C_2^2/2$ where

$$\hat{v}_2 = (v_c^2 + i_c^2 L_2/C_2)^{\frac{1}{2}} \qquad [2.16]$$

For a transformer winding L_2 is of the order of a henry and C_2 is of the order of $10^{-3}\mu$F. For a 60-MVA, 132-kV, transformer with a peak magnetising current of 18 A and $\sqrt{(L_2/C_2)} = 68\,000$ ohms, then, the additional voltage rise due to chopping would be about 1200 kV if the current were chopped at its peak value. This is beyond the withstand voltage of the system equipment which is 550 kV 1/50 wave for an earthed system (see Chapter 5). The circuit breaker must therefore be designed so that the contact gap breaks down at voltages much less than this, i.e. at voltages below the insulation level of the system. A relaxation oscillation occurs (see Fig. 2.9) as the gap lengthens with the contacts parting, until the arc is finally extinguished near normal 50-Hz zero, where the voltage rise is reasonable. Resistance switching, where linear or

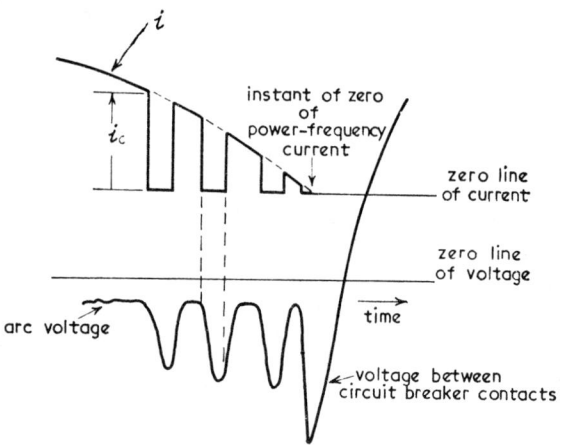

Fig. 2.9. Current chopping.

non-linear resistors are connected across the main gap by the arc itself, reduces these over-voltages (see section 2.7). Where surge diverters (see Chapter 5) are fitted at the transformers, the problem is less severe.

2.10 Interrupting the charging current of a long unloaded line with a breaker near the supply source

The circuit breaker of which one phase is shown in Fig. 2.10 is interrupting the current in a long unloaded line having a total capacitance C_2 where $C_2 \gg C_1$. When the current reaches a normal power-frequency zero, the arc is extinguished. The voltage between the breaker contacts increases from zero only very gradually, and

v_{BE} may decay only very slowly from V_m as the line capacitance gradually discharges (though if resistance switching is employed a considerable and rapid fall of voltage may take place, before the remaining voltage decays slowly, as energy from the electrostatic field is dissipated in the resistor and auxiliary arc). The final current zero could occur at a relatively small contact gap as the voltage rises relatively slowly after the zero, and a restrike may occur when this voltage has become large enough. If this restrike occurs after one half-cycle then v_{AB} is almost $2V_m$, as is shown in Fig. 2.11. When

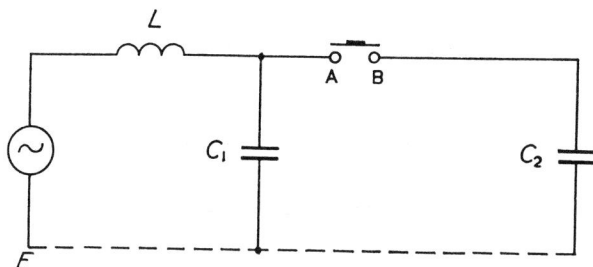

Fig. 2.10. Simplified equivalent circuit for the interruption of
the charging current of a long unloaded line.

A and B are joined by a power-frequency arc, the voltage of which is much smaller than the circuit voltage, it may be assumed that in the steady state $v_{AB} = 0$, so that v_{BE} oscillates at natural frequency about the 50-Hz voltage v_{AE}. If there were no resistance in any part of the circuit v_{BE} would overshoot to $3V_m$.

At the instant t_1, when the voltage between the contacts is at its maximum value of more than V_m, the oscillatory current passes through zero, and there are two possibilities:—

(a) a power-follow-through arc may take the path of the oscillatory current discharge. In this case extinction will again occur at a power-frequency current zero, and either the gap will by then be great enough to prevent reignition, or a repetition of the above events might occur.

or

(b) the arc may extinguish at the oscillatory current zero at t_1. If this happens the voltage v_{BE} across the line capacitance decays only slowly and one half-cycle later the voltage between the breaker contacts is again large: it would be $4V_m$ if the system had no resistances. A breakdown of the gap here could cause a further overshoot in the voltage v_{BE} as shown in Fig. 2.11, and it would be theoretically possible for extinction to occur at another zero of the oscillatory current at t_2, giving a continued sequence of voltage changes.

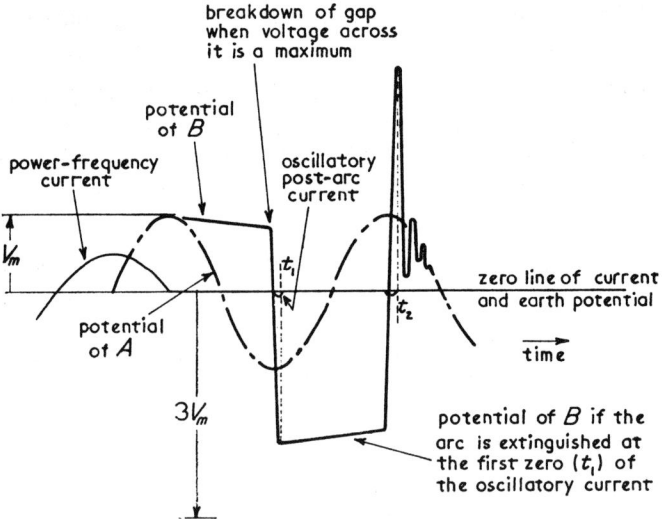

Fig. 2.11. Switching over-voltages on interrupting the charging current of a long unloaded line.

In fact damping and the increase in gap length, as the moving contact is driven rapidly away from the fixed contact, place limits on the over-voltages produced, but voltages up to about 3·5 times normal peak value have been measured in some systems due to this cause. Some breakers are designed so that only one delayed restrike can occur, and the peak voltage to earth is limited to 2·5 times normal phase voltage. The voltages at which equipment such as switchgear and transformers are tested (see Chapter 5) are high enough to allow such over-voltages on switching.

To reduce these over-voltages, many oil circuit breakers are fitted with resistance switching. Gas-blast breakers do not generally restrike when interrupting capacitor currents. Gas-blast breakers generally employ resistance switching in any case, because the fresh blast of cold un-ionised air has a much lower conductance (G in Fig. 2.2) than for the gases liberated by the high arc temperatures in an oil circuit breaker; i.e. G is generally higher in an oil breaker than in a gas-blast breaker, and thus gives some damping of the restriking voltage transient.

2.11 Interrupting the charging current of a long unloaded line with a breaker at some distance from the supply source

Switching can produce over-voltages due to surges being caused in transmission lines. A particular example of this is considered in this section and the short-line fault is considered in section 2.13. It is suggested that reference might be made to Chapter 4 before reading

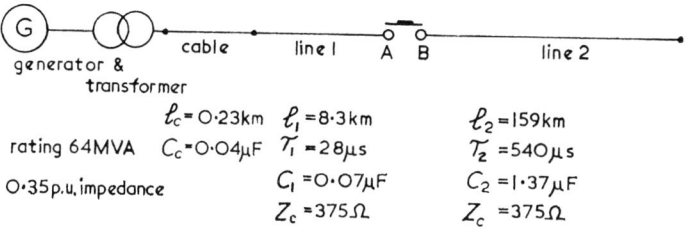

Fig. 2.12. Schematic diagram of 220-kV system with circuit breaker between two transmission lines.

this section. If a circuit breaker is situated several miles from the source of supply, as for example in the 220-kV system shown in Fig. 2.12, (see Gantenbein and Vogelsanger, 1956), the switching surges of up to $5 \times$ r.m.s. phase voltage (i.e. $3 \cdot 5 \times$ normal peak phase voltage) which can occur on opening an oil circuit breaker without resistance switching, are not caused by successive restrikes as considered in section 2.10. They are due instead to surges travelling along the transmission lines caused by the opening of the breaker, which in turn give rise to current chopping.

As in section 2.10, a restrike of the breaker gap is again assumed to occur at the instant ($t = 0$) when the supply voltage and the voltage between breaker poles are at their maximum values, which will cause the largest travelling waves of voltage and current. Since both breaker poles A and B are now connected to a transmission line, then at the instant of restrike V_{AE} and V_{BE} which were $-V_m$ and approximately V_m respectively immediately prior to the restrike, must now become equal (neglecting the arc voltage) at the mean value between them which is zero (i.e. earth potential). Thus two travelling waves move in opposite directions away from the breaker; $-V_m$ into line 2 and V_m into line 1. Since the former is a forward-travelling wave, it produces a forward-travelling wave of negative current $-I = -V_m/Z_c$. Similarly, since the latter travelling wave of voltage is a backward-

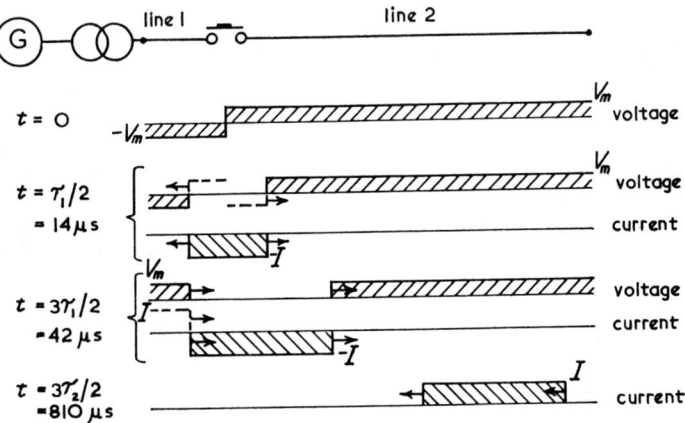

Fig. 2.13. Current and voltage surges in the transmission lines of Fig. 2.12 at various times after a restrike of the breaker gap at $t = 0$.

travelling wave in line 1 which has the same surge impedance Z_c as line 2, it produces a backward-travelling wave of current of the same value $-I$.

If the short length of cable between line 1 and the generator and transformer is neglected initially, then at $t = 28$ μs a voltage wave of V_m is reflected back into line 1, due to the very high generator inductance which constitutes a virtual open-circuit. This positive forward-travelling voltage sets up a positive forward current, which, together with the negative backward current of the same value, brings the current in line 1 to zero. It can be seen from Fig. 2.13 that the current surges in the two lines are together equivalent to a negative forward-travelling wave travelling into and along line 2 with magnitude $-I$, and with a duration which is twice the transit time τ_1 of line 1. This time τ_1 taken by a surge to travel from one end of line 1 to the other is in this case 28 μs, so that the duration of the surge current passing along line 2 is 56 μs. When the front of this current wave reaches the open end of line 2 at $t = \tau_2 = 540$ μs, this current pulse is reflected and changes sign to become $+I$.

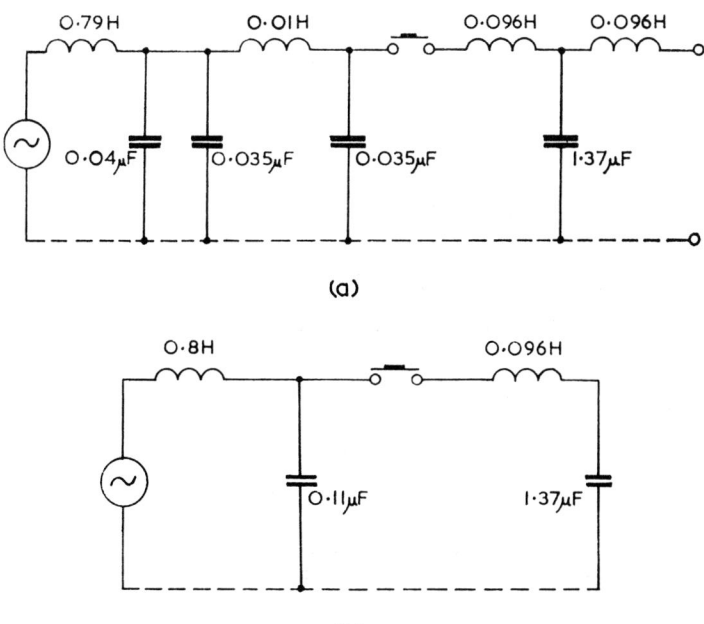

(a)

(b)

Fig. 2.14. Approximate circuit for calculation of oscillatory current.

The current wave I has a value of $I = \sqrt{2} \times 220\,000/\sqrt{3} \times 375 = 480$ A and from $t = 0$ to $t = 56$ μs a negative current of this value flows in the circuit breaker, followed by an oscillatory current. The initial amplitude of this oscillatory current may be calculated approximately as follows. (Clearly a more accurate calculation can be made but the purpose here is to encourage the student to be prepared to make quick and simple checks on the order of magnitude of quantities.) From the data in Fig. 2.12 the approximate circuit of Fig. 2.14(a) may be drawn, where in order to enable an estimate to be made more readily line 1 has been represented by a single Π section and line 2 by a single T section. The reactance of 0·79 H for generator and transformer is evaluated below. This circuit reduces approximately by inspection to that of Fig. 2.14(b) and this can be seen to be nearly equivalent to a series connection of $L = 0·896$ H and $C = 1·48$ μF.

Since the natural frequency $f_n > 50$ Hz the power frequency voltage is approximately constant at $-V_m$ during the first half-cycle of oscillatory current (see Fig. 2.11). (f_n is shown below to be 147 Hz so that the voltage will actually vary between $0·87 V_m$ and V_m.)

$$-V_m = L\ di/dt + (1/C) \int i\ dt$$

The solution for the current is shown in section 2.4 to be given by [2.9] and differentiation gives

$$L\ di/dt = 2jAL\sqrt{(1/LC)} \cos \sqrt{(1/LC)}t$$

At $t = 0$, when the breaker contact gap restrikes, the initial equation gives (since C is initially charged to V_m)

$$-V_m = L(di/dt)_{t=0} + V_m$$

$$2jAL\sqrt{(1/LC)} = -2V_m$$

$$I_m = 2jA = -2V_m\sqrt{(C/L)}$$

$$I_m \simeq -\frac{2\sqrt{2}}{\sqrt{3}}\ 220\ 000\ \sqrt{(1·48 \times 10^{-6}/0·896)}$$

$$\simeq -462\ \text{A}$$

For this system the damped oscillatory current flowing after the circuit breaker gap had broken down at $t = 0$ was in fact found (Gantenbein and Vogelsanger, 1956) to have an amplitude in the first half-cycle of 490 A.

The frequency of this current may be calculated as follows. The generator and transformer per unit impedance is 0·35 p.u. so that the combined ohmic impedance Z referred to 220 kV is (see [A1.5], p. 395 of Volume 1)

$$Z = 0·35 \times 220\ 000^2/(64 \times 10^6) = 248\ \Omega$$

This impedance is almost entirely inductive reactance, so that the total machine and transformer inductance at 50 Hz and 220 kV is about 0·79 H, which swamps that of the cable and the two lines. The total capacitance of the cable and lines is $1·48 \times 10^{-6}$ F so that the natural frequency $\simeq 1/2\pi\sqrt{(0·896 \times 1·48 \times 10^{-6})} \simeq 139$ Hz. The actual frequency was found to be about 145 Hz. (It may be noted that representing line 2 by its capacitance only in Fig. 2.14 would give closer agreement for both amplitude and frequency of oscillatory current, viz. 489 A and 147 Hz.)

The circuit breaker current in the first half-cycle after the restrike is therefore a damped sine wave of 490 A amplitude and frequency 145 Hz, with the travelling waves of current superimposed on it, as shown in Fig. 2.15. From Figs. 2.13 and 2.15 it can be seen that a

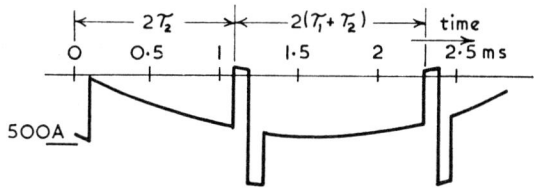

Fig. 2.15. Circuit breaker current in first half-cycle after restrike.

positive wave of current of 56 μs duration is reflected back from the open end of line 2, and reaches the breaker 1·08 ms after the restrike, i.e. at about 60° in the 147 Hz oscillatory current. The value of the oscillatory current at that instant is about -425 A, so that the surge current of 480 A is sufficient to drive the current through zero. As Fig. 2.15 shows, the current can be forced through zero four times in the first half-cycle as the current surge moves backwards and forwards.

If the effect of the cable is now considered and because it is short is taken as a lumped capacitance of 0·04 μF, then it changes effectively from a short-circuit to an open-circuit when the current surge reaches it, with a time constant for this change of $0·04 \times 375 = 15$ μs. This does not make any essential difference to the current wave shown in Fig. 2.15, though the pulses are more complex and may give additional current zeros on the second positive-going pulse.

The arc in the breaker may be extinguished at $t = 1.08$ ms when a current of about 425 A is suddenly chopped. The inductance of generator and cable at 220 kV is about 0·79 H and that of line 1 is 0·01 H, while the capacitance of the cable and line 1 is 0·11 μF. Instantaneous chopping of a current of 425 A would therefore give (see [2.16]) a voltage surge of $425\sqrt{(0.8/0.11)} = 1150$ kV. This is a theoretical maximum value obtained by assuming a restrike at the maximum voltage between breaker poles, an instantaneous chop of current, and no restrike of the breaker gap below 1150 kV. However, surges up to 625 kV ($3.47V_m$ since $V_m = 220\sqrt{2}/\sqrt{3} = 180$ kV) were recorded in the system, and resistance switching was then employed in order to reduce the switching over-voltages to $1.7V_m$ (2·4 × normal r.m.s. phase voltage).

2.12 Switching surges on closing or reclosing an unloaded line

Improvements in restriking performance of breakers, and developments in transformer core steels, have tended in recent years to reduce switching over-voltages due to interruption of line charging current and magnetising current chopping. Attention has now turned, especially at 400 kV and above, to other relatively frequent surges, such as those which may occur when an unloaded line is switched into circuit or is reclosed. If an unloaded transmission line represented by a lumped capacitance is connected to the supply at about the instant when the supply voltage on a given phase passes through its peak value V_m, then the voltage across the line capacitance of that phase will approach $2V_m$. This applies to the case where the circuit resistance is relatively small and the natural frequency of the circuit is considerably greater than the power frequency, as is generally the case in power systems (see Rudenberg, 1950).

The voltage across the line tends to reach $2V_m$ one half-period of the natural frequency transient after circuit closure, where this natural frequency is determined by the bulk (lumped) line capacitance and supply circuit inductance. In an actual transmission line, there is a surge of V_m travelling along the line when the line is energised, and, due to the infinite impedance of the open end or the high surge impedance of an unloaded transformer, the voltage at this end tends towards $2V_m$, and a reflected wave of voltage of nearly V_m returns along the line to the supply end. (It is suggested that reference should be made to Chapter 4.) There is therefore a tendency for twice normal peak phase voltage to earth to appear on the line, due either to the natural frequency oscillation or to reflected waves. The former is more likely to occur with a low MVA source and the latter is more likely with a very large MVA source.

Because the three sets of phase contacts of a breaker do not close at exactly the same instant, and there is mutual coupling between the three phases, the over-voltage at the receiving end may be further increased, and voltages of about $3V_m$ have been recorded in some power systems when a line was switched into circuit.

If a line which is in circuit but unloaded sustains a transient fault which trips the breaker at the supply end, and the breaker is then automatically reclosed, charges trapped on the line can increase the over-voltages, and values appreciably greater than $3V_m$ have occurred. This may be illustrated by referring for simplicity to a single-phase line. The circuit is similar to that shown in Fig. 2.6 which represents one phase of a 3-phase line. If a transient earth-fault occurs on one line conductor causing the breaker at the supply end to open then, since the interruption occurs at the zero of a charging current, the now healthy conductor is left charged at the peak voltage V_m. This is similar to the slowly decaying potential of point B in Fig. 2.11, and apart from energy dissipated in the switching resistor of the breaker or in shunt reactors or transformers, the line charge may take many seconds to decay. If the breaker is automatically reclosed, and if it is an air-blast breaker, which when open has the main contact closed but the isolator open, then reclosure occurs on the isolator. In these breakers the isolator gap may flash-over before contact is made, when the voltage on the other pole of the isolator (similar to the potential of point A in Figs. 2.10 and 2.11) connected to the supply becomes $-V_m$, so that the voltage across the isolator gap is $2V_m$. When an arc forms across the isolator gap following break-down under this voltage, the potential of B changes by $-2V_m$, i.e. a wave of voltage of $-2V_m$ travels down the line and at the open end reflection of almost the same value occurs, reducing the voltage there by nearly $4V_m$. The total voltage at the open end is therefore of the order of $-4V_m$ plus V_m left by the slowly-decaying line charge i.e. $-3V_m$.

Although attenuation of the surge will occur due to line losses including corona, and a transformer at the far end reduces the reflected wave since its surge impedance is finite, mutual effects due to sequential reclosing of the three phases in a 3-phase line will again increase the over-voltage. Air-blast breakers with pressure-immersed sealed heads which close on the main contact gap do not often give severe over-voltages since although with random switching closing may occur near the peak of the applied voltage, breakdown of the high pressure air before closure of the contacts is less likely to occur.

One way in which these over-voltages can be reduced in E.H.V. transmission systems is, to connect a resistor of the order of the line surge impedance (hundreds of ohms) in series with a breaker and line, and to short-circuit this by a second breaker very quickly after the first one closes or recloses. This is now being used in service with the shunt switching resistor still employed on both make and break as usual.

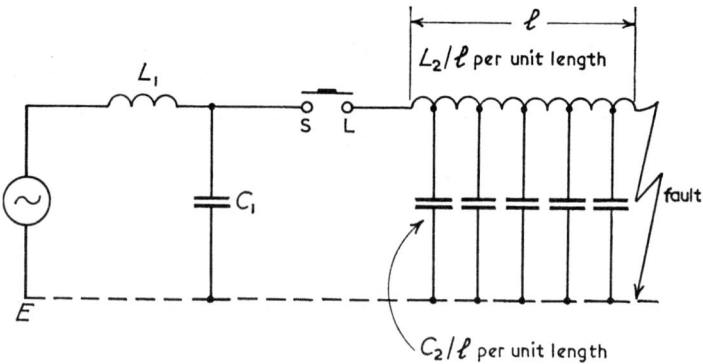

Fig. 2.16. A fault a short distance along a transmission line from a circuit breaker (Short-line fault).

2.13 Short-line or kilometric fault

If a circuit breaker clears a fault a short distance along a transmission line from the breaker, the simplified circuit is as shown in Fig. 2.16.

This represents conditions:—

(*a*) where a single line-to-earth fault occurs,

or

(b) after the first phase has cleared following a double line-to-earth fault,

or

(c) after the first two phases have cleared following a three-phase fault to earth.

If the peak phase voltage is again V_m then while the arc exists and just prior to the current zero

$$v_{LE} \simeq V_m \frac{L_2}{(L_1 + L_2)}$$

and

$$v_{SE} = v_{LE} + v_{arc}$$

Following the last current zero, v_{SE} will oscillate about the recovery voltage at the natural frequency of the source-side circuit [approximately $(1/2\pi)\sqrt{(1/L_1 C_1)}$] which is generally in the range 500 to 5000 Hz. The oscillation of v_{LE} as the line relaxes towards the uncharged state, would also be a damped sine wave if the line inductance and capacitance were considered to be lumped, but this would occur at a much higher frequency, of the order of 30 to 100 kHz. The line-side frequency f_l is in fact about 50 % higher than this and is given more closely by dividing the speed of light c by four times the length of the line, i.e. $f_l = c/4l$. This is because the velocity of a surge in an air-insulated line is virtually that of light, and the reflected waves from the two ends of the line cause the voltage waveform at a point to repeat a complete cycle after the waves have travelled four times along the line, i.e. twice in each direction (see Chapter 4). The surge velocity $u \simeq c = l/\sqrt{(L_2 C_2)}$, and the student can verify that the line-side frequency f_l is $2\pi/4$ times the value given by assuming the line inductance and capacitance to be lumped. The voltage between the breaker contacts v_{SL} would therefore contain a very high frequency component giving a very high r.r.r.v. Conditions are in fact worse than this, because the reflections occurring due to the distributed parameters L_2 and C_2, cause v_{LE} to approximate to a damped triangular waveform and this still further increases the r.r.r.v. as shown in Fig. 2.17. Fig. 2.18 shows the increased r.r.r.v. compared with a single frequency transient for a fault near the breaker.

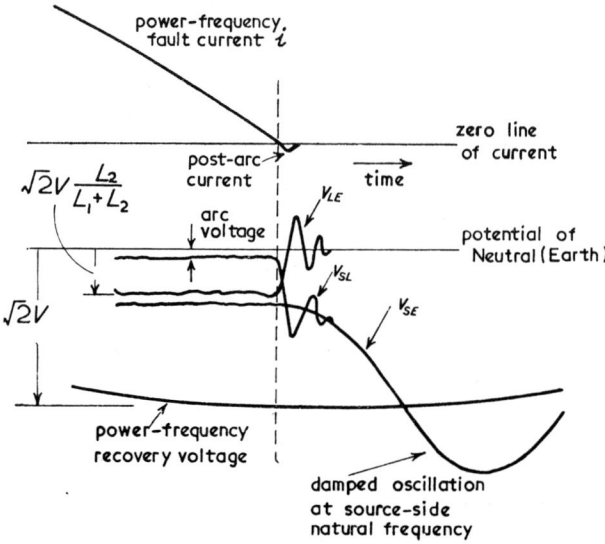

Fig. 2.17. Transient voltages at interruption of a short-line fault.

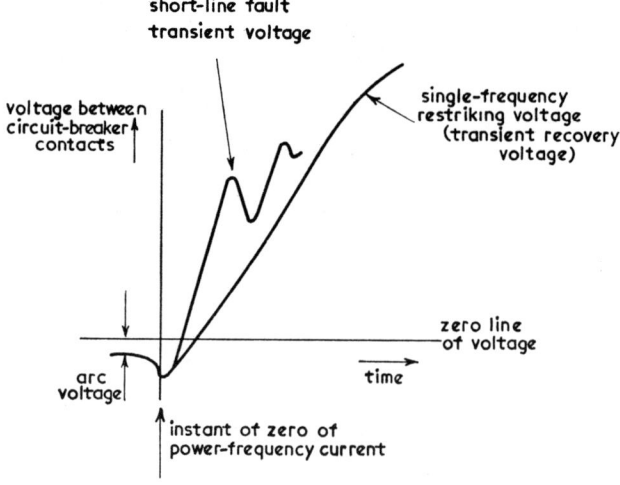

Fig. 2.18. Transient voltage between circuit breaker contacts.

The conditions are less severe for a cable fault than for a line fault since, as shown below, the r.r.r.v. of the line-side voltage is given approximately by

$$\left(\frac{dv_{LE}}{dt}\right)_{t=0} = \sqrt{2}I\omega Z_c \qquad [2.17]$$

where I is the r.m.s. current interrupted,

$$Z_c = \sqrt{\frac{L_2}{C_2}}$$

is the surge impedance of line or cable.

If resistance switching is used to place a resistance R_s in parallel with the breaker gap, then the initial rate of rise of line-side voltage is reduced to

$$\left(\frac{dv_{LE}}{dt}\right)_{t=0} = \frac{R_s Z_c}{R_s + Z_c} \cdot \sqrt{2}I\omega \qquad [2.18]$$

[2.17] and [2.18] may be verified, starting from [4.14] for a loss-free line, viz.:—

$$-\frac{\partial i}{\partial x} = \frac{C_2}{l}\frac{\partial v}{\partial t} \qquad [2.19]$$

i.e.

$$\frac{\partial v}{\partial t} = -\frac{l}{C_2}\frac{\partial i}{\partial x}$$

$$= -\frac{l}{C_2}\cdot\frac{\partial i}{\partial t}\frac{\partial t}{\partial x} = -\frac{l}{uC_2}\frac{\partial i}{\partial t}$$

where u the velocity of propagation is $l/\sqrt{(L_2 C_2)}$

i.e.

$$\frac{\partial v}{\partial t} = -\sqrt{\left(\frac{L_2}{C_2}\right)}\frac{\partial i}{\partial t} = -Z_c\frac{\partial i}{\partial t} \qquad [2.20]$$

If the current interrupted is assumed to be sinusoidal so that $i = -I_m \sin \omega t$ (the $-$ve sign denotes that the zero of time is chosen at the end of a $+$ve half-cycle of current), then

$$\frac{\partial v}{\partial t} = Z_c I_m \omega \cos \omega t$$

and

$$\left(\frac{\partial v}{\partial t}\right)_{t=0} = \sqrt{2}I\omega Z_c$$

which is [2.17].

If resistance switching is employed the circuit is as shown in Fig. 2.19. The inductive reactance of L_1 at power frequency is not likely to exceed about 1 ohm and for a typical source-side natural frequency of 5 kHz, C_1 will therefore be of the order of 1 μF. If v_{LE}

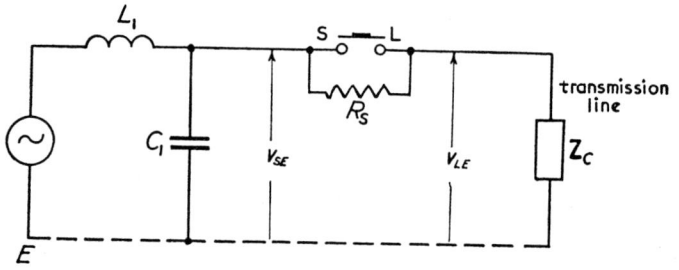

Fig. 2.19. Simple equivalent circuit for short-line fault with resistance switching.

includes components up to 100 kHz, then the reactance of C_1 initially is of the order of 1 ohm, and R_s is effectively in parallel with the line for the first few microseconds after interruption, so that [2.17] is replaced by [2.18].

[2.17] shows that if resistance switching is not employed, then for a 50-Hz line with $Z_c = 350$ ohms and an r.m.s. current of 50 kA interrupted, the r.r.r.v. would be about 8 kV/μs. This is a very high value at large interrupting capacities of about 35 000 MVA. As an illustration of the rise occurring in recent years, it may be noted that typical values for world systems in 1956 (Mortlock, 1956) ranged from 1 kV/μs at 3000 MVA to 6 kV/μs at 500 MVA. The very high values of r.r.r.v. and MVA resulting from short-line faults, may produce sufficient post-arc current to lead to thermal reignition of the arc, and impose problems in circuit breaker design and performance, particularly for gas-blast breakers above 30 kA.

From Fig. 2.17 it can be seen that since the frequency f_l of the line-side transient is very much greater than that of the source-side transient, the r.r.r.v. $dv_{SL}/dt \simeq dv_{LE}/dt$, and since v_{LE} falls as shown from $V_m L_2/(L_1+L_2)$ to zero in about a quarter of a cycle of the line-side oscillation, then

$$\left(\frac{dv_{LE}}{dt}\right)_{t=0} \simeq \frac{V_m L_2 4 f_l}{L_1+L_2}$$

and since $f_l \simeq c/4l$

$$\left(\frac{dv_{LE}}{dt}\right)_{t=0} \simeq \frac{V_m L_2 c}{l(L_1+L_2)}$$

$$V_m \simeq \sqrt{2} I \omega (L_1+L_2)$$

so that

$$\left(\frac{dv_{LE}}{dt}\right)_{t=0} \simeq \sqrt{2} I c \omega L_2/l \qquad [2.21]$$

This treatment is only very approximate in its neglect of change in line reactance with frequency and the idealised waveform of v_{LE} in the first quarter-cycle, but [2.21] is the same as [2.17]. Since the inductance per unit length of overhead transmission line (L_2/l) does not vary much with voltage and construction (see Chapter 2 of Volume 1), [2.21] does serve to indicate that the r.r.r.v. depends mainly upon the magnitude of the fault current. For this reason the r.r.r.v. is decreased and interruption is easier if the length of line between breaker and fault is increased, thus decreasing the fault current. This however is true only above a critical line-length which is usually between 0·8 and 8 km, since as the line length is decreased below this critical region the breaker is able to interrupt more current. This is largely because in a gas-blast breaker, which is the type most affected by a fault occurring within this critical, short length of line, interruption is determined within a period of a few microseconds of current zero mainly by energy balance, and the first reflection may already have occurred in the early part of this period if the fault is very near the breaker. For example, for a fault 0·4 km from the breaker, v_{LE} would reach its first peak in about one microsecond and would then fall as shown in Fig. 2.17, and thermal reignition may not then occur at the particular system fault current, but would only occur at some current beyond the fault level of the particular system.

For these reasons the short-line fault conditions are most critical when the length of the line between breaker and fault lies between two limits, and these limits are frequently about 1 and 8 km.

2.14 Present position in circuit breaking

Oil circuit breakers of multi-break type with voltage grading and resistance damping, and high-acceleration contact systems are now able to interrupt at least 25 000 MVA at 330 kV. Air-blast breakers with as many as 12 breaks in series in each phase and additional resistance-arc breaks have been built to interrupt 35 000 MVA, i.e. 50 kA at 400 kV, and may need to be designed for larger ratings, e.g. 60 and 90 kA at 1000 kV or more. Pressure-immersed breakers have air at about $2 \cdot 7$ MN/m^2 and even, in some cases, at about 7 MN/m^2, already in the breaker when it is in the closed position, and it is released through the contact gap when it is tripped. This enables shorter operating times to be achieved.

Since short-circuit testing stations, where proving tests on breakers and other equipment can be carried out directly from short-circuit alternators, have ratings of up to 5000 MVA, and it is not economical to extend them to the full breaker ratings, it has been necessary for years to adopt unit-testing of large breakers; i.e. one or more units out of those constituting one phase of the breaker are tested separately. One of the main problems, then, is to determine the correct restriking voltage to be used on test across the various units. This depends upon the voltage appearing across the first phase to clear, and the voltage distribution across that phase of the breaker. The recovery voltage may be designed to be equally shared between units, by connecting resistors across the gaps. Due to the higher frequencies in the normal restriking-voltage transient or under short-line fault conditions, the voltage distributions will then be intermediate between that due to the capacitances with no resistors, and the uniform distribution with resistance only. The units nearer the source will have the higher voltage across them if there is no resistance due.to capacitance to earth (see the analogous case of the voltage distribution across an insulator string in Chapter 4 of Volume 1). These unit tests are supplemented by tests on complete phases with full current interrupted at reduced voltage, and reduced current at full voltage.

Unit-testing places restrictions on design, in that all interrupting units must be identical and independent, and have an interrupting capacity within that of the testing station. The uneconomical extension of direct-generator and transformer installations has led to exploration of a number of circuits for synthetic testing, with current and voltage supplied from different sources. The restriking voltage is obtained from a low energy source which is a capacitor bank. This is applicable to very high voltage circuit breakers where the recovery voltage is very much higher than the arc voltage, and the current is obtained from an existing generator and transformer installation. Extensive examination of synthetic testing is being carried out, in an attempt to extend it to all aspects of proving circuit breakers. The basic conditions for equivalence between synthetic and direct tests are outlined in the next section.

Circuit breakers using sulphur hexafluoride (SF_6) are now in production up to 400 kV. This gas has certain advantages over air in circuit interruption. These are not wholly associated with its electro-negative character, but are due partly to its low dissociation temperature, high thermal conductivity and high dielectric strength, though many other factors are involved in the choice of gas.

Vacuum switches, though initially encountering certain difficulties at 10 kA, have now been made to interrupt at least 40 kA, and some switches are available for use up to 132 kV. The technique of manufacture involves the use of special zone-refined composite materials for the electrodes, and rigorous evacuation of the switch to about 10^{-8} torr (mm Hg), followed by pumping during baking at high temperature and arcing before sealing off. The vacuum switch will offer the possibility of very compact interrupters for wider ranges as research proceeds.

2.15 The conditions for equivalence between synthetic and direct tests

Fig. 2.20 shows one synthetic test circuit which has undergone extensive investigation, in which the voltage circuit produces a restriking voltage in parallel with the breaker under test. In this circuit, components are denoted by: G generator, M make switch, A auxiliary breaker, B breaker under test, S spark gap.

With A closed, the closure of M causes a 50-Hz current i to flow in B. Immediately after M closes, A and B are both opened and just before i reaches zero (instant t_1 in Fig. 2.21), the voltage circuit in which C_0 is already charged is applied to B by firing S, so that a

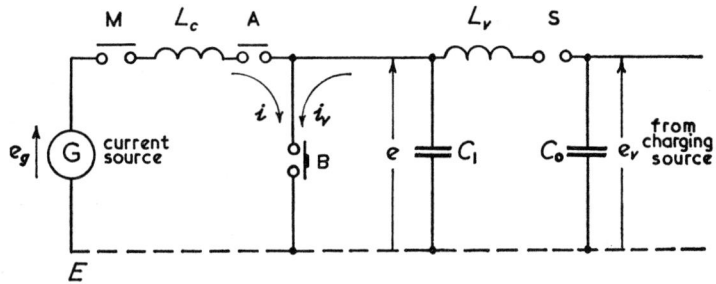

Fig. 2.20. Synthetic testing circuit.

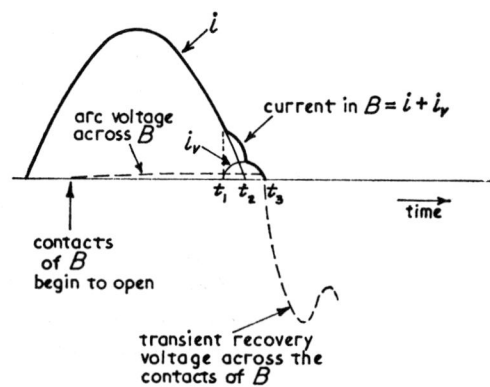

Fig. 21. Waveform sequence in synthetic testing.

current i_v also flows in B. When the breaker A brings the current i to zero at t_2, the test breaker carries the oscillatory capacitor bank current i_v until its own zero at t_3 when a restriking voltage of frequency determined by C_1 and L_v is applied to the breaker (see Fig. 2.21). If $C_0 \gg C_1$ and if the energy stored in C_0 when charged prior to the test is much greater than the energy lost in the oscillatory circuit and in the breaker when carrying i_v, then the voltage across

C_0 is almost constant at e_v during the oscillatory transient. In order that the synthetic test shall impose the same conditions as a direct test, the instantaneous currents and voltages should be the same for each. Whilst the breaker is connected to the current source in the circuit of Fig. 2.20,

$$L_c\frac{di}{dt}+v_{arc} = e_g$$

where v_{arc} the arc voltage is zero for an ideal circuit breaker, so that

$$L_c\frac{di}{dt} \simeq e_g \qquad [2.22]$$

While the breaker is connected to the voltage source and the voltage across it is e,

$$L_v\frac{di_v}{dt}+L_v\frac{d}{dt}\left(C_1\frac{de}{dt}\right)+e = e_v$$

and again for an ideal breaker the post-zero current $i_v = 0$ so that

$$L_vC_1\frac{d^2e}{dt^2}+e = e_v \qquad [2.23]$$

If in the circuit of Fig. 2.2, $R = 0$ and $G = 0$ and the source producing the voltage v is a short-circuit alternator and transformer unit, which can be used directly to test the breaker at its full rated interrupting capacity, then the corresponding equations to [2.22] and [2.23] are

$$L\frac{di}{dt} = v$$

when the breaker is closed, arc voltage can be neglected, and

$$LC\frac{d^2e}{dt^2}+e = v$$

so that the synthetic and direct tests would be equivalent for an ideal breaker if $e_g/L_c = v/L$, $e_v = v$, $L_v = L$ and $C_1 = C$.

If the losses in current and voltage circuits are small, the test plant power is then very much smaller than the interrupting capacity of the breaker. For actual breakers with finite arc voltage and post-zero current, there cannot strictly be equivalence unless all conditions are equal, viz.:—

$$e_g = v = e_v$$

$$L_c = L = L_v$$

$$C = C_1$$

but for these conditions the current and voltage circuits must have a power equal to the interrupting capacity.

Complete equivalence is not therefore possible and much research has been devoted to determining the parameters for the synthetic circuits to give a known severity at least as high as that of the direct test. Investigations have so far shown that for air-blast breakers, which do not have an arc-extinguishing effort due to energy developed during arcing, the parallel current injection circuit of Fig. 2.20 gives no appreciable differences in severity compared with direct tests up to 1200 MVA, and it is believed that this will hold at higher ratings. (See Anderson et al, 1966 and 1968.)

REFERENCES

ANDERSON, J. G. P., ELLIS, N. S., MASON, F. O., NOBLE, R. G., ORTON, L. H., REECE, M. P. & STEEL, J. G. 1966 and 1968. Synthetic testing of a.c. circuit breakers; Part 1, Methods of testing and relative severity. *Proc. I.E.E.* **113**, No. 4, 611–621; Part 2, Requirements for circuit breaker proving. *Proc. I.E.E.* **115**, No. 7, 996–1007.

BALTENSPERGER, P. & RUOSS, E. 1960. The short-line fault in high-voltage systems. *Brown Boveri Revue.* **47**, 329–339.

BOLTON, E. *et al.* 1970. Short-line fault tests on the C.E.G.B. 275-kV system. *Proc. I.E.E.* **117**, 771–784.

EDELS, H. 1962. Arc interruption and circuit breaking. *Electrical Review*, 1006–1010.

GANTENBEIN, A. & VOGELSANGER, E. 1956. New experiments concerning the disconnection of unloaded transformers and transmission lines by minimum oil-content circuit breakers. *Bulletin Oerlikon*, No. 316, 53–61.

GUILE, A. E. 1962. Interruption of short-line faults in high-voltage systems, World Power Engineering. **1**, No. 3, 22–24.

LYTHALL, R. T. 1963. *J. and P. Switchgear Book.* Johnson and Phillips Ltd., London, 6th edition.

MORTLOCK, J. R. 1956. *A.C. Switchgear.* Volume 1: a survey of requirements. Chapman & Hall, London.

RUDENBERG, R. 1950. *Transient Performance of Electric Power Systems.* McGraw-Hill, New York.

SABATH, J., SMITH, H. M. & JOHNSON, R. C. 1966. Analog computer study of switching surge transients for a 500-kV system. *I.E.E.E. Transactions on Power Apparatus and Systems*. PAS. **85**, No. 1, 1–9.

SWITCHING SURGES, 1—phase to ground voltages. 1961. *Transactions A.I.E.E.*, **80**, 240–261.

THOMAS, A. M. 1966. The switching surge strength of insulating arrangements above 100 kV. *E.R.A. Report* No. 5080.

THOMAS, A. M. 1966. Switching surges in extra high voltage transmission systems—a critical review of recent studies of amplitudes, waveforms and frequencies of occurrence. *E.R.A. Report* No. 5171.

TRENCHAM, H. 1953. *Circuit Breaking*. Butterworths Scientific Publications, London.

YOUNG, A. F. B. & ELLIS, N. S. 1963. The short-line fault. *Electrical Review*. **173**, 271–274.

Examples

1. 50 MVAr of shunt reactive compensation is provided on a 132-kV system. The installation consists of three separate reactors connected in star with the star point earthed. The capacitance of the windings can be neglected but the E.H.V. bushing on each phase has a capacitance to earth of 1000 pF. What over-voltage is produced on the reactor when, with the system unloaded, it is switched out by an air-blast breaker which chops a current of 10 A? (The source MVA is large and can be neglected in the calculation.) What is the effect of connecting a 40 kΩ resistor across the circuit breaker?

$$(349 \text{ kV}, 182 \text{ kV})$$

2. A 3-phase, 50-Hz, 200-MVA alternator is connected through a 200-MVA transformer to a 132-kV transmission line. 32 km from the generator there is a substation containing a circuit breaker to which is connected a second similar transmission line. If a single line-to-earth fault occurs 0·8 km along this second transmission line from the substation, calculate the two natural frequencies of the circuit and the initial rate of rise of the voltage across that pole of the circuit breaker contacts.

Each line conductor has inductance of 1·25 mH/km and capacitance to earth of 0·00935 μF/km. At the time of arc extinction the alternator and transformer may be assumed to have a combined per-unit inductive reactance in the faulted phase of 0·25 at rated MVA and voltage, with an effective voltage behind this reactance of 1·2 per unit.

$$(878 \text{ Hz}, 58·1 \text{ kHz}, 0·43 \text{ kV}/\mu s)$$

3. A 3-phase system consists of an alternator with its star point earthed through a 1·0-per-unit reactor, and connected through a star/star transformer A with both star points earthed, to a 16 km 132-kV transmission line which has inductance/conductor of 1·25 mH/km. At the far end of the line is a star/delta transformer B with the primary star point earthed. A 3-phase fault occurs at the secondary terminals of transformer B. Calculate the r.m.s. recovery voltage across the first phase to clear of a circuit breaker connected to the secondary terminals of transformer A. It may be assumed that resistance switching just damps out the restriking voltage transient, and that the alternator in which there is assumed to be a constant generated e.m.f. of 1·1 p.u. has reactances which may be considered to be constant during the short time of the fault. All resistances and line capacitances may be neglected and the per-unit reactances at 200 MVA and rated voltage are:—

alternator $X_1 = 0·12$ p.u.

$$X_2 = 0·15 \text{ p.u.}$$

$$X_0 = 0·06 \text{ p.u.}$$

transformer A has leakage reactance/phase of each winding of 0·04 p.u.

transformer B has leakage reactance/phase of each winding of 0·05 p.u.

The transmission line has a zero-sequence inductance which is 3·5 times the positive-sequence inductance.

(124 kV)

4. Calculate the voltage rise with respect to earth which can occur on a 66-kV, 3-phase overhead line with conductors of 1·6 cm² cross-section, with a geometric mean spacing between conductors of 1·2 m due to current being chopped in all phases at a value of 15 A.

Discuss the significance of the magnitude of the calculated voltage rise, when compared with that given in section 2.9 for chopping a transformer magnetising current. Comment upon assumptions made in the calculation.

(4·74 kV)

5. If a short-circuit occurs 0·935 km away from a circuit breaker along a transmission line having inductance 1·25 mH/km and capacitance 0·01 μF/km, deduce the line-side natural frequency and show by what factor it exceeds the value obtained by considering the line inductance and capacitance to be lumped:

(University of Leeds) (75·6 kHz)

6. A generator/transformer unit of 200 MVA, 0·45 per-unit impedance, is connected to a 275-kV transmission line 8·7 km long, having a surge impedance of 400 Ω and capacitance of 0·075 μF. This line is connected through a circuit breaker to a second line 174 km long with surge impedance of 400 Ω and capacitance of 1·5 μF.

After the circuit breaker has interrupted the charging current of the second line, the breaker gap restrikes when the voltage across it reaches its maximum value.

Calculate, noting any assumptions involved:

(i) the times taken for a surge to travel along each of the lines;

(ii) the magnitude and duration of the surge current which flows in the breaker in addition to the oscillatory current;

(iii) the peak value and frequency of the oscillatory current.

Sketch the voltage and current surges at one or two instants shortly after restrike. Sketch also the waveform of the first half-cycle of current flowing in the breaker after the restrike, and comment upon over-voltages which may arise from it.

(University of Leeds) (30 μs, 600 μs, 60 μs, 560 A, 617 A, 142·5 Hz)

Chapter 3

TRANSIENT STABILITY OF POWER SYSTEMS

3.1 Introduction

The theory of the synchronous generator (alternator) was dealt with, mainly with reference to its steady-state operation, in Volume 1, Chapter 7: the main points of concern in the present study will be summarised in section 3.2.1. This present chapter now considers the conditions necessary for the *successful* operation of a power system when changing from one steady-state operating condition to another, due to variations in its operating conditions which can be classified as either small and slow or large and sudden, where slow means a long time compared with the time constants of the field circuits of the machines, of the automatic voltage regulators and of the turbine governors: see Volume 1, sections 7.5, 7.9 and 7.10 respectively.

A synchronous power system has *steady-state stability* if, after a small slow disturbance, it can regain and maintain synchronous speed; a small slow disturbance is taken to mean normal load fluctuations, including the action of the automatic voltage regulators and turbine governors. A power system has *transient stability* if, after a large sudden disturbance, it can regain and maintain synchronous speed: a large sudden disturbance is one caused by faults and switching. In order to develop the main principles simply it is assumed that the automatic voltage regulators and turbine governors are too slow to act during the period of the analysis. *Dynamic stability* refers to the case of transient stability when the regulators and governors are fast-acting and are taken into account in the analysis. The stability limit of a system is the maximum (steady-state) power which can be transferred through the system without loss of stability. The limit depends also on the magnitude, type and location of the disturbance. The stability factor is the ratio of the stability limit to the actual load-power transfer.

It will be shown later (section 3.4.2) that all the machines in a power exporting area can be reduced to an equivalent generator (G) and, similarly, that all the machines in a power importing area can be reduced to an equivalent synchronous motor (M). The transmission system (e.g. the 400-kV Grid system in the U.K.) which connects these two areas is called the interconnector (or tie-line). The above 2-machine system can be reduced to one machine connected to an infinite busbar—a constant voltage, constant frequency system. Generally, in this chapter, resistance will be neglected relative to the inductive reactance of the system.

3.2 Power transfer: transfer reactance

3.2.1 SYNCHRONOUS GENERATOR

Fig. 3.1(a) shows the combined phasor and vector diagram for a cylindrical-rotor synchronous generator, for which no distinction is made between direct-axis (d) and quadrature-axis (q) quantities during steady-state operation. F represents the d.c. excitation m.m.f., which on open circuit would generate E_o. On load, armature reaction (equal and opposite to A) modifies the effective m.m.f. to a net m.m.f. G which creates an air-gap flux which generates E, the e.m.f. behind the stator (armature) leakage reactance X_L (or X_a as in B.S. 4296) such that $E = V + jIX_L$. If saturation effects could be ignored, armature reaction would be proportional to load current and the triangles enclosed by F and G and by E_o and E would be similar. The synchronous reactance X_d (or X_S) combines the effects of stator leakage reactance and armature reaction and would be a constant such that $E_o = V + jIX_d$. E_o behind X_d is used as the model (equivalent circuit) for steady-state power-system analysis.

If the generator is at or near rated load, the effects of saturation usually cannot be ignored and, since the effect of armature reaction on saturated iron is reduced, X_d is replaced by a smaller value, the effective synchronous reactance X_{eq} (see B.S. 2658), which is taken as the synchronous reactance of the equivalent unsaturated generator. The procedure for calculating X_{eq} involves determining the Potier reactance X_p from the zero power factor, rated current test. The Potier reactance is a form of stator leakage reactance which allows

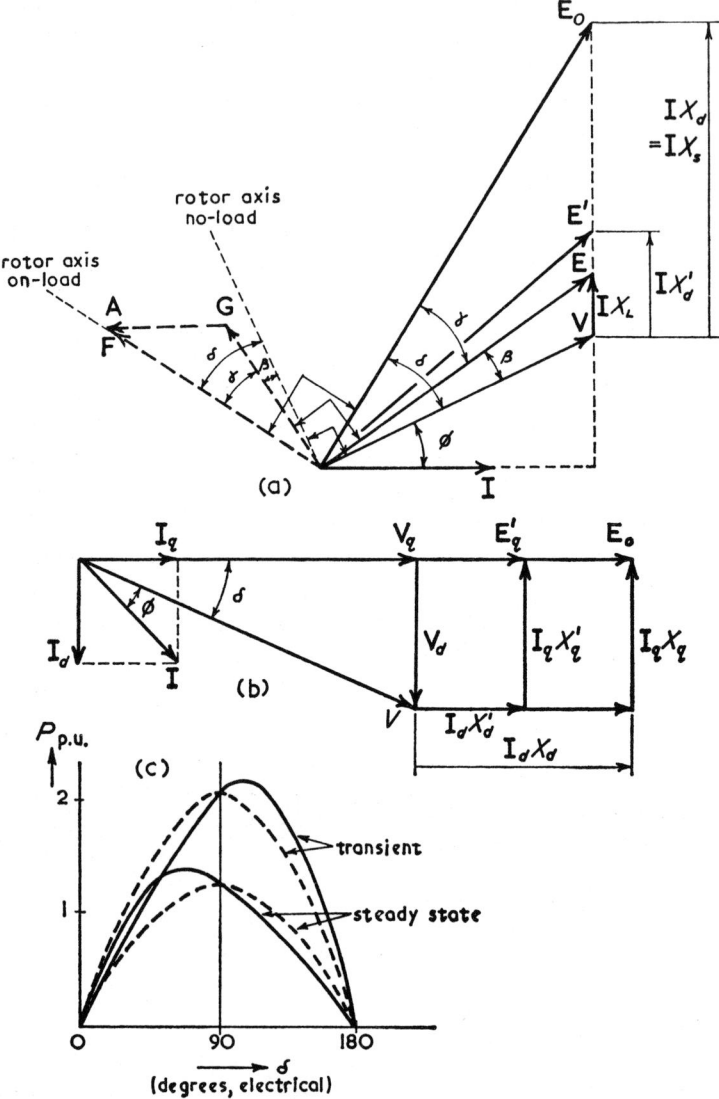

Fig. 3.1. Diagrams for a synchronous generator.
- (a) cylindrical-rotor
- (b) salient-pole type
- (c) power/angle curves for cylindrical-rotor (dotted)
  and salient-pole (solid lines).

for the increased leakage in the field winding when the generator is on load.

Referring to Fig. 3.1(a), it should be noted that the e.m.f. phasors vary sinusoidally in time at supply frequency, while the m.m.f. vectors rotate at synchronous speed in space. The superposition of the phasors and vectors on the same diagram and the particular orientation between them which gives a 90° separation between an m.m.f. and its corresponding e.m.f. is conventional and convenient but has no physical significance.

Consider now a loaded generator subjected to a sudden change in load or to an *external* fault. It was shown in Volume 1, section 7.5, that the apparent reactance of the generator increased from the subtransient value X_d'' (normally slightly greater than X_L) at the instant of the change, through the transient value X_d' to X_d during the steady-state conditions. The subtransient stage is due to the induced currents in the surface of the rotor iron which decay very rapidly (assuming there is no damper winding). The transient stage is due to the induced currents in the field winding. These induced currents delay the effect of armature reaction, on the theory that the flux linkages of two mutually-coupled circuits cannot change instantaneously. The corresponding subtransient, transient and synchronous e.m.f.s, behind the corresponding reactances are E'', E' and E_o. Fig. 3.1(a) does not show the subtransient condition as it is not of interest in the context of stability.

It must be noted that the load angle δ is always the angle between E_o and V, regardless of the e.m.f.-behind-reactance circuit being used, and is the angle through which the rotor has moved from its no-load to its on-load position.

The net mechanical power from the turbine, after supplying all losses other than the generator I^2R loss, equals the generator input power/phase P_1, where (assuming $R \neq 0$)

$$P_1 = \text{real part of } E_o I^*$$

$$= \frac{E_o^2}{Z_s} \cos \theta - \frac{VE_o}{Z_s} \cos (\theta + \delta) \qquad [3.1a]$$

where θ = impedance angle of the generator = $\tan^{-1}(X_d/R)$, and Z_s = synchronous impedance/phase = $|R + jX_d|$.

The output power of the generator, P_2, transferred to the infinite busbar is given by

$$P_2 = \text{real part of } \mathbf{VI}^*$$

$$= \frac{VE_o}{Z_s} \cos(\theta - \delta) - \frac{V^2}{Z_s} \cos \theta \qquad [3.1b]$$

When $R = 0$, the power/phase is given by

$$P = P_1 = P_2 = \frac{VE_o}{X_d} \sin \delta \qquad [3.1c]$$

This result could be obtained from Fig. 3.1, by noting that $IX_d \cos \phi = E_o \sin \delta$. If δ is the only independent variable, the output power is zero when $\delta = 0$ and a maximum when $\delta = 90°$ (or more generally when $\delta = \theta$). The steady-state loading of a generator was discussed in Volume 1, section 7.7. The power maximum, VE_o/X_d, is the steady-state stability limit.

Since δ is the angle between the rotor pole position on load and its position on no-load, then $\delta = 0$ when the generator is first synchronised (floating) on the infinite busbar. When the steam supply to the generator turbine is increased so that the generator *electrical* speed ω is temporarily greater than the synchronous speed ω_s, then

$$\delta = (\omega - \omega_s)t$$

$$\frac{d\delta}{dt} = \omega - \omega_s$$

$$\frac{d^2\delta}{dt^2} = \frac{d\omega}{dt} \qquad [3.2]$$

Thus the relative and absolute accelerations are equal and the angular velocity, in the context of this chapter, will mean the velocity relative to the synchronous speed of the infinite busbar.

Synchronising power

Assume that the generator is operating under steady-state conditions as shown in Fig. 3.1 and that the load angle δ increases by a small amount $\Delta\delta$. The increase in synchronous power output is given by

$$\Delta P = \frac{\mathrm{d}P}{\mathrm{d}\delta}(\Delta\delta) = P_r(\Delta\delta) \qquad [3.3a]$$

where

$$P_r = \frac{VE_o}{X_d}\cos\delta \qquad [3.3b]$$

is called the synchronising power coefficient. So long as $P_r > 0$, i.e. $0 \leq \delta < 90°$, an increase in the load angle δ will result in an increased output power. This power must be taken kinetically from the rotor (since the turbine governor has not been altered) so the generator slows down. Thus the synchronous generator is stable in that it inherently resists any tendency for δ to change from the value corresponding to the steady-state condition. This is not true if $\delta \geq 90°$.

Salient-pole generator

The power/load-angle formula for a salient-pole generator of negligible resistance was derived in Volume 1, p. 260, i.e.

$$P = \frac{VE_o}{X_d}\sin\delta + \frac{V^2(X_d - X_q)}{2X_dX_q}\sin 2\delta \qquad [3.4a]$$

This equation was drawn in Fig. 7.19. It should be noted that the salient-pole generator is 'stiffer' than the cylindrical-rotor generator, since the former operates at a smaller load angle for a given load. The synchronising power coefficient is

$$P_r = \frac{VE_o}{X_d}\cos\delta + \frac{V^2(X_d - X_q)}{X_dX_q}\cos 2\delta \qquad [3.4b]$$

The maximum value of P occurs when $P_r = 0$ and for this $\delta < 90°$.

Transient analysis of a salient-pole generator

The transient analysis of a loaded, salient-pole synchronous generator, having negligible stator resistance, during a short interval of time $(< T_d')$ following a sudden change in operating conditions is less complicated in terms of the fictitious direct- and quadrature-axis components of current than in terms of the actual stator phase currents. Section 7.7 in this volume discusses the transformation (by equating the effective armature reactions) of the instantaneous (time-varying) phase currents i_a, i_b and i_c into i_d and i_q, which do not vary with time (except during transient conditions), and which flow in fictitious coils which are fixed in position relative to the rotor. The detailed theory is given in textbooks such as Fitzgerald and Kingsley. The following is no more than a summary of those portions relevant to transient stability analysis.

All quantities are in per-unit form and all stator quantities are (usually) based on the rating of the generator. The base quantities for other coils (d-, and q-axis coils, field and damper coils) are then determined by transformer theory (see Volume I, p. 396). The symbol for per-unit reactance now recommended (B.S. 4296) is x, while X indicates a reactance in ohms. The per-unit values i_d and i_q are equal to the per-unit value of $\mathbf{I}_d$ and $\mathbf{I}_q$ which are phasor components of the stator current $\mathbf{I}$ as shown in Fig. 3.1(c). If the currents are in amperes then $i_d = \sqrt{2}I_d$ and $i_q = \sqrt{2}I_q$.

The transient theory of the synchronous generator assumes the theory of constant flux-linkages—that due to inductance the flux-linkages remain constant for a short time following a sudden change in operating conditions. From Fig. 3.1(b),

$$\mathbf{E}_q' = \mathbf{V}_q + \mathbf{I}_d \mathbf{X}_d' = \mathbf{V} \cos \delta + \mathbf{I}_d \mathbf{X}_d' \qquad [3.4c]$$

and it can be shown that E_q' is proportional to the total field flux-linkages and thus can be assumed constant over the transient period (Fitzgerald and Kingsley, section 5.9). The output power of a loaded generator during the transient stage is given by

$$P = VI \cos \phi = V_d I_d + V_q I_q$$

(see Volume 1, p. 404). Substituting from [3.4c] and using $V_d = V \sin \delta = I_q X_q$ gives

$$P = \frac{E_q' V \sin \delta}{X_d'} - \frac{V^2}{2}\left(\frac{1}{X_d'} - \frac{1}{X_q}\right)\sin 2\delta \qquad [3.4d]$$

This formula is similar to [3.4a] (which gives the steady-state power output), but with E_q' replacing E_o and with X_d' replacing X_d. Since

$E_q'/X_d' > E_o/X_d$ the maximum power is greater during the transient condition than during the steady state. Note also that the second term is reversed in sign since, generally, $X_d > X_q > X_d'$. Assuming that there is no field coil on the quadrature axis then $X_q' = X_q$. This is not true if a divided-winding rotor is used (see Chapter 9). Fig. 3.1(c) shows the steady-state and transient power/angle curves for a loaded, salient-pole generator. When a generator is connected to an infinite busbar, the addition of the constant interconnector reactance to the various generator reactances diminishes the difference between the two power/angle curves.

For a first approximation analysis of transient stability, transient saliency is generally neglected by omitting the second term in [*3.4d*] so that the power/angle graph becomes sinusoidal. It may be noted that this approximation is more justifiable when considering the maximum power than it is for small values of δ where the second term may in fact be of the same order as the first term. A further simplification is to use E', the e.m.f. behind the transient reactance X_d', instead of E_q'.

B.S. 4296 defines all the machine parameters necessary for the analysis of a synchronous machine, and gives methods of test for determining their values.

3.2.2 GENERAL CASE OF THE 2-PORT, 4-TERMINAL NETWORK

If the interconnector can be treated as a short line, its reactance can be added to that of the generator and the above formulae used. Clearly the variations in alternator reactance with time, from the subtransient through the transient to the synchronous state, become less significant the greater the line reactance.

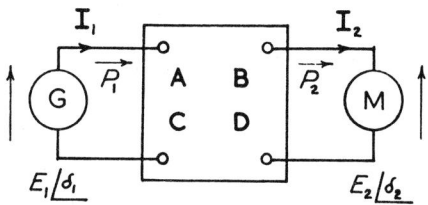

Fig. 3.2. A power system reduced to a 2-port, 4-terminal network.

Fig. 3.2 shows the power system reduced to a 2-port network. G is the equivalent generator representing the power exporting area and M is the equivalent synchronous motor representing the power importing area (or infinite busbar). The overall transfer matrix

$$\begin{bmatrix} \mathbf{A} & \mathbf{B} \\ \mathbf{C} & \mathbf{D} \end{bmatrix}$$

is obtained by multiplying the transfer matrices of G, M and the interconnector: see Volume 1, p. 49 and this volume, Appendix A.1. The equations of the 2-port network are (see Volume 1, section 1.4)

$$\mathbf{E}_1 = \mathbf{A}\mathbf{E}_2 + \mathbf{B}\mathbf{I}_2 \qquad [3.5a]$$

$$\mathbf{I}_1 = \mathbf{C}\mathbf{E}_2 + \mathbf{D}\mathbf{I}_2 \qquad [3.5b]$$

$$P_1 = \text{real part of } \mathbf{E}_1\mathbf{I}_1^*$$

$$= (DE_1^2/B)\cos(\beta - \Delta) - (E_1 E_2/B)\cos(\beta + \delta) \qquad [3.5c]$$

$$P_2 = \text{real part of } \mathbf{E}_2\mathbf{I}_2^*$$

$$= (E_1 E_2/B)\cos(\beta - \delta) - (AE_2^2/B)\cos(\beta - \alpha) \qquad [3.5d]$$

where $\mathbf{A} = A\underline{/\alpha}$, $\mathbf{B} = B\underline{/\beta}$, $\mathbf{D} = D\underline{/\Delta}$, and the overall load angle of the power system is $\delta = \overline{\delta_1 - \delta_2}$. If δ is the only independent variable, then P_2 has a maximum value when $\delta = \beta$.

For a loss-free power system (which infers no shunt loads on the interconnector) the student should be able to show that $\mathbf{A}$ is real, so that $\alpha = 0$ and $\mathbf{B}$ is a pure inductive reactance with $\beta = 90°$ (consider an equivalent-T network). Thus [3.5c and d] both reduce to

$$P = P_1 = P_2 = \frac{E_1 E_2}{B}\sin\delta \qquad [3.6]$$

B is called the short-circuit transfer reactance since $\mathbf{B} = \mathbf{E}_1/\mathbf{I}_2$ when $\mathbf{E}_2 = 0$.

The method of *mesh-current analysis* is discussed in section 8.2, using matrix methods. By way of contrast, it will now be discussed using determinants. Referring to [8.2], it should be noted that the mesh self-impedance of mesh n, Z_{nn}, is the sum of the branch impedances round mesh n and is always positive: that Z_{nm}, the mutual impedance between meshes n and m, is a branch impedance carrying both mesh currents $\mathbf{I}_n$ and $\mathbf{I}_m$, and is positive if it carries both mesh currents in the same direction: and that $\mathbf{E}_n$ is the net e.m.f. acting round mesh n and is positive if it 'generates' $\mathbf{I}_n$. It is conventional always to circulate the currents clockwise so, in general, the mutual impedances are negative. Keeping subscript 2 to indicate the output port, and referring to [8.2],

$$\mathbf{I}_2 = \frac{\begin{vmatrix} \mathbf{Z}_{11} & \mathbf{E}_1 & -\mathbf{Z}_{13} \\ -\mathbf{Z}_{21} & -\mathbf{E}_2 & -\mathbf{Z}_{23} \\ -\mathbf{Z}_{31} & 0 & \mathbf{Z}_{33} \end{vmatrix}}{\begin{vmatrix} \mathbf{Z}_{11} & -\mathbf{Z}_{12} & -\mathbf{Z}_{13} \\ -\mathbf{Z}_{21} & \mathbf{Z}_{22} & -\mathbf{Z}_{23} \\ -\mathbf{Z}_{31} & -\mathbf{Z}_{32} & \mathbf{Z}_{33} \end{vmatrix}}$$

$$= \frac{\mathbf{E}_1\Delta_{12}}{\Delta} - \frac{\mathbf{E}_2\Delta_{22}}{\Delta} \qquad [3.7a]$$

where Δ is the value of the determinant of impedance elements and Δ_{mn} is the co-factor (see Appendix A.1.9) of the impedance element $\mathbf{Z}_{mn}$. Equation [3.7a] can be rewritten as

$$\mathbf{I}_2 = \frac{\mathbf{E}_1}{\mathbf{z}_{21}} - \frac{\mathbf{E}_2}{\mathbf{z}_{22}} \qquad [3.7b]$$

where $\mathbf{z}_{21} = \Delta/\Delta_{12}$ is the short-circuit transfer impedance between meshes 1 and 2, since $\mathbf{z}_{21} = \mathbf{E}_1/\mathbf{I}_2$ when $\mathbf{E}_2 = 0$, and $\mathbf{z}_{22} = \Delta/\Delta_{22}$ is the short-circuit input impedance of mesh 2, since $\mathbf{z}_{22} = -\mathbf{E}_2/\mathbf{I}_2$ when $\mathbf{E}_1 = 0$: the negative sign is due to the choice of relative polarities for $\mathbf{E}_2$ and $\mathbf{I}_2$.

For an example, consider a power system reduced to the equivalent-T network shown in Fig. 3.3(a). Then

$$(\mathbf{z}_a+\mathbf{z}_c)\mathbf{I}_1 - (\mathbf{z}_c)\mathbf{I}_2 = \mathbf{E}_1$$

$$-(\mathbf{z}_c)\mathbf{I}_1 + (\mathbf{z}_b+\mathbf{z}_c)\mathbf{I}_2 = -\mathbf{E}_2$$

$$\Delta = \begin{vmatrix} (\mathbf{z}_a+\mathbf{z}_c) & (-\mathbf{z}_c) \\ (-\mathbf{z}_c) & (\mathbf{z}_b+\mathbf{z}_c) \end{vmatrix}$$

$$= \mathbf{z}_a\mathbf{z}_b+\mathbf{z}_b\mathbf{z}_c+\mathbf{z}_c\mathbf{z}_a$$

$$\Delta_{12} = (-1)^3(-\mathbf{z}_c) = \mathbf{z}_c$$

$$\Delta_{22} = (-1)^4(z_a + z_c) = z_a + z_c$$

$$z_{21} = (z_a z_b + z_b z_c + z_c z_a)/z_c$$

$$z_{22} = z_b + (z_a z_c/(z_a + z_c))$$

z_{22} is clearly the input impedance at port 2 when E_1 is short-circuited. Applying the star-delta transformation gives the equivalent-Π

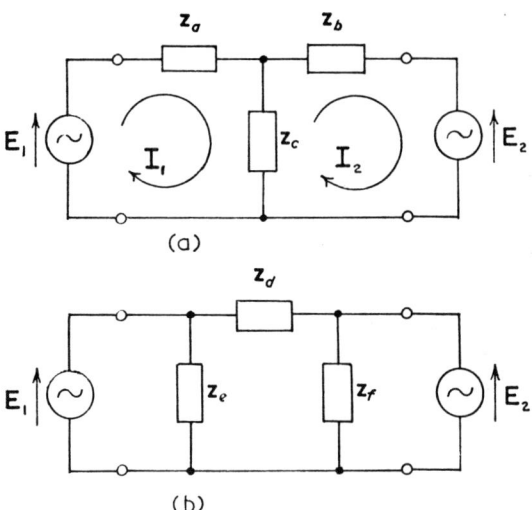

(a)

(b)

Fig. 3.3. (a) Equivalent-T and (b) equivalent-π of a power system.

circuit in Fig. 3.3(b), where $z_d = z_{21}$. The significance of z_{21} as the transfer impedance between ports 1 and 2 is apparent: it is the single equivalent impedance which directly connects the two e.m.f. sources. By interchanging the subscripts a and b, it is clear that $z_{12} = z_{21}$ (bilateral network).

If $E_1 = E_1/\delta_1$, $E_2 = E_2/\delta_2$, $\delta = \delta_1 - \delta_2$, $z_{11} = z_{11}/\theta_{11}$, $z_{12} = z_{12}/\theta_{12}$ and $z_{22} = z_{22}/\theta_{22}$, then

$$P_1 = \text{real part of } E_1 I_1^*$$
$$= (E_1^2/z_{11}) \cos \theta_{11} - (E_1 E_2/z_{12}) \cos (\theta_{12} + \delta) \qquad [3.8a]$$

$$P_2 = \text{real part of } E_2 I_2^*$$
$$= (E_1 E_2/z_{12}) \cos (\theta_{12} - \delta) - (E_2^2/z_{22}) \cos \theta_{22} \qquad [3.8b]$$

TRANSIENT STABILITY OF POWER SYSTEMS

For a loss-free power system, the impedance angles are all 90°, so that

$$P_1 = P_2 = P = \frac{E_1 E_2}{X_{12}} \sin \delta \qquad [3.9]$$

The synchronous power equations [3.1c], [3.6] and [3.9] are all of the same form so, to generalise, the equation will be written as

$$P = P_m \sin \delta \qquad [3.10a]$$

where

$$P_m = \frac{E_1 E_2}{X} \qquad [3.10b]$$

is the steady-state stability limit and X is the transfer reactance.

Transfer reactance during a fault

Fault calculations were discussed in Volume 1. Chapter 7, pp. 247–253 showed by means of worked examples how to calculate the fault currents during a 3-phase fault on a loaded power system by using Thevenin's theorem, while Chapter 8, pp. 320–323, dealt similarly with asymmetrical faults. Fault calculations usually assume that all sources of e.m.f. in the power system are equal (in p.u.) and in phase. This simplification is not valid for stability studies since the e.m.f.s of the equivalent generator (E_1) and of the equivalent motor, or infinite busbar (E_2) are not necessarily equal and are displaced by the load angle δ which is a variable. These e.m.f.s are assumed to be symmetrical, and so consist entirely of positive-sequence components. The positive-, negative- and zero-sequence networks are interconnected according to the type of fault and the complete network reduced to an equivalent delta. The transfer reactance is that which directly connects the two e.m.f. sources (i.e. z_d in Fig. 3.3). Alternatively [3.7b] can be used.

Worked example 3.1

A 132-kV, 3-phase interconnector connecting a generator-transformer set to an infinite busbar consists of two identical, parallel lines for each of which $X_1 = X_2 = X_0/2 = j0.5$ p.u. For the generator-transformer $E' = 1.2$ p.u., $X_1' = j0.2$ p.u. and $X_1' = X_2 = 4X_0$.

The infinite busbar voltage is 1 p.u. and all p.u. values are on the same base. Calculate the transfer reactance and steady-state power limit during a short-circuit fault at the midpoint of one line. Both the transformer h.v. and the infinite busbar star points are solidly earthed.

SOLUTION (see Fig. 3.4)

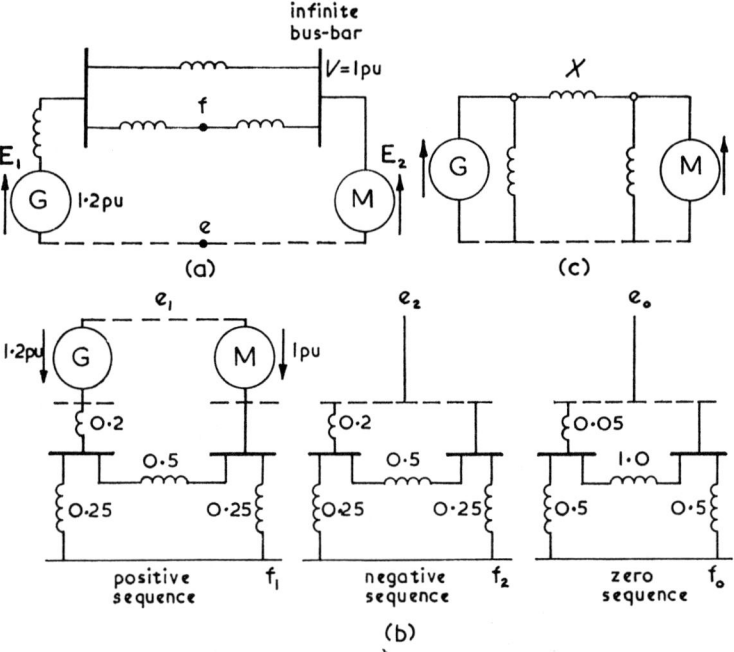

Fig. 3.4. Illustrating the calculation of the transfer reactance during a fault.

$E_1 = E' = $ the e.m.f. behind the transient reactance X_1'.

(a) 3-phase fault—connect f_1 to e_1 (Fig. 3.4b) and reduce the network to an equivalent delta (Fig. 3.4c). Hence

$$X = 1\cdot1 \text{ p.u. and } P_m = \frac{1\cdot2 \times 1}{1\cdot1} = 1\cdot091 \text{ p.u.}$$

(*b*) Earth-fault on one phase—connect all three sequence networks in series. Hence

$$X = 0{\cdot}5352 \text{ p.u. and } P_m = 2{\cdot}242 \text{ p.u.}$$

(*c*) Phase-fault across two phases—connect f_1 to f_2 and e_1 to e_2. Hence

$$X = 0{\cdot}6378 \text{ p.u. and } P_m = 1{\cdot}879 \text{ p.u.}$$

(*d*) Double-earth-fault on two phases—connect f_1 to f_2 to f_0 and e_1 to e_2 to e_0. Hence

$$X = 0{\cdot}7057 \text{ p.u. and } P_m = 1{\cdot}700 \text{ p.u.}$$

For the pre-fault conditions,

$$X = 0{\cdot}2 + (0{\cdot}5/2) = 0{\cdot}45 \text{ p.u.}$$

and

$$P_m = 2{\cdot}667 \text{ p.u.}$$

For the post-fault conditions (faulty line switched out)

$$X = 0{\cdot}7 \text{ p.u. and } P_m = 1{\cdot}714 \text{ p.u.}$$

The student should sketch the power/angle curves for all these conditions. A 3-phase fault causes the greatest reduction in power transfer and an earth-fault the least reduction. Thus, when considering the stability of a power system, it is usual to assume the worst case of a 3-phase fault.

3.3 Equal-area criterion for transient stability

If the consumers' demand for power could be assumed to increase at such a slow rate that the turbine governors could increase the generated power at the same rate, and the automatic voltage regulators could maintain the busbar voltages, then it would appear that a power system could be operated up to its steady-state stability limit, i.e. a power demand of $P_m = E_1 E_2 / X$ and a load angle $\delta = 90°$. In practice, the maximum load power must be less than P_m by a reasonable margin of safety: this will be discussed later, in section 3.7.

It will now be shown that the maximum load power is further limited by the necessity to maintain transient stability. Three cases of large, sudden changes in power-system conditions will be considered: a change in load, or in transfer reactance due to switching, or to a fault followed by the switching out of the faulty component.

3.3.1 CHANGE IN LOAD

Fig. 3.5 shows the power/angle relation for an equivalent generator (representing a power exporting area) connected to an infinite busbar. The ordinates give the synchronous power transferred along the interconnector for any given load angle δ. If the power system

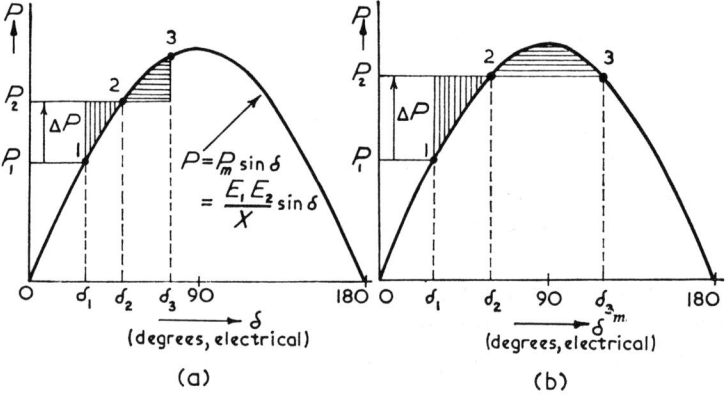

Fig. 3.5. (a) Equal area criterion for a large, sudden change in load
(b) limiting case when $\delta_{3_m} = 180 - \delta_2$.

is operating under the steady-state conditions represented by point 1 when a large local load, ΔP, within the power exporting area is switched off, then, assuming the turbine governor settings remain unaltered, there is a surplus of mechanical power input over electrical power output, equal initially to ΔP, which accelerates the turbogenerator rotor and increases the load angle δ. If $P_2 = P_1 + \Delta P$ (equals the total mechanical power input less the unchanged part of the local electrical load) then the accelerating power

$$P_a = P_2 - P_m \sin \delta \qquad [3.11]$$

decreases from ΔP, when $\delta = \delta_1$, to zero when $\delta = \delta_2$. During the time taken to change the load angle from δ_1 to δ_2 the rotor has been absorbing kinetic energy. It will be shown later (see worked example 3.3) that the change in rotor speed, relative to the synchronous speed of the infinite busbar, can be neglected for a first approximation analysis (providing stability is not lost). Thus power is proportional to torque and

$$\text{kinetic energy} = \int_{\delta_1}^{\delta_2} \text{torque} \,. \, d\delta$$
$$\propto \int \text{power} \,. \, d\delta$$
$$= \text{vertically shaded area in Fig. 3.5.}$$

When the power transfer equals P_2, then $\delta = \delta_2$ and the acceleration is zero but, since the generator now has an electrical speed slightly greater than synchronous speed, the load angle δ continues to increase. Thus the power transfer now exceeds P_2 and the rotor decelerates. The rotor swings (relative to synchronous speed) until the load angle is δ_3 and the rotor speed equals the synchronous speed. Assuming (for the moment) that all losses (e.g. resistance, eddy current damping) can be neglected, the load angle δ_3 is given by the condition that the kinetic energy gained by the rotor as it swung from δ_1 to δ_2 must equal the kinetic energy returned as it swung from δ_2 to δ_3. Thus the shaded areas in Fig. 3.5 are equal: this is the equal-area criterion for the transient stability of a power system following a large, sudden change in load. A formal proof of this criterion will be given in section 3.4.3.

For a load angle δ_3 the power transfer exceeds P_2 so the rotor continues to decelerate, sinks slightly below synchronous speed, and δ decreases. Thus the rotor will oscillate between load angles δ_1 and δ_3. In practice the oscillations will be damped due to losses and the rotor will eventually reach steady-state conditions when the load angle is δ_2.

The accelerating area is given by

$$A_{acc} = \int_{\delta_1}^{\delta_2} (P_2 - P_m \sin \delta) \, d\delta$$
$$= P_2(\delta_2 - \delta_1) - P_m(\cos \delta_1 - \cos \delta_2) \qquad [3.12a]$$

and the decelerating area by

$$A_{dec} = \int_{\delta_2}^{\delta_3} (P_m \sin \delta - P_2) \, d\delta$$
$$= P_m(\cos \delta_2 - \cos \delta_3) - P_2(\delta_3 - \delta_2) \qquad [3.12b]$$

Thus the formula for the equal-area criterion, for the case of a load change, is

$$P_m(\cos \delta_1 - \cos \delta_3) = P_2(\delta_3 - \delta_1)$$

$$\cos \delta_1 - \cos \delta_3 = (\delta_3 - \delta_1) \sin \delta_2 \qquad [3.12c]$$

where the load angles must be measured in radians.

Given δ_1 and δ_2, this equation cannot be solved for δ_3 by algebraic methods, as it is transcendental.

For a given δ_1 the maximum or limiting value of ΔP is given by $\delta_{3_m} = \pi - \delta_2$ rad (see Fig. 3.5b). Consider the generator swinging, first accelerating then decelerating, with an increasing load angle and assume that the equal-area criterion is satisfied as $\delta \to \delta_2$. It has already been shown that if δ does not exceed δ_{3_m} the generator load angle will decrease and stabilise at $\delta = \delta_2$. But if the generator swings until $\delta > \delta_{3_m}$, without having satisfied the equal-area criterion, then it is running at a speed greater than synchronous speed, so its load angle continues to increase. The power transfer is now less than P_2 so the generator continues to accelerate and hence to increase its load angle. At $\delta = \pi$ rad, the power system is 'short-circuited', i.e. E_1 and E_2 are in antiphase in time but are in phase round the closed loop between the generator and the infinite busbar. For $\pi < \delta < 2\pi$ rad the electrical power flow is negative (i.e. the machine is motoring from the infinite busbar) and adds to the mechanical power input. The acceleration could be excessive. When $\delta > 2\pi$ rad the generator is said to have pole-slipped (a pair of poles or one cycle): see end of section 3.3.3.

Substituting $\delta_{3_m} = \pi - \delta_2$ [3.12c] gives the equal-area criterion for the limiting value of a load change as

$$\cos \delta_1 + \cos \delta_2 = (\pi - \delta_1 - \delta_2) \sin \delta_2 \qquad [3.13]$$

Given δ_1, this equation is also transcendental. The equation could be solved approximately by a graphical (counting-squares) method and the answer used, if greater accuracy is necessary, to start an analytical, iterative method of solution. The above equal-area criterion can also be used to estimate the swing in the load angle of a synchronous motor when subjected to a large. sudden change in mechanical load.

3.3.2 CHANGE IN TRANSFER REACTANCE DUE TO SWITCHING

If the interconnector consists of two or more parallel lines, the switching-out of one line (assumed to occur simultaneously and instantaneously at both ends) increases the transfer reactance: see worked example 3.1. Thus the equal-area problem involves two power/angle graphs superimposed. The student should satisfy himself that the formulae of the previous section still apply (possibly with some changes in notation).

3.3.3 CHANGE IN TRANSFER REACTANCE DUE TO A FAULT

The application of the equal-area criterion requires three power/angle curves, one for the pre-fault system, another for the system during the fault and another for the system after the faulted line has been switched out. Fig. 3.6 illustrates two limiting cases of equal

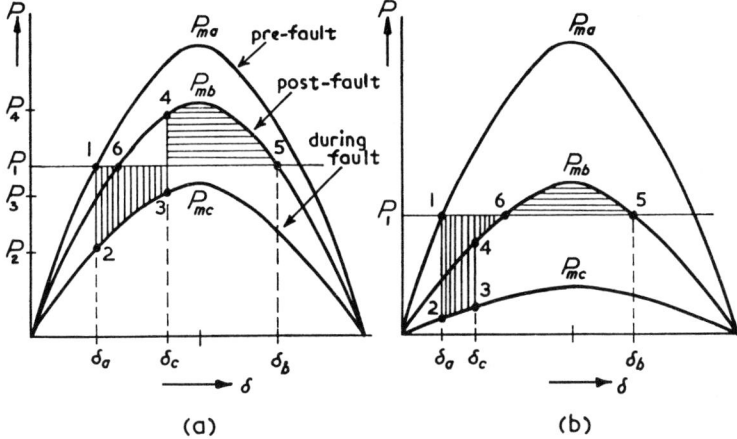

Fig. 3.6. Equal-area criterion for an interconnector fault, showing the critical switching angle δ_c for two cases.

areas. It is required to find the value of δ_c, the critical fault clearing angle, i.e. the maximum load angle at or before which the fault must be cleared in order to retain stability. The student should show that δ_c is given by

$$\cos \delta_c = \frac{P_1(\delta_b - \delta_a) + P_{mb} \cos \delta_b - P_{mc} \cos \delta_a}{P_{mb} - P_{mc}} \qquad [3.14]$$

where the symbols are defined in Fig. 3.6(a); and should consider also if [*3.14*] is applicable to Fig. 3.6(b). The two diagrams illustrate respectively fault clearance after and before the load angle corresponding to final steady-state operation.

Pole-slipping

Pole-slipping, which has already been described towards the end of section 3.3.1, occurs when the load angle δ between a generator and the infinite busbar (i.e. the remainder of the Grid system) exceeds 360°. It is liable to occur during a system fault, especially one close to the generator. It is liable also to occur as a result of interconnector switching (see section 3.3.2). Normally, if a line is to be removed from service, the load transferred by the line is reduced as much as possible before the circuit breakers are opened. If, however, a line trips out due to over-loading, the sudden transfer of the load to adjacent lines produces transients and hence power swings between groups of generators. If these swings increase in amplitude (i.e. the system is unstable), one or more generators may pole-slip. Depending on the rotor speed as δ passes through 360°, a generator may or may not resynchronise. If the slip-speed is high enough, it is possible for several pole-slips to occur.

Multiple pole-slipping is not permitted on the power system in the U.K., because during pole-slipping the excitation current produces pronounced saliency and hence considerable fluctuations in the stator current and voltage. There is also a risk that power-station auxiliaries, or feeders, may trip due to under-voltage. Thus a generator is normally tripped from the system if it pole-slips more than once in a given period of time. Pole-slipping is normally detected by a reverse-power relay which operates a timing device. The timer is designed to reset after a certain period of time, usually about 30 seconds, but if a second pole-slip is detected before the timer resets, the generator is tripped.

Problems can arise with this arrangement. Recent tests have shown that power flow in a generator does not necessarily reverse during a pole-slip. This is because induction-generator action takes place due to eddy-currents induced in the steel rotor forging. Under certain system conditions, particularly at low excitation, it is possible for machines to pole-slip very slowly, taking several minutes to complete one slip cycle. This can make protection setting adjustment very difficult. It is fortunate, however, that slow pole-slips produce much smaller system disturbances than do fast pole-slips.

Pole-slipping also produces undesirable effects in the excitation system. During a pole-slip, very large currents are induced in the rotor winding as the load angle approaches 180°. For angles greater than 180° the rotor current decreases rapidly to maintain an m.m.f. balance between the rotor and stator windings. With modern rectifier-excitation systems, the rotor current cannot reverse, thus when the rotor current decreases to zero and attempts to reverse, the m.m.f. balance can no longer be maintained and a rapid flux change takes place. This results in a large reverse voltage being developed across the rectifiers which may break down if not adequately rated. It is thus important to be able to predict reverse induced voltages under pole-slip conditions. As the steel rotor body acts as a damper winding whose induced currents help to maintain the m.m.f. balance, the induced voltages are less than would be the case if the rotor body currents did not flow. It is thus also important to be able to predict the effects of rotor body currents.

3.4 Mechanics of angular motion

The equations of angular motion are summarised below.

Inertia	$= \text{mass} \times (\text{radius of gyration})^2$
Momentum	$= \text{inertia} \times \text{velocity}$
Torque	$= \text{rate of change of momentum}$
	$= \text{inertia} \times \text{acceleration}$
Energy	$= \text{torque} \times \text{displacement}$
Kinetic energy	$= \frac{1}{2} \times \text{inertia} \times (\text{velocity})^2$
	$= \frac{1}{2} \times \text{momentum} \times \text{velocity}$
Power	$= \text{torque} \times \text{velocity}$
	$= (\text{inertia} \times \text{velocity}) \times \text{acceleration}$
	$= \text{momentum} \times \text{acceleration}$ [3.15]

3.4.1 MOMENTUM (M) AND INERTIA CONSTANT (H)

In [3.15], it is convenient to electrical engineers to quote a unit for the momentum, M, which is consistent with the units of accelerating power, P_a (W, or more usually MW) and the angular acceleration, $d^2\delta/dt^2$ (rad/s²). Thus a unit for M is the MWs²/rad, or the MJs/rad, and [3.15] can be written as

$$P_a = M \frac{d^2\delta}{dt^2} \qquad [3.16]$$

$$\left[MW = \frac{MWs^2}{rad} \cdot \frac{rad}{s^2} \right]$$

For a given torque, the corresponding power varies directly as the velocity: so too does the momentum of a given rotating mass. In the context of this chapter, the changes in speed are small (providing stability is not lost) so both P_a and M are the values corresponding to synchronous speed. The load angle δ can be measured in terms of the radian or degree and both can be either mechanical or electrical, but both M and δ must be measured in the same angular unit. If the unit of δ is changed from radian to degree its numerical value increases by the factor $(360/2\pi)$ while that of M decreases by the same factor. Thus the numerical value of P_a is independent (as it must be) of the choice of angular unit. Similarly, the conversion factor to be used in changing between mechanical and electrical angular measure is

$$1° \text{ mechanical} = p° \text{ electrical} \qquad [3.17]$$

where p is the pole pairs of the machine. The frequency, f Hz, and the rotor velocity, n rev/s, are related by

$$f = pn \qquad [3.18]$$

If the kinetic energy of the rotor at synchronous speed is denoted by ($K.E.$), then its momentum at synchronous speed is given by

$$M = 2(K.E.)/\omega \qquad [3.19]$$

where ω, the synchronous speed, is given by $\omega = 2\pi n$ mechanical rad/s $= 2\pi f$ electrical rad/s $= 360f$ electrical deg/s. The units used throughout [3.19] must be consistent.

Even for generator sets of similar design, e.g. 3000 r.p.m. steam-turbine sets, the kinetic energy and momentum vary over a wide range with variation in rating. So, instead of momentum, the *inertia constant* (or stored-energy constant) is often used as an alternative and is defined as

$$H = \frac{(K.E.)}{S} \frac{MJ}{MVA} \ (= s) \qquad [3.20]$$

where S is the rating of the generator in MVA. The values of H in seconds vary over quite a small range, e.g. 1500 r.p.m. steam-turbine sets, from about 10 for small sets to about 6 for larger sets (100 MW) at present in use; 3000 r.p.m. sets from about 6 for small sets to about 3 for large sets (500 MW); while for water-turbine sets, H increases with rating from about 2 to 4 depending on speed.

M and H are related by

$$(K.E.) = HS = \tfrac{1}{2}M\omega \quad MJ \qquad [3.21]$$

Thus, for a given set, M and H are directly proportional, and

$$M = HS/180f \quad MWs^2/\text{elec. deg.} \qquad [3.22a]$$

The equation for the accelerating power, [3.16], is rewritten as

$$\frac{d^2\delta}{dt^2} = \left(\frac{180f}{HS}\right)P_a \quad \text{elec. deg./s}^2 \qquad [3.22b]$$

where P_a is in MW, and S in MVA.

If the accelerating power is in per unit (p.u.) then $P_{apu} = P_a/S$ and

$$\frac{d^2\delta}{dt^2} = \left(\frac{180f}{H}\right)P_{apu} \quad \text{elec. deg./s}^2 \qquad [3.23]$$

If the accelerating power is in p.u. and the synchronous speed in electrical radians per second, then

$$\frac{d^2\delta}{dt^2} = \left(\frac{\pi f}{H}\right)P_{apu} \quad \text{elec. rad./s}^2 \qquad [3.24]$$

For each case above, the factor in brackets equals $1/M$. Equation [3.23] is the one normally used and in section 3.5 it will be integrated twice to obtain the load angle δ as a function of time t.

3.4.2 REDUCTION OF A POWER SYSTEM TO ONE MACHINE CONNECTED TO AN INFINITE BUSBAR

If several generating sets are operating in parallel at one end of an interconnector, and the transfer reactance between them can be neglected relative to the reactance of the interconnector, then these sets can be replaced with a single equivalent set (subscript e) by equating the total kinetic energy. Thus

$$H_e S_e = H_1 S_1 + H_2 S_2 + \ldots$$

If the several ratings are replaced by an arbitrary base rating S_b, with a corresponding change in H to H_b such that, for example, $H_{1b} S_b = H_1 S_1$, then

$$H_{eb} S_b = H_{1b} S_b + H_{2b} S_b + \ldots$$
$$H_{eb} = H_{1b} + H_{2b} + \ldots \qquad [3.25]$$

Thus the inertia constants based on their individual thermal ratings are changed to an arbitrary base rating, using an inverse proportion, and then added.

If [3.25] is applied to all the sets at the power-exporting end of an interconnector and then to all the sets at the power-importing end, the power system is reduced to an equivalent generator and motor connected by the interconnector—the two-machine system. This system will now be reduced to one equivalent generator connected to an infinite busbar. Assume that, due to an increase in the transfer reactance of the interconnector, the power transfer decreases by ΔP. Then from [3.16] the acceleration of the generator is $\Delta P / M_g$ and the deceleration of the motor is $\Delta P / M_m$. The acceleration of the generator relative to the motor is the algebraic difference, and this last must equal the acceleration of the equivalent single generator relative to the infinite busbar. Thus, after cancelling ΔP,

$$\frac{1}{M_e} = \frac{1}{M_g} + \frac{1}{M_m} \qquad [3.26a]$$

From [3.22a] $M' = M/S_b = H/180f$, so

$$\frac{1}{H_e} = \frac{1}{H_g} + \frac{1}{H_m} \qquad [3.26b]$$

where all the H values are on the common base, S_b.

Thus a complex power system can be reduced to an equivalent generator connected to an infinite busbar, which was the type of system assumed at the beginning of this chapter.

Worked example 3.2

A generator A is rated at 50 Hz, 60 MW, 75 MVA, 1500 r.p.m., and has an inertia constant $H = 7 \cdot 5$ MJ/MVA. The corresponding data for another generator B is 50 Hz, 120 MW, 133·3 MVA, 3000 r.p.m., 4·5 MJ/MVA. (a) If these two generators operate in parallel in a power station, calculate H for the equivalent generator on a base of 100 MVA. (b) If the power station is connected to another power station which has two of each type of generator, calculate H for the equivalent generator connected to an infinite busbar.

SOLUTION

Base 100 MVA.

(a) $H_A = 7 \cdot 5(75/100) = 5 \cdot 625$ MJ/MVA
   (i.e. 562·5 MJ per 100 MVA)
   $H_B = 4 \cdot 5(133 \cdot 3/100) = 5 \cdot 999$ MJ/MVA
   For the equivalent generator
   $H_e = 5 \cdot 625 + 5 \cdot 999 = 11 \cdot 624$ MJ/MVA.

(b) H_e for the second power station will be twice that for the first, i.e. 23·248 MJ/MVA. So for the equivalent generator connected to the infinite busbar

$$\frac{1}{H_e} = \frac{1}{11 \cdot 624} + \frac{1}{23 \cdot 248}$$

$$H_e = 7 \cdot 749 \text{ MJ/MVA.}$$

3.4.3 PROOF OF THE EQUAL-AREA CRITERION

From [3.16], and referring to Fig. 3.5,

$$\frac{d^2\delta}{dt^2} = \frac{P_a}{M}$$

Multiplying through by $d\delta/dt$,

$$\frac{d\delta}{dt} \cdot \frac{d^2\delta}{dt^2} = \frac{1}{2}\frac{d}{dt}\left(\frac{d\delta}{dt}\right)^2 = \frac{P_a}{M} \cdot \frac{d\delta}{dt}$$

$$\frac{d\delta}{dt} = \left[\int \frac{2P_a}{M} \cdot \frac{d\delta}{dt} \cdot dt\right]^{\frac{1}{2}}$$

$$= \left[\frac{2}{M}\int P_a \cdot d\delta\right]^{\frac{1}{2}}$$

The load angle δ will have a turning (maximum) value when $d\delta/dt = 0$, i.e. when

$$\int_{\delta_1}^{\delta_3} P_a \cdot d\delta = 0$$

Thus the rotor swings from δ_1 to a maximum value of δ_3, at which time the relative velocity is zero. The two shaded areas must be equal and of opposite sign.

3.5 Swing curve (load angle/time curve)

Referring to Fig. 3.6(a), for example, the power difference between the input to and output from the generator set is given by

$$\Delta P = P_1 - P_m \sin \delta \qquad [3.27]$$

where P_1 is the initial steady-state mechanical input power to the generator which is assumed to be constant throughout this analysis because of the relatively long time-constant of the governor and of the entrapped steam. $P_m \sin \delta$ is the electrical output of the generator and equals the power transfer along the interconnector. P_m will have the appropriate value P_{ma}, P_{mb} or P_{mc}, and is given, in general terms, by [3.10].

The equation of motion of the rotor (neglecting damping) is

$$M\frac{d^2\delta}{dt^2} + P_m \sin \delta = P_1 \qquad [3.28]$$

Equation [3.28] is non-linear and is normally solved approximately by the following step-by-step or iterative procedure, in which the various quantities are assumed to remain constant over a short interval of time Δt (usually about 0·05 s): see Fig. 3.7. When the sudden, large disturbance occurs at $t = 0$, the difference power ΔP is known, since $\delta = \delta_a$ (Fig. 3.6a), and if this were taken to be the accelerating power, the resulting change in angle could be estimated, and the procedure repeated. Clearly this would result in

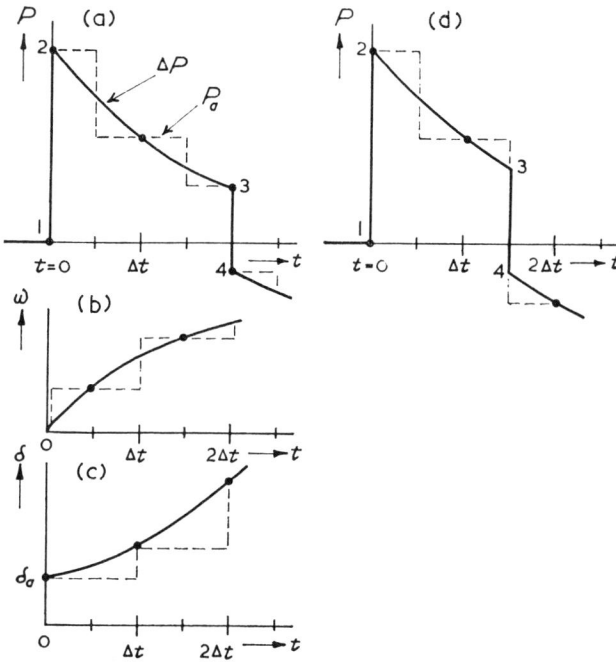

Fig. 3.7. Step-by-step solution of the swing curve, showing the discontinuity in the accelerating power at (a) the end, and (d) the middle of a time interval. The numbered points correspond with those in Fig. 3.6a.

the assumed accelerating power being always greater than the actual accelerating power. Therefore the following assumptions are made.

(a) The accelerating power P_a to be assumed at the beginning of any interval is assumed to have existed from the middle of the previous interval to the middle of the present interval. This results in a stepped assumed power graph whose average approximates to the actual power. For the first interval the accelerating power equals

$P_1 - P_2$ for only half the interval and is zero for the previous half-interval. Thus the accelerating power P_a to be used for the first interval is $(P_1 - P_2)/2$, i.e. the average of the difference powers immediately before and after the discontinuity in ΔP. Similarly, if the switching angle δ_c (Fig. 3.6) occurs at the end of a time interval, the average value $(P_3 + P_4)/2$ is assumed for the next interval (see Fig. 3.7a). If, however, the discontinuity occurs at the middle of a time interval, no corrections are necessary; see Fig. 3.7(d).

(b) Allowing for discontinuities as above, [3.28] can be rewritten, using the units in [3.23], as

$$\frac{d^2\delta}{dt^2} = \left(\frac{180f}{H}\right)P_{apu} \simeq \frac{\Delta\omega}{\Delta t}$$

$$\Delta\omega = \left(\frac{180f\,\Delta t}{H}\right)P_{apu} \qquad [3.29]$$

Equation [3.29] gives the change in velocity $\Delta\omega$, during the time interval Δt. It is assumed that $\Delta\omega$ occurs at the beginning of the interval, and the iteration starts with $\omega = 0$ (relative to synchronous speed) at $t = 0$. The estimated velocity for any interval is assumed to be correct at the middle of that interval.

(c) The velocity during any interval is

$$\omega = \frac{d\delta}{dt} \simeq \frac{\Delta\delta}{\Delta t}$$

$$\Delta\delta = (\Delta t)\omega \qquad [3.30]$$

Equation [3.30] gives the change in load angle during the interval and the iteration starts with $\delta = \delta_a$ at $t = 0$.

The estimated load angle for any interval is assumed to be correct at the end of that interval. Table 3.1 shows a convenient layout for the iteration (if the number of iterations is large).

An alternative method of iteration, which omits the calculation of the rotor velocity, is as follows. If n is the interval number, i.e. from times $(n-1)\Delta t$ to $(n)\Delta t$, then

$$\Delta\omega_{n-1} = \omega_{n-\frac{1}{2}} - \omega_{n-1\frac{1}{2}} = \left(\frac{180f\,.\,\Delta t}{H}\right)P_{apu(n-1)}$$

$$\Delta\delta_n = \delta_n - \delta_{n-1} = \omega_{n-\frac{1}{2}}\,.\,\Delta t$$

TABLE 3.1. *Swing curve calculation for worked example 3.3*

time s	pre-fault 0·0	fault 0·1	0·2	switch 0·2	post-fault 0·3	0·4	0·5
δ deg.	30·0	42·0	72·59	72·59	72·59	99·29	106·77
sin δ		0·5	0·6691	0·9542	0·9542	0·9869	0·9575
power input p.u.	1·0 ↳			1·0	1·0	1·0	1·0
power transfer	1·0	0·4	0·5353	0·7634	1·4313	1·4804	1·4363
ΔP	0·0 ↳	0·6	0·2366	0·2366 ↳	−0·4313		−0·4363
		2)0·6			0·2366		
		0·0			2) −0·1947		
P_a		0·3	0·4647		−0·0973	−0·4804	
$\Delta\omega = 400 P_{a\,pu}$ deg./s		120·0	185·9		−38·92	−192·16	−174·52
initial ω		0·0	120·0		305·9	266·96	74·80
ω during interval		120·0	305·9		266·98	74·80	−99·72
$\Delta\delta = 0\cdot1\omega$ deg.		12·0	30·59		26·70	7·48	−9·97
initial δ		30·0	42·0		72·59	99·29	106·77
final δ		42·0	72·59		99·29	106·77	96·80

$$\Delta\delta_{n-1} = \delta_{n-1} - \delta_{n-2} = \omega_{n-1\frac{1}{2}} \cdot \Delta t$$

$$\Delta\delta_n - \Delta\delta_{n-1} = (\omega_{n-\frac{1}{2}} - \omega_{n-1\frac{1}{2}})\Delta t$$

$$\Delta\delta_n = \Delta\delta_{n-1} + \left(\frac{180f(\Delta t)^2}{H}\right)P_{apu(n-1)} \qquad [3.31]$$

The change in load angle at the end of any interval is equal to the change in angle due to the accelerating power at the beginning of that interval, plus the change in load angle in the previous interval.

Worked example 3.3

An equivalent generator connected to a 50-Hz infinite busbar has an inertia constant of 2·25 MJ/MVA. The steady-state power limits before, during a fault, and after fault clearance are respectively 2·0, 0·8 and 1·5 p.u. Calculate the critical fault clearing angle if the initial load is 1·0 p.u. Estimate the swing curve using a time interval of 0·1 s. The fault is cleared at $t = 0·2$ s.

SOLUTION

Referring to Fig. 3.6(a), and with angles measured in electrical degrees

$$\sin \delta_a = \tfrac{1}{2}, \ \delta_a = 30°$$

$$\sin \delta_b = 1/1·5, \ \delta_b = 41·81°$$

$$\cos \delta_b = 0·7454$$

$$\delta_b = 180 - 41·81 = 138·19°, \ \cos \delta_b = -0·7454$$

Substituting in [3.14] gives the critical fault clearing angle as $\delta_c = 83·69°$.

From [3.23], since power is in per-unit, the acceleration

$$\frac{d^2\delta}{dt^2} = \left(\frac{180f}{H}\right)P_{apu} = \left(\frac{180 \times 50}{2·25}\right)P_{apu} = 4000\,P_{apu}$$

$$\Delta\omega = (4000 \times 0·1)P_{apu} = 400P_{apu} \quad \text{elec. deg./s}$$

$$\Delta\delta = 0·1\omega \quad \text{elec. deg.}$$

Power transfer during the fault = 0·8 sin δ p.u.

Power transfer after the fault = 1·5 sin δ p.u.

S:e Table 3.1. If the results are plotted it will be seen that the generator is transiently stable, with a maximum load angle of about 107 elec. deg., and a maximum velocity rise of about 1·7% of synchronous speed (18 000 elec. deg./s). Also the half-period is about 0·39 s, giving an oscillation frequency of about 1·28 Hz.

The alternative method is as follows.

$$\frac{180f(\Delta t)^2}{H} = \frac{180 \times 50 \times 0·01}{2·25} = 40$$

$$n = 1. \quad \Delta\delta = 40P_{apu} = 40 \times 0·3 = 12°$$

$$\delta = 30 + 12 = 42°$$

$$n = 2. \quad \Delta\delta = 12 + (40 \times 0·4647)$$

$$= 12 + 18·59 = 30·59°$$

$$\delta = 42 + 30·59 = 72·59°$$

3.6 Small oscillations

Referring to Fig. 3.5(a), assume that the generator set is operating at point 1 under steady-state conditions when the mechanical power input is suddenly increased by an amount ΔP so small that the sine curve can be assumed to be a straight line over the range of oscillation $\delta_3 - \delta_1$. Consider the resulting instantaneous change in load angle $(\Delta\delta) = \delta - \delta_1$. The change in electrical power transmitted along the interconnector is given by

$$\frac{dP}{d\delta}(\Delta\delta) = P_r(\Delta\delta) \qquad [3.32a]$$

where P_r, the synchronising power coefficient, is obtained by differentiating [3.10], i.e.

$$P_r = \frac{dP}{d\delta} = \frac{E_1 E_2}{X} \cos \delta \qquad [3.32b]$$

and is, in this context, a constant corresponding to its value when $\delta = \delta_1$. The equation of motion of the rotor is

$$M\frac{d^2\delta}{dt^2} + P_r(\Delta\delta) = \Delta P \qquad [3.33]$$

The transient solution is obtained by putting $\Delta P = 0$ and substituting $p = d/dt$, $p^2 = d^2/dt^2$.

$$Mp^2 + P_r = 0$$

$$p = \pm (-P_r/M)^{\frac{1}{2}} = \pm j\omega_n \qquad [3.34]$$

where $\omega_n = (P_r/M)^{\frac{1}{2}}$ is the undamped natural frequency of oscillation in rad/s, or $f_n = \omega_n/2\pi$ Hz, and the period is $T_n = 2\pi/\omega_n$ s/cycle. The student could check this by showing that the complete solution of [3.33] is

$$\delta = \delta_2 - (\delta_2 - \delta_1) \cos \omega_n t \qquad [3.35]$$

So far, in this chapter, all losses and the consequent damping of the oscillations, have been neglected. Let K_d be the damping coefficient, i.e. the damping power per unit angular velocity. Then

$$M\frac{d^2\delta}{dt^2} + K_d\frac{d\delta}{dt} + P_r(\Delta\delta) = \Delta P \qquad [3.36]$$

The damping considered here is the electrical damping due to eddy-currents in the generator rotor due to the slip speed: it is not the mechanical friction, due to the absolute speed, which is assumed to be deducted from the turbine power.

The characteristic equation corresponding to the transient solution of [3.36] is

$$Mp^2 + K_d p + P_r = 0 \qquad [3.37a]$$

$$p = \{-K_d \pm (K_d^2 - 4MP_r)^{\frac{1}{2}}\}/2M$$

$$= -(K_d/2M) \pm j\left(\frac{4MP_r - K_d^2}{4M^2}\right)^{\frac{1}{2}} \qquad [3.37b]$$

The motion will be under-, critically- or over-damped depending on whether the last radical is positive, zero or negative. The damping coefficient for critical damping, K_{dc}, is given by

$$K_{dc} = 2(MP_r)^{\frac{1}{2}} \qquad [3.38]$$

The under-damped frequency of oscillation is given by

$$\omega_d^2 = (P_r/M) - (K_d/2M)^2$$

$$= \omega_n^2\left[1 - \left(\frac{K_d}{2(MP_r)^{\frac{1}{2}}}\right)^2\right]$$

$$\omega_d = \omega_n(1 - \zeta^2)^{\frac{1}{2}} \qquad [3.39]$$

where $\zeta = K_d/K_{dc} = K_d/(2(MP_r)^{\frac{1}{2}})$ is called the damping ratio: the three cases above correspond to ζ less than, equal to, and greater than unity.

Dividing [3.37a] through by M shows that only two of the three coefficients can be specified independently, and it is usual to specify them indirectly in terms of ω_n and ζ.

Thus

$$K_d/M = (2\zeta(MP_r)^{\frac{1}{2}})/M = 2\zeta\omega_n$$

$$P_r/M = \omega_n^2$$

and the complementary function of [3.36] is often written as

$$\frac{\mathrm{d}^2\delta}{\mathrm{d}t^2} + 2\zeta\omega_n\frac{\mathrm{d}\delta}{\mathrm{d}t} + \omega_n^2(\Delta\delta) = 0$$

Since δ_1 is a constant, it should be noted that

$$\mathrm{d}(\Delta\delta)/\mathrm{d}t = \mathrm{d}\delta/\mathrm{d}t \quad \text{and} \quad \mathrm{d}^2(\Delta\delta)/\mathrm{d}t^2 = \mathrm{d}^2\delta/\mathrm{d}t^2.$$

The theory of small oscillations can be applied to a number of problems including that of a change in the load on a synchronous motor.

3.7 Steady-state stability

The steady-state stability limit of a power system, reduced to an equivalent generator connected to an infinite busbar, is given from [3.10b] by $P_m = E_1E_2/X$. This simple calculation assumes that the turbine governors can increase the mechanical power input at the same rate as that of the slowly increasing electrical load, and that the voltage regulators and excitation systems can maintain a constant system voltage. In a practical calculation for a complex power system, the fall in system voltage as the load rises must be assumed to operate the transformer tapchangers. When these have reached their limits, the system voltage will continue to fall as the load rises. A reasonable assumption is that the steady-state stability limit has been reached when the system voltage has fallen to about 0·8 p.u. Any actual system load must be less than this theoretical limit by a reasonable margin of safety because, for example, the statutory lower limit of consumers' voltage is 0·94 p.u.

For any given system load (e.g. the estimated peak load several years ahead), a load flow analysis would be performed as discussed in Chapter 8. At this load the system must operate with steady-state stability, which could be determined in two ways. (*a*) If a generator's load angle is assumed to increase by say 5°, the synchronising power will cause the generator to return, with damped oscillations, to its original load angle. A typical calculation might assume an increase in the angular separation of the two generators which already have the largest angular separation. A simple test that the motion is a damped oscillation is that the synchronising power coefficient, $P_r = (E_1 E_2 / X) \cos \delta$, of all the generators is positive: see [*3.6*]. (*b*) If a small increase in load is assumed then all the generators must respond with an increased electrical output, i.e. $P_r = \mathrm{d}P/\mathrm{d}\delta$ is positive for all the generators. The choice of the load change, in magnitude and location, should be a worst-case choice, i.e. the one most likely to cause instability. Thus both cases lead to the same criteria.

Given, for example, a 3-generator power system interconnected by a ring transmission system, the transfer reactance between any pair of generators can be determined by the methods outlined in section 3.2.2. Using [*3.9*] with assumed values of load angle, the power equations can be written down and these, when differentiated with respect to load angle, give the differential powers in the form $\mathrm{d}P = (\mathrm{d}P/\mathrm{d}\delta) \, \mathrm{d}\delta$. Solving these equations gives the value of P_r for each generator, and each must be positive if the system is to be stable for the assumed load conditions.

Power systems are subjected continually to fluctuations in load and these must result, at worst, in damped oscillations converging to a final steady-state condition. The initial redistribution of power, following on a large (say 10%), sudden increase in load, is determined by the values of the synchronising power coefficients, P_r, and subsequently by the inertia constants, H, and finally by the speed droops of the turbine governors.

3.8 Methods of improving power-system stability

From Fig. 3.5 and [*3.10*] it can be seen that the problem of maintaining power-system stability is easier if $P_m = (E_1 E_2 / X)$ is large, since for any given load, δ_1 is small, and for any given increase in load the increase in load angle $\delta_2 - \delta_1$ is small. In this sense, salient-pole generators are more stable than round-rotor generators (see section 3.2.1). P_m can be increased by raising the system voltage above its nominal value and by reducing the transfer reactance. Thus the inter-connectors (e.g. the Grid), (a) should have a low series inductive reactance by using bundle conductors (Volume 1, p. 93) with mini-mum spacing between phases and by using series capacitors (Volume 1, p. 59), (b) should consist of several parallel lines and closed rings, and should use auto-transformers (which have a lower leakage reactance than the double-wound type). Stability can be improved by reducing the total fault clearance time (protection plus circuit breaker times) so that the fault is cleared at a load angle less than the critical fault-clearing angle, δ_c in Fig. 3.6. The transfer reactance should still be low after the faulty component has been disconnected from the system. If auto-reclosing circuit breakers (section 1.4.2) are used, and the fault does not clear in the first dead time, the effect of a reclosure would be to reduce the synchronising power and so tend to cause instability. On the Grid system most auto-reclosures are of the delayed, single-shot type, i.e. the circuit breaker at one end is reclosed and must remain closed, to check that the fault has cleared, before the other circuit breaker is closed. Some very long trans-mission lines incorporate, at intervals of about 150 km, very fast *automatic* shunt reactance compensation equipment (see Volume 1, Fig. 1.21) to maintain the system voltage at the installation. Each section between such installations could be operated at a load angle of about 60°, with an overall load angle of about 150°. (This is not yet applicable to the U.K. Grid system.)

The sections following discuss other methods of improving stability.

3.8.1 RESONANT LINKS

The use of current-limiting reactors to reduce a fault current was discussed in Volume 1, 8.2.2. These reactors increase the voltage regulation due to load current and also increase the transfer reactance thus worsening the problem of maintaining power system stability. The use of series capacitors to reduce the overall reactance of a line was mentioned in Volume 1, 1.5.1. A problem arising from their use is the excessive voltage across them when carrying fault current.

A resonant link combines these two ideas into one piece of equipment, and three stages of their development are shown in Fig. 3.8. The link is mainly used in industrial installations, and for linking these installations to the public supply networks. As experience is gained and ratings increased, the links could be installed in the public supply networks.

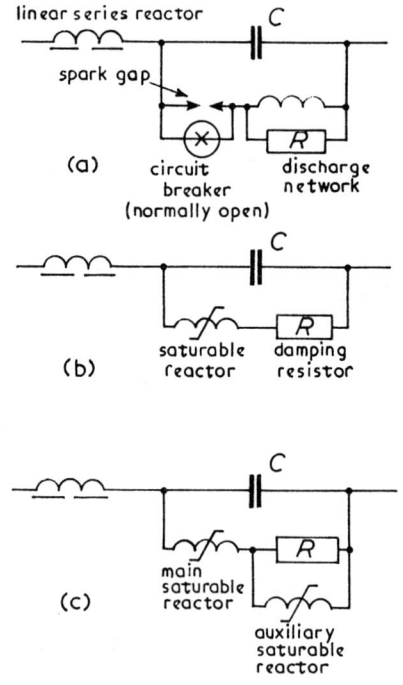

Fig. 3.8. Resonant links; (a) spark gap (b) single saturable reactor (c) double saturable reactor (short circuit limiting coupler—SLC).

Fig. 3.8(a) shows a simple spark gap link in which the capacitor C is tuned approximately to series resonance, at system frequency, with the linear inductor. The voltage across C, when carrying current, appears across the spark gap which is set so as not to flashover for any normal load current, including peak load current due to starting heavy motor loads etc. Thus the resonant link presents almost zero impedance to all load currents. When a system fault occurs, the spark gap flashes-over, C is partially short-circuited, and the overall impedance of the link is raised to limit the fault current almost instantaneously. The discharge network limits and damps the discharge current from C. This discharge current is detected by a relay which closes the circuit breaker, thus allowing the spark gap to de-ionise and reset. When the system fault is cleared by the appropriate circuit breaker, the link circuit breaker is opened manually. But the sudden insertion of C back into the link causes a voltage transient which is liable to restrike the spark gap, so that, in practice, the system load flow must be re-arranged to reduce the link current to about half rated current. This delay in re-setting the link could adversely effect the system stability immediately after fault clearance. The spark gap link is relatively cheap but can only be used when system security and stability are not of major importance.

Fig. 3.8(b) shows the link circuit breaker replaced by an a.c. saturable reactor (reference Friedlander) whose knee-point voltage is greater than the voltage across C for all load currents. Thus the normal losses in the damping resistor are negligible. When a system fault occurs, the reactor saturates and partially short-circuits C so that the overall link impedance is mainly resistive. The damping resistor is large and expensive and its presence adversely affects system stability. The problems of the single saturable reactor link are largely overcome by the double saturable reactor link, often called a short-circuit limiting coupler (SLC): see Fig. 3.8(c). For all normal operating conditions, the shunt circuit across C has a very high impedance so that the link has almost negligible impedance. When a system fault occurs, the main reactor saturates in a fraction of a cycle and passes a large current through the damping resistor R, which damps the d.c. transient current and reduces the fault current asymmetry. The voltage across R then causes the auxiliary reactor to saturate and bypass R. In this condition the link is designed so that the parallel L, C circuit has an overall inductive reactance which adds to that of the linear inductor to limit the fault current, while minimising the heat loss in (and hence cost of) R during the remainder of the fault period. When the fault is cleared the auxiliary reactor de-saturates and re-inserts R into the link to damp the transient as C is re-inserted. The link reverts to the resonant condition in one or two cycles, thus improving stability during the post-fault period.

The changes in link impedance are completely automatic and almost instantaneous.

Resonant links are used where the normal load current through them is small and where a large reduction of fault level is required when they operate. Such a situation is between busbar sections or the connection between an industrial (private) generation plant and the public supply: in such cases the in-feed and the load on each section are equalised as far as possible thus minimising the power transfer through the link. Another application is the installation of new plant to operate in parallel with existing plant: the link is installed with and as part of the new plant so that the fault levels at the existing circuit breakers are not increased above their ratings.

The presence of a resonant link between an industrial generating station and the public supply system restricts the flow of fault current either way and also restricts the flow of synchronising power: in this sense it is not a method of improving system stability during a fault. However, if the rating of the link is comparable with the rating of the industrial plant, ample post-fault synchronising power is available for no-load check synchronisation and power swings of up to twice normal full load. Resonant links can maintain synchronism between two supplies providing fast protection times are maintained on the industrial generation side. If, however, fault clearance times are not short enough to avoid instability, pole-slipping can occur. The impedance across the link becomes high during the actual pole-slip, and providing thermal ratings are not exceeded time is available for the governors to restore stable conditions (see section 3.8.3). The operation of the resonant link under these conditions has the great advantage that synchronising power is available immediately after the pole-slip without any interruption of the supply.

As the high-voltage a.c. power system in the U.K. (the Grid) increases in size and fault level, two methods are available for reinforcing the system without significantly increasing its fault level: one is the high-voltage d.c. (h.v.d.c.) link (see Chapter 6) and the other is the resonant link. The resonant link is the cheaper but, to date, none has been installed. Both these links, when interconnecting two points of one synchronous system, must do so in parallel with a.c. interconnectors to provide the necessary synchronising power during system disturbances.

3.8.2 EXCITATION AND AUTOMATIC VOLTAGE REGULATORS

The satisfactory operation of the synchronous generators of a complex power system at high load angles and during transient conditions is very much dependent on the source of excitation for the generators and on the automatic voltage regulators (A.V.R.). These will now be discussed on the assumption that the student is familiar with the basic theory of control systems.

Voltage regulators: There are several requirements of a good voltage regulator and these conflict to a certain extent. A practical voltage regulator is therefore a compromise between these conflicting requirements, which are as follows: regulation, open-circuit response, steady-state stability and transient response.

(*a*) Regulation: The requirements of voltage control of the transmission system in the U.K. are very stringent, a maximum variation in generator terminal voltage of $\pm$ 0·5% being specified. Variations in the output of a voltage regulator are due to both system and operational causes. System variations arise due to temperature variations in the component values, drift in amplifiers and voltage supplies, and non-linearities in such components as magnetic amplifiers or servos. These variations can be eliminated by good design and careful selection of component tolerances. Operational variations arise due to changing load conditions, which require different generator field voltages.

A basic voltage regulator functions by comparing the rectified and smoothed terminal voltage V_t of the generator with a pre-set d.c. reference signal V_r. The error between these values is amplified, and the amplified voltage is applied to the field winding of the generator. The amplifier may consist of magnetic amplifiers or thyristor amplifiers, or a d.c. pilot exciter feeding into a main d.c. exciter or an a.c. exciter with rectifiers.

The excitation equation under steady-state conditions is of the form

$$V_f = \mu(V_r - V_t)$$

where V_f is the voltage applied to the rotor field winding and μ is the overall gain of the excitation system. It is useful to use per-unit values for the voltages; thus $V_t = 1$ p.u. = rated generator voltage, and $V_f = 1$ p.u. = field voltage to give rated generator voltage on open-circuit. If the machine is excited with a reference voltage V_r, then $V_f = V_t$ on open-circuit, so

$$V_t = \mu(V_r - V_t)$$

$$V_t = V_r\left(\frac{\mu}{1+\mu}\right)$$

Thus V_t is slightly less than the reference value and if μ is very large, $V_t = V_r$ approximately.

When the machine is on rated load the field voltage is several times the open-circuit value—typically 3 p.u. Under these conditions the excitation equation gives

$$3 = \mu(V_r - V_t)$$

$$V_t = V_r - (3/\mu)$$

Thus from no-load to full-load the terminal voltage has drooped by a fraction $(3/\mu)$. A typical specification is that this droop must not exceed 1% (a range of $\pm\ 0.5\%$) so $\mu = 300$. Thus by selecting a sufficiently high value of gain it is possible to maintain the operational voltage variation within prescribed limits. As discussed later, high gain may have disadvantages.

(b) Open-circuit response: The behaviour of the voltage regulator should be such that under open-circuit conditions the response is well damped and not prone to oscillations which may place high voltages on the stator winding insulation.

(c) Steady-state stability: A good voltage regulator should enable a generator to operate stably at load angles well in excess of 90° (see Volume 1, Fig. 7.21).

Consider the simple case of a single generator supplying a static load. The equivalent circuit (model) can be represented by the field voltage V_f in per-unit behind a reactance and V_t across the load. If a fixed field voltage is used the current drawn by the load is proportional to the terminal voltage. As load current is increased due to a lowering of the load impedance, the terminal voltage drops due to the synchronous reactance X_s.

If a voltage regulator is employed which controls the field voltage, the variations in terminal voltage are very much reduced. The effect is the same as having a fixed voltage supply and a much lower reactance. It can be shown that a perfect voltage regulator effectively reduces the machine reactance from its synchronous value to its transient value. The machine is thus operating in a dynamic manner and, even under steady-state conditions, can operate on the transient stability curve stably at load angles of up to about 135° (see Volume 1, sections 7.8 and 7.9).

A practical voltage regulator system cannot given an instantaneous response as it contains time delay elements, particularly the exciter field, so this theoretical maximum cannot always be achieved. In

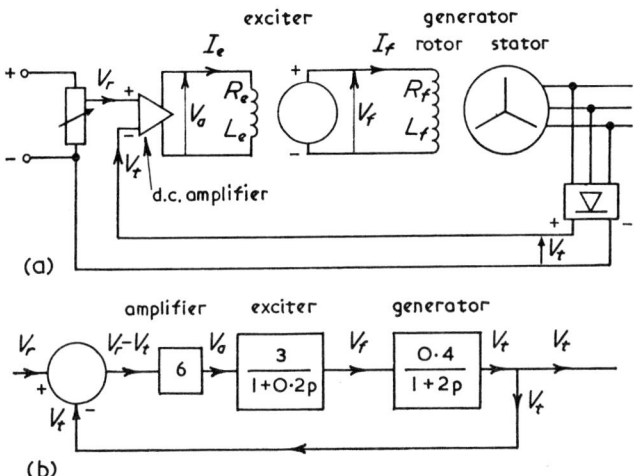

Fig. 3.9. Basic form of voltage regulator and exciter; (a) circuit diagram (b) block diagram of the control loop.

particular, too high a value of regulator gain μ can lead to instability even at relatively low load angles. For this reason techniques have been developed to analyse the steady-state stability of generators fitted with voltage regulators.

Fig. 3.9(a) shows a basic form of voltage regulator, with simple negative feed-back, supplying the field of a d.c. exciter. It is required to develop formulae for the output/input ratio, or transfer function, of each component and hence of the overall excitation system. For the d.c. amplifier (assuming no time delay)

$$v_a = G_a(v_r - v_t)$$

where G_a is the amplifier gain. For the exciter field circuit

$$v_a = R_e i_e + L_e(di_e/dt)$$

or, in operational form where p replaces d/dt,

$$v_a = (R_e + L_e p)i_e = R_e(1 + T_e p)i_e$$

where $T_e = L_e/R_e$ = the time constant of the exciter field circuit. Thus

$$\frac{i_e}{v_a} = \frac{(1/R_e)}{1 + T_e p}$$

For the exciter, neglecting saturation,

$$v_f = k_e i_e$$

where k_e = the slope of the magnetisation curve. Thus, for the exciter as a single unit, its overall transfer function is

$$\frac{v_f}{v_a} = \frac{v_f}{i_e} \cdot \frac{i_e}{v_a} = \frac{G_e}{1 + T_e p}$$

where $G_e = k_e/R_e$ = exciter gain.

Similarly, for the synchronous generator

$$\frac{v_t}{v_f} = \frac{k_g/R_f}{1 + T_f p} = \frac{G_g}{1 + T_f p}$$

where k_g = the slope of the magnetisation curve, R_f = the resistance of, and $T_f = L_f/R_f$ = the time constant of the rotor field circuit and $G_g = k_g/R_f$ = the gain of the generator.

Since the components are in cascade, multiplication of their output/input ratios gives the forward-loop transfer function as

$$\frac{v_t}{v_r - v_t} = \frac{G}{(1 + T_e p)(1 + T_f p)} \qquad [3.40a]$$

where G = the overall gain = the product of the component gains. Thus the overall transfer function is

$$\frac{v_t}{v_r} = \frac{G}{(1 + T_e p)(1 + T_f p) + G} \qquad [3.40b]$$

It should be noted that each of the linear factors in p in the denominator arises from a component in which there is a time delay between a change in the input and the resulting change in the output.

Fig. 3.9(b) shows the control-system block diagram: it shows the flow of the control signals (not the flow of power). Substituting the assumed typical data gives

$$\frac{v_t}{v_r} = \frac{18}{p^2 + 5 \cdot 5p + 20 \cdot 5} \qquad [3.40c]$$

The denominator gives the characteristic equation (see section 3.6) as

$$p^2 + 5 \cdot 5p + 20 \cdot 5 = 0 \qquad [3.40d]$$

which is in the form

$$p^2 + 2\zeta\omega_n p + \omega_n^2 = 0$$

The natural frequency of oscillation of the control system is $\omega_n = (20 \cdot 5)^{\frac{1}{2}} = 4 \cdot 52$ rad/s or $4 \cdot 52/2\pi = 0 \cdot 72$ Hz, and the damping ratio is $\zeta = 5 \cdot 5/(2 \times 4 \cdot 52) = 0 \cdot 61$, which indicates slight under-damping.

The roots of the characteristic equation, which give the poles of the transfer function, are

$$p = -2 \cdot 75 \pm j3 \cdot 59$$

The damped frequency of oscillation is $3 \cdot 59$ rad/s or $0 \cdot 573$ Hz, and the time constant is $1/2 \cdot 75 = 0 \cdot 363$ s.

A linear or a quadratic characteristic equation gives roots which have a negative real part, consequently the control system will be stable. A cubic, or higher, characteristic equation may have at least one root with a non-negative real part in which case the control system will be unstable.

Fig. 3.10 shows a block diagram of a typical voltage regulator of the type used on many older machines. Two static magnetic amplifiers are cascaded, and the output supplies the field of a d.c. exciter, or a.c. exciter and rectifiers. The exciter output feeds the field of the main generator. Each stage of amplification is characterised by a gain and a time constant, i.e. is of the form

$$\frac{G}{1 + Tp}$$

For example, for the first stage magnetic amplifier, $G = 20$ and

$T = 0.05$ s, so for a 1 V step input the output would rise exponentially to 20 V: also the rise time to 20 V would be 0·05 s if the initial rate of rise had been maintained.

To improve the overall response a stabilising loop which feeds back a fraction of the derivative (hence p in the numerator) of the output is used. The complete voltage regulator is made up of a series of such amplifiers and stabilising loops. The generator is much simplified in this example and is characterised by a variable gain $G(\delta)$ which is a

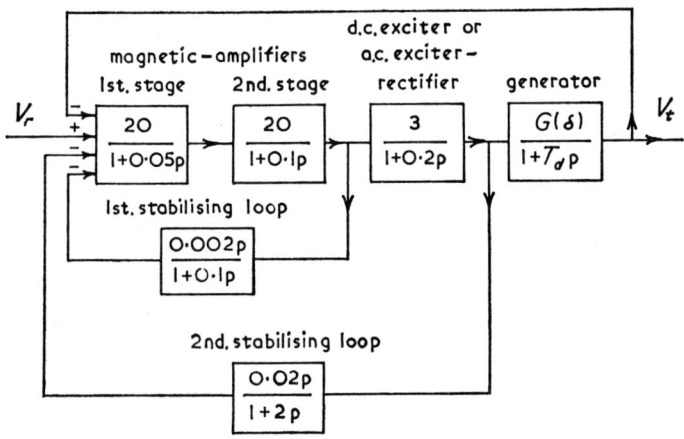

Fig. 3.10. Block diagram for regulators types 4 and 5 (see list near end of Section 3.8.2).

function of load angle and load, and a short-circuit field time constant, T_d, typically 1–2 seconds.

The stability of such a system may be analysed by the Routh–Hurwitz criterion. The overall transfer function of the system is derived in the form

$$\frac{v_t}{v_r} = \frac{B_0 \mathrm{p}^n + B_1 \mathrm{p}^{n-1} + B_2 \mathrm{p}^{n-2} + \ldots + B_n}{A_0 \mathrm{p}^m + A_1 \mathrm{p}^{m-1} + A_2 \mathrm{p}^{m-2} + \ldots + A_m} \qquad [3.41]$$

where the coefficients are functions of the machine and voltage regulator parameters. The denominator of this function is the characteristic equation, and the Routh–Hurwitz criterion is applied to the coefficients of this equation. In this way the maximum permissible load angle as a function of load and regulator parameters may be plotted.

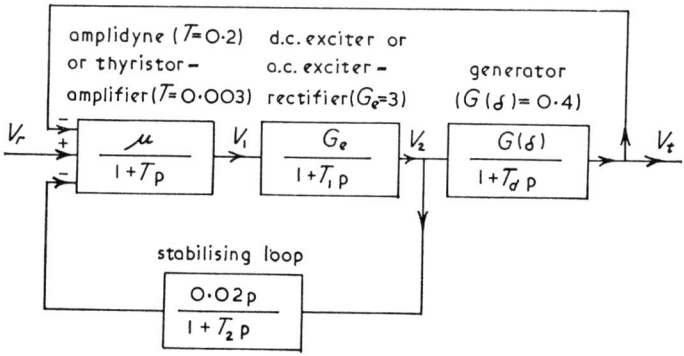

Fig. 3.11. Block diagram for regulators types 3, 6, and 7.

For example, for the regulator shown in Fig. 3.11 the transfer function is obtained as follows:—

$$v_1 = \left\{ v_r - v_t - \left(\frac{0.02p}{1+T_2p} \right) v_2 \right\} \frac{\mu}{1+Tp}$$

$$v_2 = \left\{ \frac{G_e}{1+T_1p} \right\} v_1$$

$$v_t = \left\{ \frac{G(\delta)}{1+T_dp} \right\} v_2$$

Hence, after some algebraic reductions,

$$\frac{v_t}{v_r} = \frac{\mu G_e G(\delta)(1+T_2p)}{(1+Tp)(1+T_1p)(1+T_2p)(1+T_dp) + \mu G_e G(\delta)(1+T_2p) + \\ + 0.02\mu G_e p(1+T_dp)}$$

$$[3.42]$$

The root of the numerator gives the zero of the transfer function, and the roots of the denominator, after reduction to a polynomial

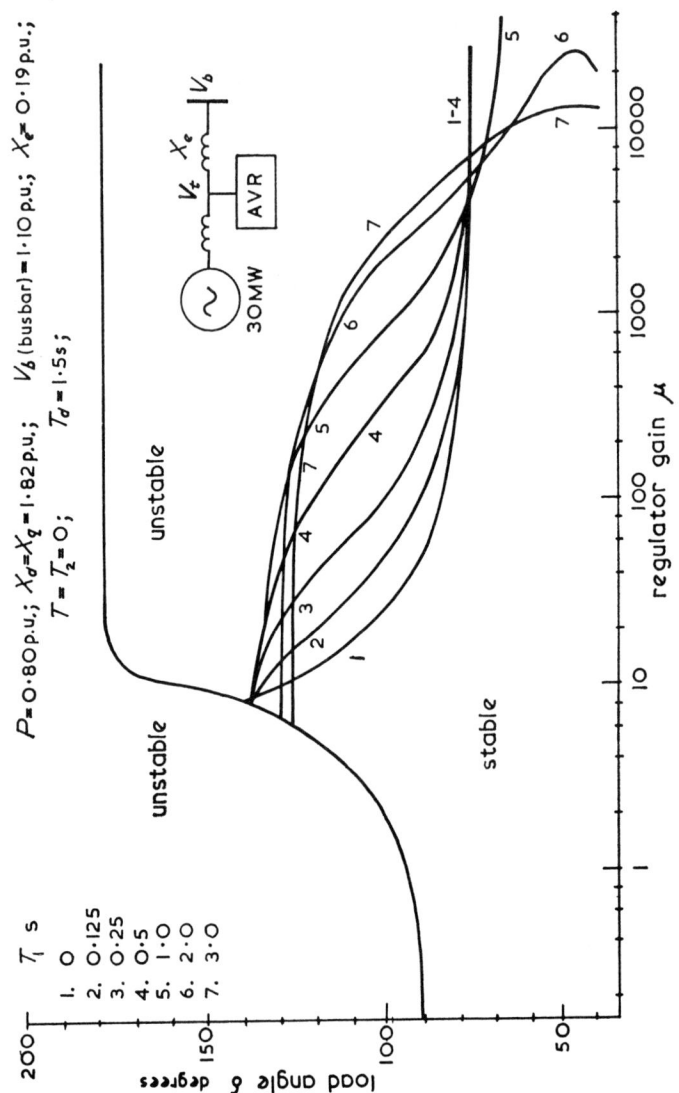

Fig. 3.12. Alternator stability loci with voltage regulator.

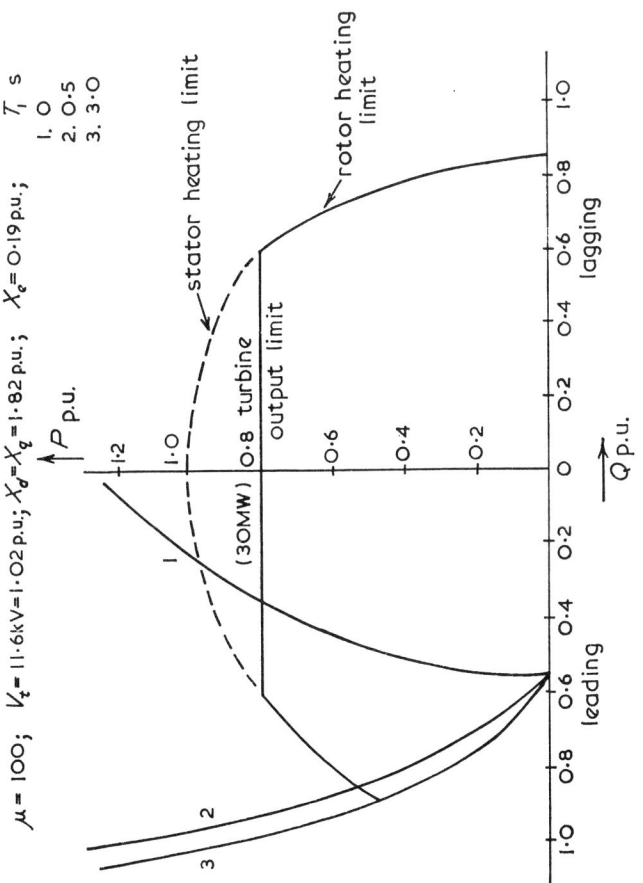

Fig. 3.13. Alternator performance chart.

in p, give the poles: hence the stability of the control system can be investigated.

The stability boundaries for a simplified form of this regulator are shown in Fig. 3.12, for $T = 0$, $T_2 = 0$, $T_d = 1 \cdot 5$ s and for different values of T_1. A 30-MW alternator at full load is considered. It is seen that for high regulator gains a high regulator time constant T_1 is required to maintain stability at high load angles. Angles of 130°–140° can only be achieved with low time constants if the regulator gain is low, typically 5–20. By doing similar studies at different loads it is possible to plot the stability limit on the machine operating chart, Fig. 3.13, which shows plots for a regulator gain of 100, and shows how a small regulator time constant greatly extends the stability boundary.

(d) Transient response: Under system fault conditions it is desirable to be able rapidly to increase excitation, so that the maximum synchronising power exists when the fault is cleared, to prevent loss of synchronism.

For example, in Fig. 3.14, curve (a) shows the synchronous power/load angle relationship for prefault conditions with the generator operating at point 1. During the fault (curve (b)) the rotor accelerates. With no excitation control, when the fault is cleared, the generator operating point will return to curve (a) at point (3) (assuming no change in the transfer reactance). The forward momentum, even whilst now decelerating, may take the operating point beyond point 4, at which point the rotor begins to accelerate again, and synchronism may be lost. If, however, forced excitation control is used, the operating point is point 5 on curve (c), and the limiting swing is given by point 6. The decelerating power is much greater and the chances of losing synchronism are reduced. The ability of the excitation system to force rapidly so that a recovery curve such as (c) is obtained, depends very much on the delays in the regulator circuit. It is thus very desirable to reduce these to a minimum, for optimum transient response.

Types of voltage regulator: The previous sections have shown how regulator design is influenced by the performance requirements, and how this leads to conflict, i.e. delays are useful for steady-state stability but undesirable for transient stability, whilst high gain is good for regulation but undesirable for steady-state stability. The development of regulators over the years has been aimed at reducing these conflicts. Below are listed the developments in excitation systems, roughly in chronological order.

1. D.C. exciters with manual field control.
2. D.C. exciters with electro-mechanical regulators.
3. D.C. exciters with amplidyne control.
4. D.C. exciters with magnetic amplifier control.
5. A.C. exciters with magnetic amplifier control and static diode rectifiers.

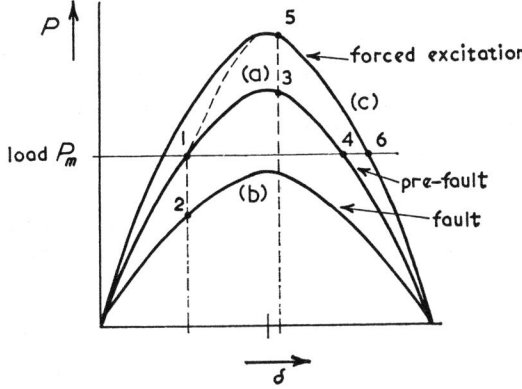

Fig. 3.14. Illustrating the effect of field forcing.

6. A.C. exciters with thyristor field control and static diode rectifiers.
7. A.C. exciters with thyristor field control and rotating diode rectifiers.
8. Static thyristor exciters.
9. A.C. exciters with rotating thyristors.

Manual control of the field was rapidly superseded by the use of electro-mechanical regulators which saw many years' service, but did not improve dynamic stability, only regulation. Amplidyne and magnetic amplifier regulators for d.c. exciters were virtually contemporaneous, but suffered from time lags and required stabilising loops (Figs. 3.10 and 3.11).

On 500-MW and 660-MW alternators in the U.K., a.c. exciters and static rectifiers have been substituted for d.c. exciters to obtain several advantages. The exciter time constant is reduced from about 3 s to about 0·1 s, the commutation difficulties with a sizeable machine running at 3000 r.p.m. are eliminated and the overall length of the turbo-alternator is reduced. More recently a system with a rotating armature exciter has been developed for 660-MW sets, which is brushless and has shaft-mounted diodes.

The advent of the power thyristor has enabled the exciter field time constant to be eliminated, and transformer-fed static-thyristor

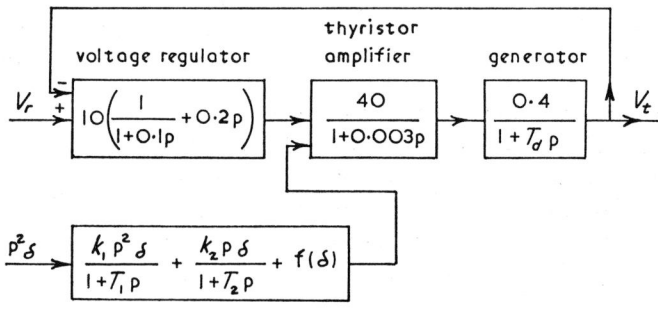

Fig. 3.15. Block diagram for regulators types 8 and 9.

rectifier systems are becoming increasingly popular. Such systems have the advantage that the inherent delays are very small and the voltage regulator can be designed to give optimum performance. One such system is shown in Fig. 3.15. A high gain is used to maintain good regulation, and a delay is incorporated in the voltage channel to ensure good steady-state performance with the high gain. Good transient performance is obtained by using derivatives of the load angle δ, which feed into the thyristor amplifier independently of the voltage channel. Thus signals other than voltage can be used to advantage with a thyristor excitation system, as they are not distorted by delays in an exciter.

The logical extension of this system is to use shaft-mounted thyristors and a completely brushless arrangement. Work on this development has been proceeding for some time and a trial installation is already being made on a 60-MW alternator.

3.8.3 TURBINE GOVERNORS

In the past, governors have been used to maintain a degree of steady-state stability of power systems under changing load conditions, and to prevent overspeeding during emergency trip conditions (load rejection).

Variations in the load on a power system produce frequency variations which have to be limited by altering the generation connected to the system. If the load increases suddenly, frequency will drop unless increased power is injected into the system. Too rapid a drop in frequency can result in instability and system failure. Frequency changes are too rapid to be controlled by manual variations of generation alone, so the governor of each turbo-alternator is given a 'droop' characteristic which automatically varies generated power as the system frequency changes.

A 4% droop is typical. Thus a 4% drop in system frequency will open the governor of an alternator from the no-load to the full-load position. This droop may be represented by the equation

$$P_G = P_S - \frac{25(f - f_0)P_R}{f_0} \qquad [3.43]$$

where P_G is the steam-flow power input,

P_R is the rated output power,

P_S is the setting of the governor speeder gear,

f and f_0 are the instantaneous and base system frequency, respectively.

Fig. 3.16 shows a diagrammatic representation of a turbine and its governor. The steam flow from the boiler is controlled by the governor valve servo which has a delay T_s, such that the steam flow into the H.P. cylinder P_H is

$$P_H = \frac{P_G}{1 + T_s \mathrm{p}}$$

A fraction AP_H of the mechanical output power is generated in the H.P. section, whence the steam flows into the reheater, which constitutes a delay of the order of 10–15 seconds, T_r. The intermediate cylinder steam flow is thus

$$P_I = \frac{P_H}{1+T_r\mathrm{p}}$$

and a fraction BP_I of the output power is generated.

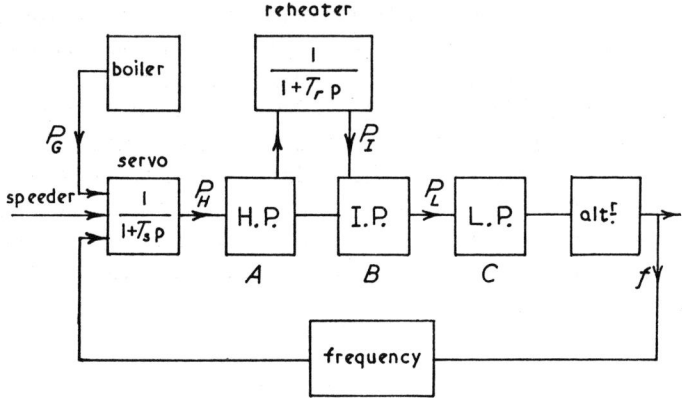

Fig. 3.16 Block diagram for a turbo-generator set.

The steam flows straight into the low pressure section from the I.P. section, thus $P_L = P_I$. The remaining mechanical output power is generated in the L.P. section, CP_L, such that $A+B+C = 1$.

The total mechanical output power

$$\begin{aligned}
P_m &= AP_H+BP_I+CP_L \\
&= AP_H+(B+C)P_I \\
&= AP_H+(1-A)P_I \\
&= AP_H+\left(\frac{1-A}{1+T_r\mathrm{p}}\right)P_H \\
&= P_H\left(\frac{1+AT_r\mathrm{p}}{1+T_r\mathrm{p}}\right) \\
&= \frac{P_G(1+AT_r\mathrm{p})}{(1+T_r\mathrm{p})(1+T_s\mathrm{p})} \\
&= \frac{1+AT_r\mathrm{p}}{(1+T_r\mathrm{p})(1+T_s\mathrm{p})}\left[P_S-\frac{25(f-f_0)}{f_0}P_R\right]
\end{aligned}$$

$$[3.44]$$

This equation shows how the mechanical power will respond to a change in speeder gear setting, or in system frequency.

A more recent development in governor design has been the electronic governor, in which the governor valve is driven electrically and controlled by electronic signals, thus reducing the servo time constant, T_s, and improving the response. In this way acceleration signals as well as speed signals may be incorporated in the governor control, which can improve transient stability by shutting off steam during a fault. Unfortunately about 50% of the mechanical power is produced in the H.P. cylinder ($A = 0.5$), and although this can be controlled very rapidly, the I.P. and L.P. power changes are delayed by the stored energy in the reheater limb, so full control of steam power is extremely difficult. One way of overcoming this problem is to govern the interceptor valves in the reheater loop, and the latest designs of machines with electronic governors incorporate this feature.

REFERENCES

BROWN, et al. 1970. Effects of excitation, turbine energy control and transmission on transient stability. *I.E.E.E. Trans.* **PAS-89,** 1247–1251.

COLES, H. E. 1965. Effects of prime-mover governing and voltage regulation on turboalternator performance. *Proc. I.E.E.* **112,** 1395–1405.

COLES, H. E. 1968. Dynamic performance of a turboalternator utilising 3-term governor control and voltage regulation. *Proc. I.E.E.* **115,** 266–279.

DEMELLO, F. P. & CONCORDIA, C. 1969. Concepts of synchronous machine stability as affected by excitation control. *I.E.E.E. Trans.* **PAS-88,** 316–329.

DINELEY, J. L. & KENNEDY, M. W. 1964. Influence of governors on power-system transient stability. *Proc. I.E.E.* **111,** 98–106.

FITZGERALD, A. E. & KINGSLEY, C. 1961. *Electric Machinery.* McGraw-Hill.

FRIEDLANDER, E. 1966. Static network stabilisation. *G.E.C. Journal* **33,** 58.

GOVE, R. M. 1965. Geometric construction of the stability limits of synchronous machines. *Proc. I.E.E.* **112,** 977–985.

HO, K. W. 1964. Graphical method for solving transient stability problems. *Proc. I.E.E.* **111,** 335–342.

HUMPAGE, W. D. & STOTT, B. 1964. Effects of autoreclosing circuit breakers on transient stability of e.h.v. transmission systems. *Proc. I.E.E.* **111,** 1287–1298.

JACOVIDES, L. J. & ADKINS, B. 1966. Effect of excitation regulation on synchronous-machine stability. *Proc. I.E.E.* **113,** 1021–1034.

KABRIEL, B. J. 1970. Choosing power-system voltage-regulator parameters by use of standard forms. *Proc. I.E.E.* **117,** 1809–1814.

KABRIEL, B. J. & EVANS, F. J. 1969. Simple formulae for voltage-regulator gains for best alternator stability. *Proc. I.E.E.* **116,** 1907–1914.

KANEFF, S. & AKHTAR, M. Y. 1970. Influence of synchronous-machine rotor angular-velocity variations in transient-stability studies. *Proc. I.E.E.* **117**, 1675–1682.

KIMBARK, E. W. 1969. Improvement of power system stability by changes in the network. *I.E.E.E. Trans.* **PAS-88**, 773–781.

LOKAY, H. E. & BOLGER, R. L. 1965. Effect of turbine generator representation in system stability studies. *I.E.E.E. Trans.* **PAS-84**, 933–942.

O'KELLY, D. 1970. Steady-state power/rotor-angle characteristic for synchronous machine including hysteresis. *Proc. I.E.E.* **117**, 1683–1690.

OLIVE, D. W. 1966. New techniques for the calculation of dynamic stability. *I.E.E.E. Trans.* **PAS-85**, 767–777.

PATCHETT, G. N. 1970. *Automatic Voltage Regulators and Stabilizers.* Pitman.

SHACKSHAFT, G. 1970. Effect of oscillatory torques on the movement of generator rotors. *Proc. I.E.E.* **117**, 1969–1974.

STROEV, V. A. & SREEDHRAN, R. 1967. Steady-state stability of alternators as affected by voltage regulators. *Proc. I.E.E.* **114**, 939–945.

SURANA, S. L. & HARIHARAN, M. V. 1968. Transient response and transient stability of power systems. *Proc. I.E.E.* **115**, 114–120.

B.S. 2658: Guide to terms used in a.c. power system studies.

B.S. 4296: Methods of test for determining synchronous machine quantities.

Examples

1.　Two generator-transformer sets are interconnected by a 132-kV, single-circuit interconnector. The total (pre-fault) system reactances are $X'_1 = X_2 = X_0/2 = j30\,\Omega$/phase, and the e.m.f.s of both sources are 132 kV and can be assumed to remain constant. Both sources are solidly earthed and the system resistance can be neglected. If a fault occurs 2/3rds from the sending end, calculate the transfer reactances during each of the following faults: earth-fault, phase-fault, double-earth-fault, 3-phase fault. Sketch the corresponding power/angle graphs, including that for the pre-fault condition, and comment on them.

(L.P.) (40, 60, 75, ∞ Ω/ph)

2.　Three generator-transformer sets A, B and C have internal reactances of 0·3, 0·4 and 0·5 p.u., respectively. The ring interconnectors AB, BC and CA have reactances of 0·6, 0·7 and 0·8 p.u., respectively. All these per-unit values are on the same base. Calculate the short-circuit transfer reactances between the pairs of generators A–B, B–C and C–A.

(L.P.) (1·542, 2·237, 1·971 p.u.)

3. A 3-phase generator-transformer set is connected to an infinite busbar via duplicate lines (two identical lines in parallel) and the steady-state stability limit is 200 MW. This limit is reduced to 70 MW during a 3-phase short-circuit midway along one line. Calculate the p.u. reactance (neglect resistance), on a base of 100 MVA, of the generator-transformer and of each line, and the power limit when the faulty line is isolated. What reasonable assumption must be made regarding the system e.m.f.s.?

(L.P.) (0·4286, 0·1429 p.u., 175 MW)

4. An importing power area has a total demand of 25 MW from an infinite busbar via an interconnector. The steady-state power limit is 80 MW. Estimate, using the equal-area criterion, the maximum additional area load that could be suddenly switched on without the system losing stability. Make and state any necessary assumptions. Investigate an iterative procedure for improving the accuracy of the estimated answer.

(L.P.) (40·5 MW)

5. A large cylindrical-rotor generator is delivering power over a transmission system to an infinite busbar when a fault occurs on the system. The transfer reactances between the generator and the busbar before, during, and after the fault are 0·4, 1·1 and 0·5p.u. respectively. Resistance is negligible. Calculate the critical clearing angle if, at the time of the fault, the load transfer was 0·8 p.u. Sketch the power/angle curves and show the equality of the accelerating and decelerating areas.

(L.P.) (116°)

6. A 150-MVA generator-transformer unit having an overall reactance of 0·3 per unit is delivering 150 MW to infinite busbars over a double-circuit 220-kV line having an impedance per phase per circuit of j100 Ω. A 3-phase fault occurs midway along one of the transmission lines. Calculate the maximum angle of swing that the generator may achieve before the fault is cleared without loss of stability.

(I.E.E.) (90°)

7. The transfer reactances between a generator and an infinite busbar operating at 132 kV, before, during and after a fault on the interconnector are:

Before the occurrence of the fault   140 Ω per phase
During the fault   385 Ω per phase
After clearance of the fault   175 Ω per phase

If the fault is cleared when the generator rotor has advanced 80 electrical degrees from its steady position before the fault, determine the maximum load that could be carried without the fault causing instability.

(I.E.E.) (52·4 MW)

8. A 2-pole, 50 Hz, 11·5-kV, turbo-alternator has a rating of 60 MW, power factor 0·85 lagging. Its rotor has a moment of inertia of 8800 kgm^2. Calculate its inertia constant in MJ/MVA and its momentum in MJs/electrical degree.

(L.P.) (6·15, 0·0483)

9. A power station A has six identical generator sets each rated at 60 MVA and each having an inertia constant of 6 MJ/MVA: the corresponding data for a second station B is 5 sets each of 200 MVA and 2 MJ/MVA. If these two stations are close together at one end of a long tie-line (interconnector), calculate the inertia constant of the single equivalent set on a base of 100 MVA.

(L.P.) (41·6 MJ/MVA)

10. A 4-pole, 50-Hz turbo-alternator is rated at 45 MW, power factor 0·8 lagging, and has an inertia of 25 000 kgm^2. It is connected via a transmission system to another set whose corresponding data is 2-pole, 50-Hz, 60-MW, 0·85 lagging, 9000 kgm^2. Calculate the inertia constant of each set on its own rating, and that of the single equivalent set connected to an infinite busbar and on a base rating of 100 MVA.

(L.P.) (5·48, 6·29, 2·07 MJ/MVA)

11. A turbo-alternator set has a momentum of 0·01 MJs/electrical degree and an output of 50 MW to an infinite busbar via a lossless interconnector system whose steady-state power limit is 100 MW.

When a section of the interconnector is switched out the power limit is reduced to 60 MW. Estimate the relative angular displacement of the alternator, with respect to the infinite busbar, at the end of the first two intervals of 50 ms immediately following switching.

(L.P.) (32·5, 39·5°)

12. A single machine rated at 50 Hz, 500 MVA, is excited to 455 kV (on the 400-kV level and behind its transient reactance), has an inertia constant of 2 MJ/MVA, is loaded to 450 MW, and is connected to a 400-kV infinite busbar via an interconnector. The transfer reactance of the system is 10% prior to a fault, 20% during a fault and 15% after fault clearance, all on 100 MVA. Estimate the rotor angle at the end of the first three intervals each of 0·075 second, assuming the fault is cleared at $t = 0·15$ s.

(L.P.) (29·0, 43·52, 57·72°)

13. A 50-Hz, 50-MVA generator has a momentum of 0·05 MWs^2/ electrical degree and is loaded to 0·8 per unit. It is connected to an infinite busbar such that the steady-state power limits before, during and after a fault are 1·6, 0·2, and 1·2 per unit respectively. Estimate the load angle at the end of each of the first four intervals of 0·1 s assuming the fault is cleared at $t = 0·25$ s. All the per-unit values are on a base rating of 50 MVA.

(L.P.) (33·5, 43·9, 60·91, 75·43°)

Chapter 4

TRAVELLING WAVES IN TRANSMISSION LINES

4.1 Introduction

In 'lumped' circuit theory, it is assumed that changes in voltage and current due to a disturbance in one part of a network occur simultaneously with those in all other parts. This does not, of course, imply that the changes are completed at once, but merely that the changes begin to take place at the same instant as the disturbance begins, and that after a transient period a steady-state condition is reached. This is equivalent to assuming that the size of the circuit is negligibly small or that the disturbance moves through the circuit with an infinite velocity. Since the latter cannot exceed the speed of light $c = 3 \times 10^8$ ms^{-1}, the time taken for a disturbance to propagate along a transmission line is significant, and the representation of a line by a 'lumped' circuit is inadequate except for certain limited purposes.

In section 1.3.2 of Volume 1, the case of a single conductor with an infinite earth plane nearby was considered, and the distributed-parameter line differential equations were solved and various supply-frequency conditions were examined. Forward (or incident) and backward (or reflected) waves were referred to briefly in section 1.3.6. In order to develop further the discussion of travelling waves it is necessary to review as briefly as possible in the next section this same case of one conductor and an earth plane.

4.2 Single conductor line with nearby earth plane

In the circuit of Fig. 4.1, Z_{11} and Y_{11} represent the series self-impedance and shunt admittance per unit length of the single conductor. V_{10} is the voltage between the conductor at a given point and the earth plane, so that V_{10} represents the r.m.s. phase voltage of a particular frequency on a completely balanced three-phase line.

From Fig. 4.1,

$$\frac{dV_{10}}{dx} = -Z_{11}I_1 \qquad [4.1]$$

$$\frac{dI_1}{dx} = -Y_{11}V_{10} \qquad [4.2]$$

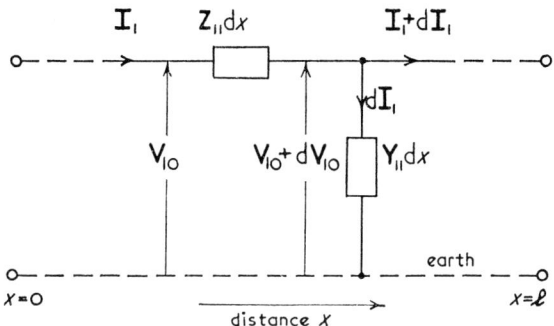

Fig. 4.1. Line element.

so that

$$\frac{d^2V_{10}}{dx^2} = Y_{11}Z_{11}V_{10} \qquad [4.3]$$

$$\frac{d^2I_1}{dx^2} = Y_{11}Z_{11}I_1 \qquad [4.4]$$

The solution of [4.3] is

$$V_{10} = a\,e^{\gamma_{11}x} + b\,e^{-\gamma_{11}x} \qquad [4.5]$$

where

$$\gamma_{11} = \sqrt{(Y_{11}Z_{11})} = \alpha_{11} + j\beta_{11} \qquad [4.6]$$

a and b are arbitrary constants which, as shown in section 1.3.6 of Volume 1, are phasors for a sinusoidal single-frequency voltage applied to the line, and the two terms of [4.5] represent respectively a backward-travelling wave $V_{b1}e^{\gamma_{11}x}$ and a forward-travelling wave $V_{f1}e^{-\gamma_{11}x}$ (note that a forward-travelling wave is here taken to travel in the direction of increasing x whereas in sections 1.3.2 and 1.3.6 of Volume 1 the reverse convention was adopted for convenience).

[4.5] may be re-written using an arbitrary constant T_{11} (though in the present case it is unity) as

$$V_{10} = T_{11}(V_{f1} e^{-\gamma_{11}x} + V_{b1} e^{\gamma_{11}x}).$$ [4.7]

From [4.1]

$$I_1 = Z_{11}^{-1}\gamma_{11}T_{11}(V_{f1} e^{-\gamma_{11}x} - V_{b1} e^{\gamma_{11}x})$$ [4.8]

since T_{11} is 1 here and γ_{11} is defined by [4.6]

$$I_1 = \frac{1}{Z_{c11}} (V_{f1} e^{-\gamma_{11}x} - V_{b1} e^{\gamma_{11}x})$$ [4.9]

where

$$Z_{c11} = \sqrt{\left(\frac{Z_{11}}{Y_{11}}\right)}$$ [4.10]

and Z_{c11} is the characteristic (or surge) self-impedance of conductor 1. If the line losses can be neglected Z_{c11} becomes $Z_c = \sqrt{(L/C)}$.

If R_1, L_1 and C_1 are the line series resistance, inductance and capacitance to earth per unit length, and the shunt conductance is negligibly small then

$$\gamma_{11} = [(R_1 + j\omega L_1)(j\omega C_1)]^{\frac{1}{2}}$$
$$= j\omega(L_1 C_1)^{\frac{1}{2}}(1 + R_1/j\omega L_1)^{\frac{1}{2}}$$
$$\simeq j\omega(L_1 C_1)^{\frac{1}{2}}(1 + R_1/2j\omega L_1)$$

and from [4.6]

$$\alpha_{11} \simeq \frac{R_1}{2}\sqrt{\frac{C_1}{L_1}} \simeq \frac{R_1}{2Z_c}$$

Due, however, to the high voltages on lines during surge conditions causing very significant corona losses and currents which may flow in earth paths, the simple expression given above for attenuation coefficient has only a very limited use. Transmission lines are subjected to non-sinusoidal disturbances such as occur due to lightning or to switching (see Chapters 5 and 2), but it is not possible to obtain a general solution to [4.3] for such arbitrary waveshapes. The lightning voltage waveshape may be approximately represented for standard test purposes as the difference of two exponential com-

ponents which gives a wave of the type shown in Fig. 4.2, e.g. the
wave usually applied in the impulse testing of equipment is the
1/50 wave, where $t_1 = 1\ \mu s$ and $t_2 = 50\ \mu s$.

4.2.1 LOSS-FREE LINE

When a surge voltage travels along a line the instantaneous voltage v
at any time t and position x in the line may be written as some
function F of t and x denoted by

$$v = F(t + Ax) \qquad [4.11]$$

If for simplicity a loss-free line is considered for the present purpose,
then changing from r.m.s. values of sinusoidal currents and voltages
to instantaneous values, [4.3] becomes

$$\frac{\partial^2 v}{\partial x^2} = L_1 C_1 \frac{\partial^2 v}{\partial t^2} \qquad [4.12]$$

where L_1 and C_1 are the inductance and capacitance per metre.
Since the velocity of propagation u of the surge along the line is
(see section 1.3.6 of Volume 1) $u = \omega/\beta_{11}$ for each component
frequency of the surge $\omega/2\pi$ and the phase-change coefficient
$\beta_{11} = \omega\sqrt{(L_1 C_1)}$ for a loss-free line (see section 1.3.8 of Volume 1)
then

$$\frac{\partial^2 v}{\partial x^2} = \frac{1}{u^2} \frac{\partial^2 v}{\partial t^2} \qquad [4.13]$$

For the general surge voltage of [4.11], [4.13] gives

$$A^2 F''(t + Ax) = \frac{1}{u^2} F''(t + Ax)$$

so that $A = \pm \dfrac{1}{u}$

$$v = F\left(t \pm \frac{x}{u}\right) \text{or } v = F(x \pm ut)$$

are solutions to [4.13].

   If a voltage $v_f = f(x - ut)$ of the type illustrated in Fig. 4.2 is
impressed on the line of Fig. 4.1 at the point $x = 0$ at time $t = 0$, and,

by time t_1, has moved a distance x_3 along the line, then the voltage at positions intermediate between $x = 0$ and $x = x_3$ will be as indicated in Fig. 4.3. Since only a loss-free line is being considered at the moment, the voltage at some intermediate position x_1 will be

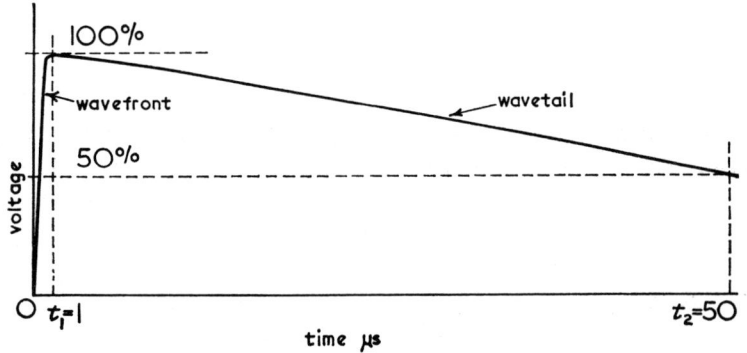

Fig. 4.2. 1/50 impulse voltage waveform.

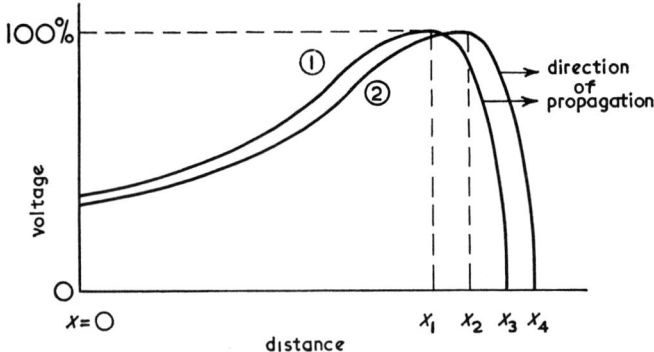

Fig. 4.3. Voltage distribution along a line due to a forward-travelling wave,

    1   at instant t_1
    2   at instant t_2,   where $t_2 > t_1$.

100% (see Fig. 4.3). At a time $t_2 > t_1$ when the wavefront has reached x_4 and the peak voltage occurs at a position x_2, this distance x_2 must be given by $(x_1 - ut_1) = (x_2 - ut_2)$ which shows that $x_2 > x_1$ as illustrated by Fig. 4.3, and this component is, therefore, a forward-travelling wave. Similarly $v_b = b(x+ut)$ represents a backward-travelling wave moving in the direction of decreasing x back towards the source of the original disturbance, and caused by some discontinuity (see section 4.4).

For a loss-free line, [4.2] becomes for instantaneous conditions

$$-\frac{\partial i}{\partial x} = C_1 \frac{\partial v}{\partial t}$$ [4.14]

and for a forward-travelling wave $v_f = f(x-ut)$ which causes a forward-travelling wave of current i_f

$$\frac{\partial i_f}{\partial x} = C_1 u f'(x-ut)$$

so that

$$i_f = C_1 u f(x-ut) = C_1 u v_f$$

and since

$$u = \frac{1}{\sqrt{(L_1 C_1)}}$$

$$i_f = v_f / Z_c$$ [4.15]

where $Z_c = \sqrt{(L_1/C_1)}$ is the characteristic (self) impedance (or surge impedance, for which the symbol Z_0 is often used but which is avoided here because of possible confusion with zero-sequence impedance). In general the characteristic impedance is given by [4.10] but for a loss-free (or distortion-free) line it is a pure resistance Z_c. Applying [4.14] to a backward-travelling voltage wave $v_b = b(x+ut)$ with a resulting backward-travelling wave of current i_b gives

$$\frac{\partial i_b}{\partial x} = -C_1 u b'(x+ut)$$

so that

$$i_b = -C_1 u b(x+ut) = -C_1 u v_b$$

$$i_b = -v_b / Z_c$$ [4.16]

The complete solution of [4.13] is, therefore, an instantaneous voltage v of the conductor with respect to earth which is given by

$$v = v_f + v_b = f(x-ut) + b(x+ut)$$ [4.17]

and the instantaneous current i flowing in the conductor is

$$i = i_f + i_b = \frac{1}{Z_c}(v_f - v_b) \qquad [4.18]$$

The significance of the negative sign in [4.16] and in the more general equation [4.9] is that the backward-travelling wave of current is flowing in the negative x direction in the line conductor when its backward-travelling voltage v_b is positive (i.e. where v_b makes the conductor positive with respect to earth). Conversely i_b flows in the positive x direction through the conductor if v_b is negative. Thus if both forward and backward waves of voltage are positive, the backward wave of current differs from the forward one in two respects, viz. in its direction of flow in the conductor as well as in the direction in which the current wave propagates along the line. Positive forward and backward waves of voltage which raise the potential of the line conductor above that of earth are shown in Fig. 4.4(a) and (b) respectively, with their corresponding current waves. Similarly Fig. 4.4(c) and (d) show currents set up by forward- and backward-travelling voltage waves which are negative so that the line potential is reduced below that of earth.

The characteristic (self) impedance $Z_c = \sqrt{(L/C)}$ for the loss-free transmission line is a resistive approximation to the surge self-impedance operator given by [4.10] which is a function of frequency

$$\sqrt{\frac{\mathbf{Z}_{11}}{\mathbf{Y}_{11}}} = \sqrt{\left(\frac{R_1 + L_1 s}{G_1 + C_1 s}\right)} = \sqrt{\frac{L_1}{C_1}}\sqrt{\left(\frac{s + \alpha_{11} + \beta_{11}}{s + \alpha_{11} - \beta_{11}}\right)} = \mathbf{Z}_{c11}(s)$$

where s is the complex variable $\sigma + j\omega$. For a line with some power losses, the waveshapes of voltage and current are no longer exactly the same. Furthermore, the equivalent circuit of Fig. 4.1 only represents conditions in each phase of a three-phase line if the circuit is completely balanced (see section 7.3.1). The inductance and capacitance of a three-phase transmission line are not, however, completely balanced because of the asymmetric spacing of the line conductors (see Chapters 2 and 3 of Volume 1), so that instead of a single resistive characteristic impedance Z_c, there is strictly a set of complex self- and mutual impedances for each of the component frequencies which make up a given disturbance on a line. For many purposes the simple approximation of an average resistive characteristic self-impedance Z_c which is employed in section 4.4 is sufficient. In recent years, however, problems have arisen for which this approximation is unsatisfactory, for example in power-line carrier communication protection of long lines (see section 1.3),

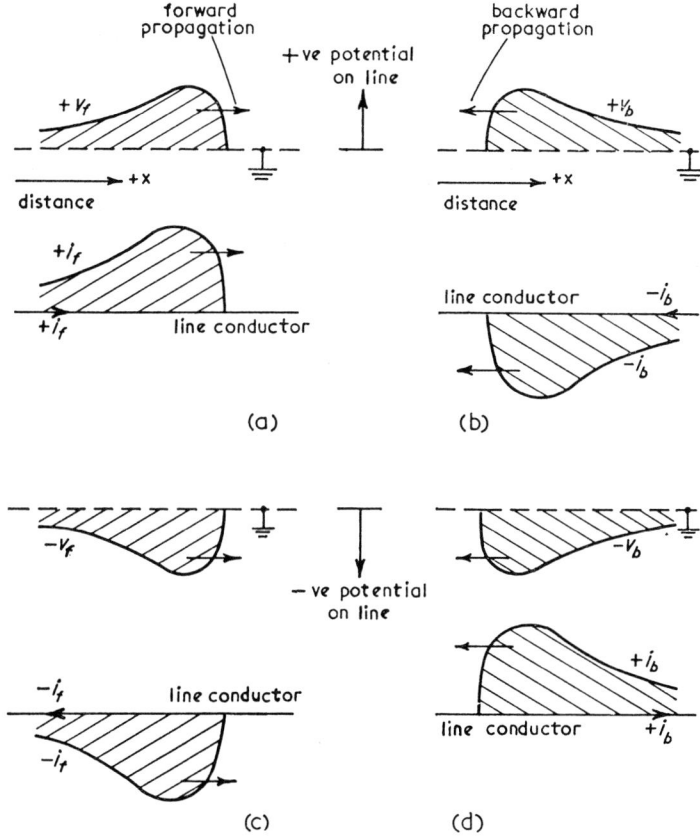

Fig. 4.4. Schematic representation of polarities and directions of voltage and current surges.

fault location (see section 4.4.2), switching out unloaded lines (see sections 2.10 and 2.11), and recovery voltages of circuit breakers under short-line fault conditions (see section 2.13). For such purposes, it has been shown by Wedepohl (1963) that matrix methods can be used to formulate equations which may then be solved by digital computer. Some introduction to the work of Wedepohl in the application of matrix methods is given in section 4.3 for the simple case of two conductors with nearby earth plane, after first introducing the problem in the next section.

4.2.2 TWO CONDUCTORS WITH NEARBY EARTH PLANE

In place of the single conductor of Fig. 4.1 there are now two conductors 1 and 2 with self-impedance Z_{11} and Z_{22} respectively per unit length, and mutual impedance per unit length between them of Z_{12}; $Z_{12} = Z_{21}$ for a bilateral system and this will be due mainly to mutual inductance, though power transfer resistances can arise which represent power transfer between one phase and another, due to asymmetrical spacing (see section 2.9 of Volume 1). Due to these series parameters for an element of line, the two equations corresponding to [4.1] are

$$\frac{dV_1}{dx} = -Z_{11}I_1 - Z_{12}I_2 \qquad [4.19]$$

$$\frac{dV_2}{dx} = -Z_{12}I_1 - Z_{22}I_2 \qquad [4.20]$$

where V_1 and V_2 are written in place of V_{10} and V_{20} as the r.m.s. voltages of a particular frequency between the respective line and earth.

Similarly there are shunt admittances per unit length between each line and earth and between the two lines. The latter is denoted by Y_{12} but if the symbols for the former were Y_{11} and Y_{22} then the equation corresponding to [4.2] would appear as $dI_1/dx = -Y_{11}V_1 - Y_{12}V_{12} = -(Y_{11}+Y_{12})V_1 + Y_{12}V_2$. In order to give equations which correspond with [4.19] and [4.20], it is more convenient to define the shunt admittance of line 1 to earth as $(Y_{11}-Y_{12})$ instead of Y_{11}, and that between line 2 and earth as $(Y_{22}-Y_{12})$ instead of Y_{22}. The equations corresponding to [4.2] are then

$$\frac{dI_1}{dx} = -Y_{11}V_1 + Y_{12}V_2 \qquad [4.21]$$

$$\frac{dI_2}{dx} = Y_{12}V_1 - Y_{22}V_2 \qquad [4.22]$$

Differentiation and substitution of these four equations gives

$$\frac{d^2V_1}{dx^2} = P_{11}V_1 + P_{12}V_2 \qquad [4.23]$$

$$\frac{d^2V_2}{dx^2} = P_{21}V_1 + P_{22}V_2 \qquad [4.24]$$

$$\frac{d^2I_1}{dx^2} = P_{11}I_1 + P_{21}I_2 \qquad [4.25]$$

$$\frac{d^2I_2}{dx^2} = P_{12}I_1 + P_{22}I_2 \qquad [4.26]$$

where

$$P_{11} = (Z_{11}Y_{11} - Z_{12}Y_{12}) \qquad [4.27]$$

$$P_{12} = -(Z_{11}Y_{12} - Z_{12}Y_{22}) \qquad [4.28]$$

$$P_{21} = (Z_{12}Y_{11} - Z_{22}Y_{12}) \qquad [4.29]$$

$$P_{22} = -(Z_{12}Y_{12} - Z_{22}Y_{22}) \qquad [4.30]$$

By writing [4.23] to [4.26] in operational form these equations may be solved, but the solutions are more conveniently obtained by matrix algebra as indicated in section 4.3, since the latter can readily be extended to deal with digital computer calculations involving larger numbers of asymmetrically-spaced conductors, as occur with bundle-conductor and double-circuit three-phase transmission lines.

4.3 Matrix methods applied to two conductors with nearby earth plane

[4.19] to [4.26] can be written in matrix form as

$$\left[\frac{dV}{dx}\right] = -[Z][I] \qquad [4.31]$$

$$\left[\frac{dI}{dx}\right] = -[Y][V] \qquad [4.32]$$

$$\left[\frac{d^2V}{dx^2}\right] = [Z][Y][V] = [P][V] \qquad [4.33]$$

$$\left[\frac{d^2I}{dx^2}\right] = [Y][Z][I] = [P]_t[I] \qquad [4.34]$$

If a component-voltage matrix $[V_c]$ is used such that

$$[V] = [T][V_c] \qquad [4.35]$$

then $[T]$ is a linear transformation matrix (similar to the symmetrical component transformation matrix $[T_s]$—see section 7.6) which for the two-conductor case has four elements, and which for the single-conductor case had only one element which was unity, so that for a single conductor $V_{c1} = V_{10}$ is given by

$$V_{c1} = V_{f1}\, e^{-\gamma_{11}x} + V_{b1}\, e^{\gamma_{11}x} \qquad [4.36]$$

In the present two-conductor case $[V_c]$ will (like $[V]$) be a column vector, and it has two components V_{c1} and V_{c2}. From $[4.33]$ and $[4.35]$

$$\left[\frac{d^2 V_c}{dt^2}\right] = [T]^{-1}[P][T][V_c] = [\gamma^2][V_c] \qquad [4.37]$$

if

$$[\gamma^2] = [T]^{-1}[P][T]$$

The transformation matrix $[T]$ may be chosen so as to make $[\gamma^2]$ a diagonal matrix, in which case

$$\frac{d^2}{dx^2}\begin{bmatrix} V_{c1} \\ V_{c2} \end{bmatrix} = \begin{bmatrix} \gamma_1^2 & 0 \\ 0 & \gamma_2^2 \end{bmatrix}\begin{bmatrix} V_{c1} \\ V_{c2} \end{bmatrix} \qquad [4.38]$$

so that

$$\frac{d^2 V_{c1}}{dx^2} = \gamma_1^2 V_{c1} \qquad [4.39]$$

$$\frac{d^2 V_{c2}}{dx^2} = \gamma_2^2 V_{c2} \qquad [4.40]$$

$[4.39]$ and $[4.40]$ are then exactly the same as $[4.3]$ so that

$$V_{c1} = V_{f1}\, e^{-\gamma_1 x} + V_{b1}\, e^{\gamma_1 x} \qquad [4.41]$$

$$V_{c2} = V_{f2}\, e^{-\gamma_2 x} + V_{b2}\, e^{\gamma_2 x} \qquad [4.42]$$

which are similar to $[4.36]$ which applies to the single-conductor case. Since $[T]^{-1}[P][T] = [\gamma^2]$, premultiplying both sides by $[T]$ gives

$$[P][T] = [T][\gamma^2] \qquad [4.43]$$

Expanding [4.43]

$$\begin{bmatrix} \mathbf{P}_{11} & \mathbf{P}_{12} \\ \mathbf{P}_{21} & \mathbf{P}_{22} \end{bmatrix}\begin{bmatrix} \mathbf{T}_{11} & \mathbf{T}_{12} \\ \mathbf{T}_{21} & \mathbf{T}_{22} \end{bmatrix} = \begin{bmatrix} \mathbf{T}_{11} & \mathbf{T}_{12} \\ \mathbf{T}_{21} & \mathbf{T}_{22} \end{bmatrix}\begin{bmatrix} \gamma_1^2 & 0 \\ 0 & \gamma_2^2 \end{bmatrix}$$

$$\begin{bmatrix} (\mathbf{P}_{11}\mathbf{T}_{11}+\mathbf{P}_{12}\mathbf{T}_{21}) & (\mathbf{P}_{11}\mathbf{T}_{12}+\mathbf{P}_{12}\mathbf{T}_{22}) \\ (\mathbf{P}_{21}\mathbf{T}_{11}+\mathbf{P}_{22}\mathbf{T}_{21}) & (\mathbf{P}_{21}\mathbf{T}_{12}+\mathbf{P}_{22}\mathbf{T}_{22}) \end{bmatrix} = \begin{bmatrix} \mathbf{T}_{11}\gamma_1^2 & \mathbf{T}_{12}\gamma_2^2 \\ \mathbf{T}_{21}\gamma_1^2 & \mathbf{T}_{22}\gamma_2^2 \end{bmatrix}$$

Equating corresponding elements

$$\left.\begin{array}{l} \mathbf{P}_{11}\mathbf{T}_{11}+\mathbf{P}_{12}\mathbf{T}_{21} = \mathbf{T}_{11}\gamma_1^2 \\ \mathbf{P}_{21}\mathbf{T}_{11}+\mathbf{P}_{22}\mathbf{T}_{21} = \mathbf{T}_{21}\gamma_1^2 \end{array}\right\} \qquad [4.44]$$

$$\left.\begin{array}{l} \mathbf{P}_{11}\mathbf{T}_{12}+\mathbf{P}_{12}\mathbf{T}_{22} = \mathbf{T}_{12}\gamma_2^2 \\ \mathbf{P}_{21}\mathbf{T}_{12}+\mathbf{P}_{22}\mathbf{T}_{22} = \mathbf{T}_{22}\gamma_2^2 \end{array}\right\} \qquad [4.45]$$

[4.44] and [4.45] are both contained within

$$\left.\begin{array}{l} \mathbf{P}_{11}\mathbf{T}_{11}+\mathbf{P}_{12}\mathbf{T}_{21} = \mathbf{T}_{11}\gamma^2 \\ \mathbf{P}_{21}\mathbf{T}_{11}+\mathbf{P}_{22}\mathbf{T}_{21} = \mathbf{T}_{21}\gamma^2 \end{array}\right\} \qquad [4.46]$$

[4.46] corresponds to [A.1.24] and [A.1.25] in Appendix 1 except that it is one order less. The corresponding quantities are [T] and [X], [P] and [A], and γ^2 and λ. (λ being a scalar which may be either a real or complex number, can equally pre-multiply or post-multiply [X] in [A.1.25].)

Thus the two values of γ^2 required, viz. γ_1^2 and γ_2^2 are the eigenvalues of the matrix [P] as shown in Appendix A.1.10. Applying [A.1.28] gives

$$|\mathbf{P}-\gamma^2| = 0 \qquad [4.47]$$

so that from [4.46]

$$\begin{vmatrix} (\mathbf{P}_{11}-\gamma^2) & \mathbf{P}_{12} \\ \mathbf{P}_{21} & (\mathbf{P}_{22}-\gamma^2) \end{vmatrix} = 0$$

$$(\mathbf{P}_{11}-\gamma^2)(\mathbf{P}_{22}-\gamma^2)-\mathbf{P}_{12}\mathbf{P}_{21} = 0$$

This quadratic in γ^2 gives

$$\left.\begin{array}{l} \gamma_1^2 = \tfrac{1}{2}(\mathbf{P}_{11}+\mathbf{P}_{22}+\{(\mathbf{P}_{11}-\mathbf{P}_{22})^2+4\mathbf{P}_{12}\mathbf{P}_{21}\}^{\frac{1}{2}}) \\ \gamma_2^2 = \tfrac{1}{2}(\mathbf{P}_{11}+\mathbf{P}_{22}-\{(\mathbf{P}_{11}-\mathbf{P}_{22})^2+4\mathbf{P}_{12}\mathbf{P}_{21}\}^{\frac{1}{2}}) \end{array}\right\} \qquad [4.48]$$

Thus [4.48] enables the values of the two propagation coefficients appearing in [4.41] and [4.42] to be calculated from the system impedances and admittances using [4,27] to [4.30]. The next stage is to calculate the elements of the transformation matrix [T]. [4.44] and [4.45] which may be written in matrix form as

$$[P - \gamma_i^2][T_i] = 0 \qquad [4.49]$$

where $[T_i]$ is a column vector the elements of which are those of column i of [T], show that T_{11} and T_{21} are mutually dependent and that T_{12} and T_{22} are mutually dependent. For each of these pairs of elements it is necessary for one to be specified, e.g. if T_{11} and T_{22} are specified then from [4.44]

$$T_{21} = (\gamma_1^2 - P_{11})\frac{T_{11}}{P_{12}} \qquad [4.50]$$

and from [4.45]

$$T_{12} = (\gamma_2^2 - P_{22})\frac{T_{22}}{P_{21}} \qquad [4.51]$$

The values of γ_1^2 and γ_2^2 given by [4.48] can now be substituted into [4.50] and [4.51], and these in turn can be substituted into [4.35] using the component voltages of [4.41] and [4.42] to evaluate the voltages to earth of the two lines which are given by

$$V_1 = T_{11}(V_{f1}\,e^{-\gamma_1 x} + V_{b1}\,e^{\gamma_1 x}) + T_{12}(V_{f2}\,e^{-\gamma_2 x} + V_{b2}\,e^{\gamma_2 x})$$

$$[4.52]$$

$$V_2 = T_{21}(V_{f1}\,e^{-\gamma_1 x} + V_{b1}\,e^{\gamma_1 x}) + T_{22}(V_{f2}\,e^{-\gamma_2 x} + V_{b2}\,e^{\gamma_2 x})$$

$$[4.53]$$

The arbitrary constants V_{f1}, V_{b1}, V_{f2} and V_{b2} depend upon the waveshapes of the disturbances in any particular case, and there is no general solution to the wave equation for arbitrary waveshapes. All of the equations given here apply only to components of one single frequency.

The line currents can be determined by re-writing [4.31] as

$$[I] = -[Z]^{-1}\left[\frac{dV}{dx}\right] \qquad [4.54]$$

Substituting values of dV_1/dx and dV_2/dx obtained from [4.52] and [4.53] into [4.54] gives

$$\begin{bmatrix} I_1 \\ I_2 \end{bmatrix} = -\frac{1}{|Z|} \begin{bmatrix} Z_{22} & -Z_{12} \\ -Z_{12} & Z_{11} \end{bmatrix}$$

$$\times \begin{bmatrix} T_{11}\gamma_1(-V_{f1}\,e^{-\gamma_1 x}+V_{b1}\,e^{\gamma_1 x})+T_{12}\gamma_2(-V_{f2}\,e^{-\gamma_2 x} \\ \hspace{6cm} +V_{b2}\,e^{\gamma_2 x}) \\ T_{21}\gamma_1(-V_{f1}\,e^{-\gamma_1 x}+V_{b1}\,e^{\gamma_1 x})+T_{22}\gamma_2(-V_{f2}\,e^{-\gamma_2 x} \\ \hspace{6cm} +V_{b2}\,e^{\gamma_2 x}) \end{bmatrix}$$

$$I_1 = \frac{1}{|Z|}\{\gamma_1(Z_{22}T_{11}-Z_{12}T_{21})(V_{f1}\,e^{-\gamma_1 x}-V_{b1}\,e^{\gamma_1 x})$$
$$+\gamma_2(Z_{22}T_{12}-Z_{12}T_{22})(V_{f2}\,e^{-\gamma_2 x}-V_{b2}\,e^{\gamma_2 x})\} \quad [4.55]$$

$$I_2 = \frac{1}{|Z|}\{\gamma_1(Z_{11}T_{21}-Z_{12}T_{11})(V_{f1}\,e^{-\gamma_1 x}-V_{b1}\,e^{\gamma_1 x})$$
$$+\gamma_2(Z_{11}T_{22}-Z_{12}T_{12})(V_{f2}\,e^{-\gamma_2 x}-V_{b2}\,e^{\gamma_2 x})\} \quad [4.56]$$

where

$$|Z| = Z_{11}Z_{22}-Z_{12}^2$$

[4.52] and [4.53] show that the voltage to earth of each of the two line conductors contains one forward- and one backward-travelling wave, both with propagation coefficient γ_1, and a second pair of forward- and backward-travelling waves with propagation coefficient γ_2. Similarly [4.55] and [4.56] show that each line current contains a pair of forward and backward waves associated with γ_1 and a second pair with γ_2.

The impedance $Z_{1\gamma1}$ relating the voltage and current components for conductor 1 and for the γ_1 forward and backward waves is (from [4.52] and [4.55]) given by

$$Z_{1\gamma1} = \frac{T_{11}|Z|}{\gamma_1(Z_{22}T_{11}-Z_{12}T_{21})} = \frac{|Z|}{\gamma_1(Z_{22}-Z_{12}T_{21}/T_{11})}$$

$$[4.57]$$

Three other impedances relate component currents and voltages of one conductor for the two-conductor case, viz. for conductor 1 and the γ_2 waves

$$Z_{1\gamma 2} = \frac{T_{12}|Z|}{\gamma_2(Z_{22}T_{12} - Z_{12}T_{22})} = \frac{|Z|}{\gamma_2(Z_{22} - Z_{12}T_{22}/T_{12})}$$

[4.58]

for conductor 2 and the γ_1 waves

$$Z_{2\gamma 1} = \frac{T_{21}|Z|}{\gamma_1(Z_{11}T_{21} - Z_{12}T_{11})} = \frac{|Z|}{\gamma_2(Z_{11} - Z_{12}T_{11}/T_{21})}$$

[4.59]

for conductor 2 and the γ_2 waves

$$Z_{2\gamma 2} = \frac{T_{22}|Z|}{\gamma_2(Z_{11}T_{22} - Z_{12}T_{12})} = \frac{|Z|}{\gamma_2(Z_{11} - Z_{12}T_{12}/T_{22})}$$

[4.60]

Each of these four impedances is a constant for a given frequency component of the disturbance since [4.48] and [4.27] to [4.30] show that they depend only upon the system impedances and admittances. Although these impedances are similar to surge impedances, the usual definition of the latter is that they are that set of terminal impedances which give a reflection factor of zero (see next section), and the matrix of self- and mutual surge impedances for multi-conductor systems can be shown to be $[T][\gamma]^{-1}[T]^{-1}[Z]$. It is, however, beyond the scope of this book to develop this topic further, and, for the further application of matrix methods to three-phase lines, reference should be made to the original paper by Wedepohl (1963) and to his subsequent work.

4.4 Effects of various terminations

In this section in which the reflections occurring at various terminations are considered, it is necessary to make the simplifying assumptions that

(a) The line is a single-circuit 3-phase line with one conductor per phase;

(b) The line has no losses;

(c) The line is completely balanced.

It is thus possible to use a single surge (self) impedance $Z_c = \sqrt{(L_1/C_1)}$ which is purely resistive, and only one set of forward- and backward-travelling waves in one line conductor need to be considered (which would have only a single propagation coefficient for each component frequency).

For further simplicity, unit function voltage will be considered, where the voltage imposed at $x = 0$ is zero until $t = 0$ when it rises to V and stays constant during the period of interest which may be several transit times of a particular line. Transit time $\tau = l/u$ (where l = line length) and is about 3·3 μs/km. Impulse voltages applied during equipment testing in Britain have 1 μs front time (t_1 in Fig. 4.3), and 50 μs to half peak value; lightning voltages may have front times between 1 and 5 μs and can fall to near zero in some tens of microseconds; and many switching surges have front times of one or more hundred microseconds. Due to line series resistance, corona (see section 3.8 of Volume 1), the resistance of earth paths and line shunt conductance, surges will in fact undergo appreciable attenuation. For these reasons the assumption of unit function voltage is far from the truth, but since there is no general solution to the wave equation for an arbitrary waveshape, some assumption of waveshape is required for the present purpose and the rectangular one allows a simple examination of various terminations. This particular case of a zero wavefront time and the voltage remaining at peak value instead of having a falling tail, has the further advantage of being pessimistic in its stressing of insulation at both front and tail.

4.4.1 GENERAL TERMINATION

If a line were infinitely long the ratio of voltage to current at any point in it would be equal to the line characteristic impedance (as shown in section 1.3.5 of Volume 1), and a disturbance at one end would cause a forward-travelling wave of voltage and a forward-travelling wave of current and there could be no reflections. Consider a line of finite length l. At $x = 0$ and $t = 0$, a disturbing voltage $V(t)$ is applied and there is a series impedance $Z_1(s)$ connecting the source of the disturbance to the line. At $x = l$ there is a shunt impedance $Z_2(s)$ connected between the line conductor and earth.

The series and shunt impedances $Z_1(s)$ and $Z_2(s)$ and the characteristic impedance operator $Z_{c11}(s)$ (see section 4.2.1), which are strictly functions of frequency so that they are not identical for each component of the surge, are all assumed to be linear and are therefore replaced by Z_1, Z_2 and Z_c. The voltages and currents are all functions of time but are written as V rather than $V(t)$ etc. for convenience.

At $t = 0$ and $x = 0$ a forward-travelling wave $v_{f1} = V - i_1 Z_1$ sets off along the line. The forward-travelling wave reaching the terminating impedance Z_2 at $x = l$ is $v_{f1} e^{-\gamma l}$. This termination causes a backward-travelling wave v_{b1} to be reflected such that the total voltage of the line with respect to earth at $x = l$, $= v_l$ where

$$v_l = v_{f1} e^{-\gamma l} + v_{b1} \qquad [4.61]$$

and

$$v_l - i_l Z_2 = 0 \qquad [4.62]$$

where the line current at $x = l$ is given by [4.9] as

$$i_l = 1/Z_c(v_{f1} e^{-\gamma l} - v_{b1}) \qquad [4.63]$$

Substituting from [4.61] and [4.63] into [4.62]

$$v_{b1}\left(1 + \frac{Z_2}{Z_c}\right) = v_{f1} e^{-\gamma l}\left(\frac{Z_2}{Z_c} - 1\right)$$

$$v_{b1} = \left(\frac{Z_2 - Z_c}{Z_2 + Z_c}\right) v_{f1} e^{-\gamma l} \qquad [4.64]$$

[4.64] shows that the backward wave reflected from the terminating impedance Z_2 is K_r times the forward (incident) wave reaching the termination where

termination reflection factor $K_r = \left(\dfrac{Z_2 - Z_c}{Z_2 + Z_c}\right) \qquad [4.65]$

[4.65] shows the well-known effects of voltage doubling at the open-circuited end of a line where $Z_2 = \infty$ and $K_r = 1$, and zero voltage must occur at a short-circuit where $Z_2 = 0$ and the reflected voltage is equal and opposite to the incident voltage.

This first reflection occurs at $t = \tau$ where the line transit time is $\tau = l/u$. At $t = \tau$, the backward-travelling wave v_{b1} sets off towards the point $x = 0$ where the original disturbance occurred, and at $t = 2\tau$ it reaches this point where the line is closed by the impedance $\mathbf{Z}_1$ connected between line and earth. When the backward wave of initial value v_{b1} reaches $\mathbf{Z}_1$ it has undergone attenuation and phase change as it travels a distance l so that it is now $v_{b1} e^{\gamma(-l)} = v_{b1} e^{-\gamma l}$. The incident wave which is this backward-travelling wave will cause a reflection which is a second forward-travelling wave v_{f2} to set off from $x = 0$ towards $x = l$. If the voltage between line and earth is v then at the time $t = 2\tau$, v will have three components:

(a) The original forward wave v_{f1}

(b) The first backward wave $v_{b1} e^{-\gamma l}$

(c) The second forward wave v_{f2}

and the source voltage V is the sum of these voltages on the line and the voltage drop in the impedance $\mathbf{Z}_1$

$$V = v_{f1} + v_{f2} + v_{b1} e^{-\gamma l} + i_2 \mathbf{Z}_1 \qquad [4.66]$$

and $v_{f1} = V - i_1 \mathbf{Z}_1$ where $i_1 = V_{f1}/Z_c$ is the original line current which was entirely composed of a forward-travelling wave of current leaving $x = 0$ at $t = 0$.

Substituting these values for v_{f1} and i_1 into [4.66] gives

$$-v_{f1} \frac{\mathbf{Z}_1}{Z_c} + v_{f2} + v_{b1} e^{-\gamma l} + i_2 \mathbf{Z}_1 = 0 \qquad [4.67]$$

The line current i_2 at $x = 0$ and $t = 2\tau$ is given by

$$i_2 = 1/Z_c(v_{f1} + v_{f2} - v_{b1} e^{-\gamma l}) \qquad [4.68]$$

Substituting from [4.68] into [4.67] gives

$$v_{f2} = \left(\frac{\mathbf{Z}_1 - Z_c}{\mathbf{Z}_1 + Z_c}\right) v_{b1} e^{-\gamma l} \qquad [4.69]$$

[4.69] is exactly similar to [4.64] where in this case the forward wave reflected from the source impedance $\mathbf{Z}_1$ is K_s times the backward (incident) wave where

source reflection factor $K_s = \left(\dfrac{\mathbf{Z}_1 - Z_c}{\mathbf{Z}_1 + Z_c}\right) \qquad [4.70]$

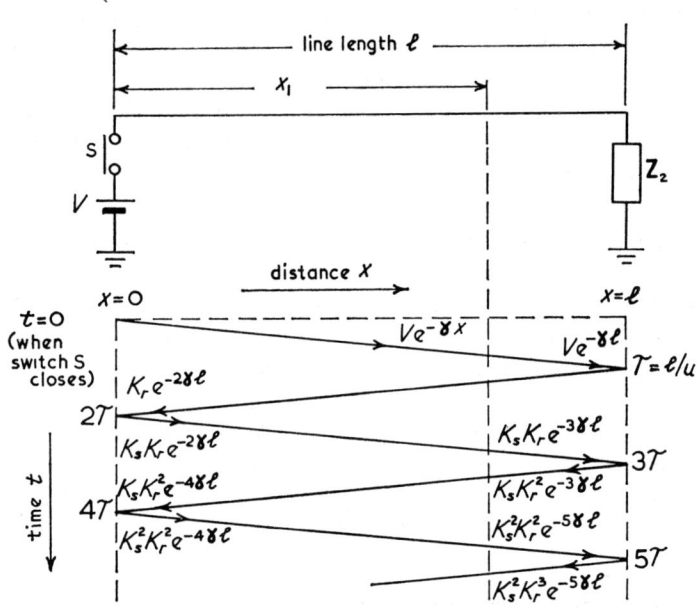

Fig. 4.5. Successive voltage reflections along a line illustrated on a (Bewley) lattice diagram.

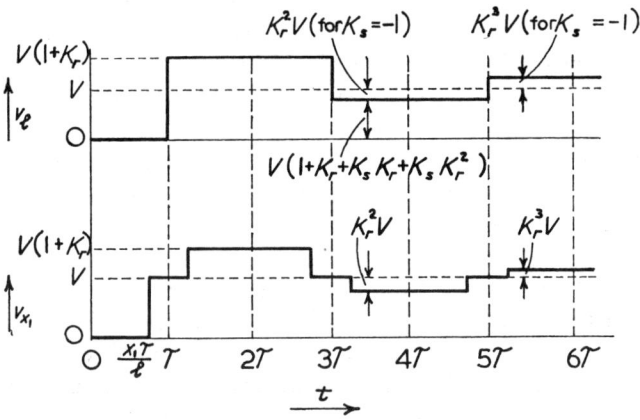

Fig. 4.6. Time variation of voltage at two points in the line of Fig. 4.5.

The voltages at any point in the line at any instant of time can be displayed by means of a (Bewley) lattice diagram which enables any number of successive reflections to be considered.

If for simplicity in the above analysis the voltage disturbance applied at $t = 0$ is the unit function voltage of magnitude V with $Z_1 = 0$ (so that $K_s = -1$), then Fig. 4.5 enables the voltage at any time or place in the line to be evaluated by inspection and super-position; e.g. at $x = x_1$, $v_{x1} = 0$ for $0 < t < (x_1/l)\tau$, $v_{x1} = V e^{-\gamma x_1}$ for $(x_1/l)\tau < t < (\tau+(l-x_1)\tau/l)$ and so on. The voltage/time waveforms at $x = x_1$ and $x = l$ built up from Fig. 4.5 in this way are displayed in Fig. 4.6, with the simplification for convenience of $\gamma \to 0$, i.e. phase change as well as attenuation has been neglected. In Fig. 4.6, K_r is taken as positive which indicates that $Z_2 > Z_c$. Fig. 4.6 shows that since the voltage across Z_2 differs from the applied surge voltage V by K_rV, $-K_r^2V$, K_r^3V, etc. on successive reflections (for the given assumptions) and $K_r < 1$, then the voltage across Z_2 is approaching the steady-state voltage V.

4.4.2 RESISTIVE TERMINATION; JUNCTIONS OF LINES AND CABLES

Since the characteristic impedance Z_c of a balanced transmission line or cable which is assumed to be without losses is $\sqrt{(L/C)}$ then it is a pure resistance Z_c. The reflection coefficient at a purely resistive termination R between line conductor and earth is given by [4.65] as $K_r = (R-Z_c)/(R+Z_c)$ which is a real number. This result may alternatively be deduced directly since the voltage across the resistor $= v_f + K_r v_f = iR$ where $i = v_f/Z_c - K_r v_f/Z_c$. A Thevenin equivalent circuit consisting of a source of voltage twice the applied surge, i.e. $2V$, with internal series-resistance Z_c switched into the terminating resistor R may be used for this condition. With $R = \infty$ the voltage at the open end is $2V$. If a circuit breaker happens to connect a long unloaded line to the supply at the instant of peak voltage, then apart from a slight reduction due to line losses, the voltage at the open end tends to reach twice the peak value of the phase voltage (see section 2.12). With a line on short-circuit for which $R = 0$ the current is $2V/Z_c$, and in general the current in R which is also the current transmitted into a second line or cable if one is connected there, is $2V/(R+Z_c)$. The voltage wave transmitted into a second line or cable of characteristic impedance R is the voltage across $R, = 2VR/(R+Z_c)$, which could have been deduced from [4.64] since it is the sum of the incident and reflected waves. Thus

$$\text{reflection coefficient} = (R-Z_c)/(R+Z_c) \qquad [4.71]$$

$$\text{transmission coefficient} = 2R/(R+Z_c) \qquad [4.72)$$

For the three important cases of short-circuit, open-circuit and matched termination ($R = Z_c$), the voltage and current waves are as shown in Fig. 4.7 a short time after the surge has reached the

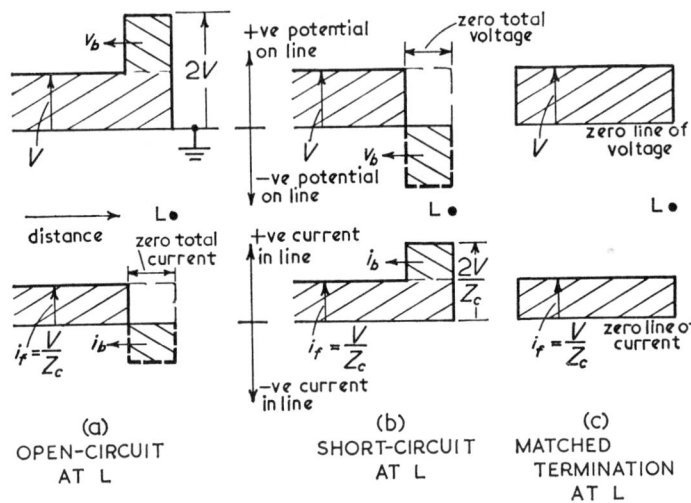

Fig. 4.7. Voltage and current reflections for three types of end termination.

discontinuity. Since the characteristic impedance of an E.H.V. cable is about 30 to 50 Ω whereas that for a line is frequently about 250 to 350 Ω, the voltage transmitted from a line into a cable is of the order $40/(40+300)$, i.e. about $\frac{1}{8}$ of the surge voltage travelling along the line. For this reason a transmission line is sometimes connected to a substation by a cable of perhaps less than one mile length, so that lightning or switching surges travelling along the line are much attenuated by the cable, and are less likely to flashover or damage apparatus in the substation.

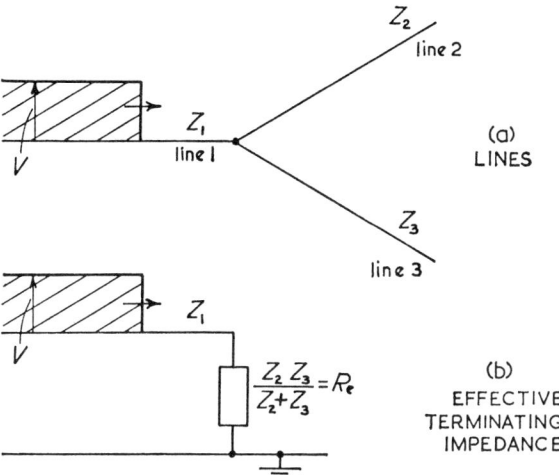

Fig. 4.8. Just before a surge reaches the junction of a line
with two other lines.

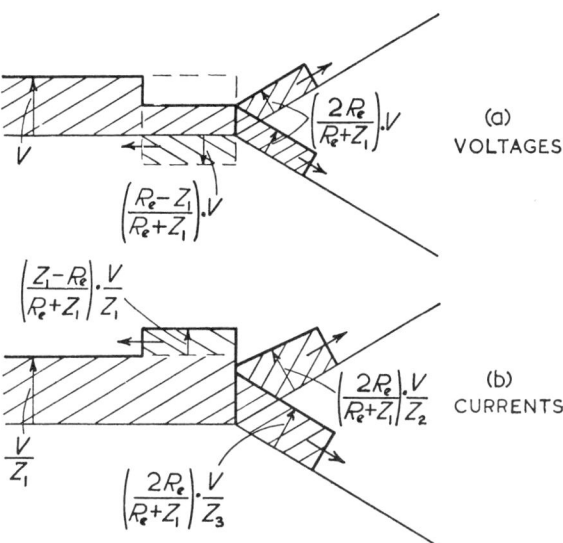

Fig. 4.9. Just after a surge reaches the junction of a line
with two other lines.

If a surge V travelling along a line of surge impedance Z_1 reaches the junction with two lines having surge impedances of Z_2 and Z_3 (Fig. 4.8), then this, as far as initial conditions at the junction are concerned, is equivalent to a terminating resistor of $Z_2Z_3/(Z_2+Z_3)$ $= R_e$. If $R_e < Z_1$, the voltage wave reflected back along the first line will be negative as shown in Fig. 4.9.

By applying voltage pulses which may have a duration appreciably less than the line transit time, at one end of a line on which a fault has occurred, the sign of the surges reflected back to the same end indicates whether it is a shunt or series fault (broken conductor), and the time taken for the incident wave to reach the fault and the reflected wave to return shows how far away the fault is.

4.4.3 REACTIVE TERMINATIONS

When a loss-free line is terminated by an impedance which contains inductive or capacitive elements, the resulting voltages can be obtained by Laplace transformation. In simple cases, of say a single capacitor or inductor, the impedances of these elements may be written as $1/Cs$ and Ls respectively, and the voltages and currents will vary exponentially. If a unit function surge of voltage V is switched on to a line which is terminated by an initially uncharged shunt capacitor of C farads, then the Laplace transform of the voltage reflected is from [4.65]

$$\mathscr{L}[v_b] = \left(\frac{(1/Cs) - Z_c}{(1/Cs) + Z_c}\right)\frac{V}{s}$$

$$= \left[1 - \frac{2s}{s + (1/CZ_c)}\right]\frac{V}{s} \qquad [4.73]$$

From standard tables of inverse transforms

$$v_b = V(1 - 2\exp(-t/CZ_c)) \qquad [4.74]$$

It should be noted that t is here measured from the time the surge eaches the capacitor.

The voltages and currents are illustrated in Fig. 4.10. The currents follow from [4.15], [4.16] and [4.74]. The distributions of voltages and current shown in Fig. 4.10 and the time variation of voltage given by [4.74] might have been deduced from the fact that the applied surge is assumed to be of infinitely steep wavefront so that it contains a range of frequencies extending to infinity. The capacitor therefore acts initially as a short-circuit so that the reflected wave is negative and brings the voltage at that point to zero. The capacitor then charges up with a time constant of Z_cC until, when completely charged, it constitutes an open-circuit and the voltage is then $2V$.

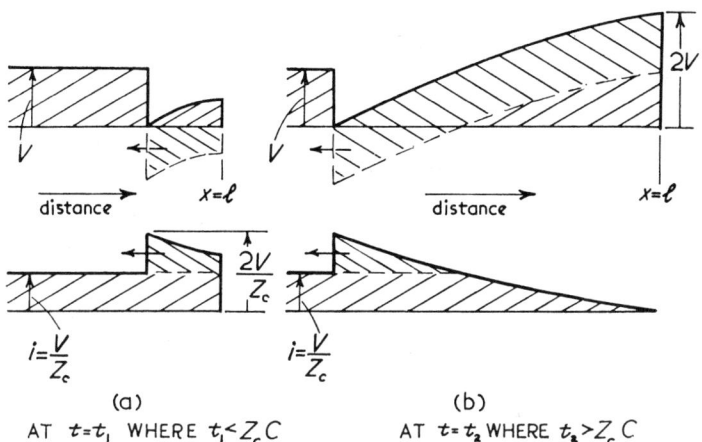

(a)   (b)
AT $t=t_1$ WHERE $t_1 < Z_cC$ AT $t=t_2$ WHERE $t_2 > Z_cC$

Fig. 4.10. Voltages and currents at a capacitive line termination.

The corresponding case of a shunt inductor of L Henry gives the transform of the reflected voltage as

$$\mathscr{L}[v_b] = \frac{(Ls - Z_c)}{(Ls + Z_c)}\frac{V}{s}$$

$$= \left[1 - \frac{2(Z_c/L)}{s + Z_c/L}\right]\frac{V}{s} \qquad [4.75]$$

and from standard inverse transforms

$$v_b = V(-1 + 2\exp\left[-(Z_ct/L)\right]) \qquad [4.76]$$

The voltage and current distributions are as shown in Fig. 4.11.

The voltages and currents for this case are the duals of those for the capacitor termination. At the instant the surge arrives the inductor appears as an open-circuit so that the voltage doubles. The current increases exponentially from zero in L until finally it corresponds to a short-circuit if the resistance of the inductor is negligible.

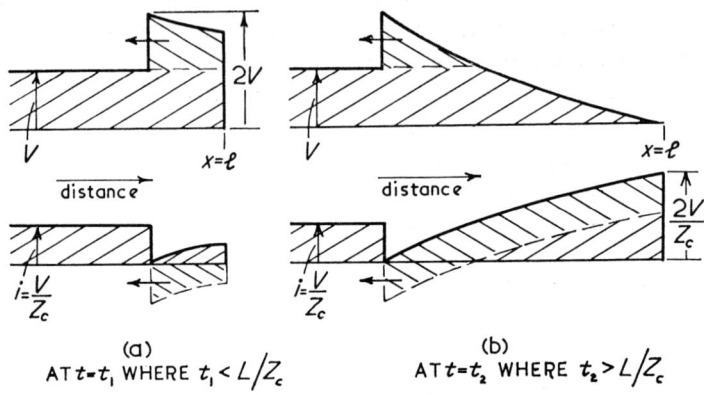

Fig. 4.11. Voltages and currents at an inductive line termination.

REFERENCES

Bewley, L. V. 1963. *Travelling Waves on Transmission Systems.* Dover, New York.

Wedepohl, L. M. 1963. Application of matrix methods to the solution of travelling-wave phenomena in polyphase systems. *Proc. I.E.E.*, **110**, No. 12, 2200–2212.

Examples

1. Calculate the approximate surge impedance of a single-phase line for which the effective ratio of conductor spacing/radius = 60.

(H.N.C.) (490 Ω)

2. A transmission line has a capacitance of 0·00778 μF/km and an inductance of 0·933 mH/km. This overhead line is connected to an underground cable having a capacitance of 0·187 μF/km and induc-

tance of 0·155 mH/km. Calculate the rise of voltage produced at the junction of the line and cable by a voltage surge of crest value 50 kV along the cable.

(H.N.C.) (92·3 kV)

3. Find in exponential form, an approximate expression for a 0·5/5 μs wave form. Assume that the exponential which governs the front of the wave if negligible when it has fallen to 0·002.

(H.N.C.) $(v = V(e^{-0.139t} - e^{-12.4t}))$

4. A cable 1 km long and of surge impedance 80 Ω is connected between a long transmission line and a substation. The transformer in the latter causes the effective termination of the cable to be a resistance of 10 000 Ω. A voltage surge of 90 kV appears in the line and travels towards the cable and substation. The velocity of the wave in the cable is 10^{10} cm/s. What is the voltage at the transformer at the time when the second rise of voltage occurs there, and how long after the 90-kV surge first reaches the junction between line and cable does this occur?

(University of Rangoon) (98·5 kV, 30 μs)

5. The ends of two long transmission lines A and B are connected by a cable C 1·5 km long. The lines have capacitance 10 pF/m and inductance 1.6×10^{-6} H/m, and the cable has capacitance 89 pF/m and inductance 5×10^{-7} H/m. A rectangular voltage wave of magnitude 10 kV and of long duration, travels along line A towards the cable. Find the magnitude of the second voltage step occurring at the junction of the cable and line B. What will be the voltage at the junction of line A and the cable, 20 μs after the initial surge reaches this point?

(University of Leeds) (7·8 kV, 6·79 kV)

6. A long line is connected to a cable of length 1 km, the far end of which is isolated. The line has capacitance 0·114 pF/cm and inductance 0.978×10^{-8} H/cm, and the cable has capacitance 1·36 pF/cm and inductance 0.75×10^{-8} H/cm.

A rectangular surge of magnitude 10 kV travels along the line towards the cable. Calculate the voltage on the cable at its mid-point

38 μs after the arrival of the initial surge at the junction of cable and line.

(12·8 kV)

7.   A 500-kV surge on a long overhead line of characteristic impedance 400 Ω, arrives at a point where the line continues into a cable AB of length 1 km having a total inductance of 264 μH and a total capacitance of 0·165 μF. At the far end of the cable, connection is made to a transformer of characteristic impedance 1000 Ω. The surge has negligible rise-time and its amplitude may be considered to remain constant at 500 kV for a time longer than the transit times involved here.

Plot against time, a curve of the voltage at the junction A of the cable for 26·4 μs after the arrival at this junction of the original surge, using a ladder diagram to show the reflections along the cable.

(L.C.T.) (91 kV up to 13·2 μs, 243·75 kV from 13·2 to
26·4 μs rising to 359·25 kV at 26·4 μs)

8.   A rectangular surge of 100 kV and 20 μs duration travels along a line of surge impedance 500 Ω and 100 000 metres long with a velocity of 3×10^8 m/s, towards the junction of the line with a cable. The cable, 400 metres long, has a surge impedance of 50 Ω and the surge velocity in it is 2×10^8 m/s. At the far end of the cable is a transformer which may be regarded as a parallel combination of a capacitor of 100 $\mu\mu$F and an inductor of 0·5 H, connected between the cable core and earth.

Calculate the approximate voltage across the transformer 6 μs after the surge travelling along the line reaches its junction with the cable.

If the same surge travelling along a similar line reached the point where the line terminated in a capacitor of 2×10^{-2} μF connected between the line and earth, calculate the maximum voltage appearing across the capacitor.

(University of Leeds) (66·2 kV, 174·3 kV)

9.   A 20-km-long cable connects two overhead lines A and B of surge impedance 400 Ω. The inductance and capacitance of the cable are 0·01 H and 1·78 μF.

A 10-kV rectangular surge of 350 μs duration is approaching from line A. Calculate the total voltage at the junction point of the cable and the line B, 410 μs and 500 μs after the initial surge reaches the junction of the line A and the cable.

What is the effective permittivity of the cable dielectric?

(University of Leeds) (7·81 kV, 2·49 kV, 4)

Chapter 5

INSULATION CO-ORDINATION

5.1 Introduction

Insulation co-ordination consists of the correlation of the insulation strengths of the various components of a high-voltage power system, so as to minimise damage and loss of supply caused by over-voltages. The supply authorities in their planning of the system and specification of over-voltage tests of equipment, endeavour as far as is economically reasonable to minimise supply interruptions, and steps taken towards this are:

(i) to ensure that the system insulation will withstand all normal working stresses and most of the abnormal ones;

(ii) to discharge or divert over-voltages which exceed the withstand strength of the apparatus;

(iii) to ensure that breakdowns occur by external flashover rather than by internal failure of equipment such as puncture or breakdown of solid or liquid dielectrics;

(iv) to control the points at which breakdowns will occur in the light of the relative importance of the various items of equipment.

The problem is one of balancing the cost of specifying insulation able to withstand higher voltages without flashover or damage, as well as the cost of providing more or better protective devices, against the possible damage to equipment and the number and severity of supply interruptions which may be estimated to be likely to occur. A subject which depends so heavily upon engineering judgement and experience of the many probabilities and other statistical features, and of the economics of various alternatives, does not readily lend itself to brief treatment in a textbook, but an attempt has been made to outline some of the chief points. The conditions discussed in this chapter apply particularly to the British grid system, and do not apply to all other cases; tropical areas pose some special problems (see, for example, Golde 1956) and so do d.c. systems (see section 5.7).

5.2 Over-voltages

Over-voltages in a power system may be either at system frequency or they may be due to transient surges with higher frequency components. System-frequency over-voltages are commented upon briefly in the next section and transient over-voltages due to lightning or switching in the following one. The latter consist mainly of transient recovery-voltages of natural-frequency oscillating about system-frequency voltages when circuit breakers open or close, and although these are not dealt with in this chapter because they have been discussed in some detail in Chapter 2, the significance of these relatively long wavefront surges in the breakdown of long gaps in E.H.V. systems above about 300 kV is indicated in section 5.4. Some discussion of the relatively short wavefront surges caused by lightning which are likely to exceed switching surges in distribution or lower voltage transmission systems is given in section 5.2.3.

System equipment has always been subjected to power-frequency over-voltage tests generally of 1 minute duration, and Table 5.2 in section 5.3 shows voltage levels between about 1·5 and 2 times rated voltage which are applied to power transformers. For many years major equipment has been subjected to standard impulse voltage tests which have peak voltages in excess of the peak power-frequency test voltage as shown in Tables 5.1 and 5.2 and these are aimed at ensuring adequate insulation to earth to withstand lightning surges. In recent years considerable work has been in progress to determine the ability of E.H.V. system insulation to withstand switching surges.

5.2.1 SYSTEM-FREQUENCY OVER-VOLTAGES

Under steady-state conditions the system-frequency voltage may exceed the rated voltage in some parts of a system so that equipment must be insulated for the highest system voltage, e.g. 300 kV for the 275-kV system. Larger increases in system-frequency voltage may occur for times of the order of a few seconds or less, in such conditions as loss of load on a generator (up to about 20% for a steam turbine and 30% for a water turbine), or opening a circuit breaker at the receiving end of a long line (up to 90% under particularly unfavourable conditions). Harmonic distortion can sometimes increase the peak stresses applied to equipment. If an earth fault occurs on one phase the voltages to earth of the other two phases can exceed the normal 58% of the line (phase-phase) voltage until the fault is cleared, if, for example, the system neutral is not earthed or is earthed through an arc suppression (Petersen) coil. The reactance of an arc suppression coil is of such a value as to produce resonance with the system capacitance during an earth fault on a single phase, so that the capacitance current through the fault is neutralised and the voltage across the fault path after current zero in it rises relatively slowly. Due to this, a flashover between one phase and earth which is the most common fault, and may be due to lightning flashover of an insulator string, will generally be self-extinguishing after the lightning current has passed, so that a power-follow-through arc which would cause breaker operation is avoided. Where a Petersen coil is used or the neutral is not earthed, it is possible for the voltages to earth on two phases to reach 100% of the line voltage due to an earth fault on one phase or to unbalanced leakage to earth. System neutrals are rarely left isolated (except sometimes at low voltages) because of over-voltages such as those which occur if an arc to earth on one phase is unstable (arcing earth), where conditions could arise similar to those in section 2.10 (see Fig. 2.11). If the neutral is earthed either solidly or through a resistor or reactor, the voltage to earth of the sound phases depends upon the ratios of zero-sequence resistance R_0 and reactance X_0 to the system positive-sequence reactance X_1. If $1 < X_0/X_1 < 3$ and $R_0/X_1 < 1$, the maximum voltage to earth during earth fault on one phase does not exceed 80% of the line voltage and the system is said to be effectively earthed, whereas systems where the voltage can exceed this value are said to be non-effectively earthed.

5.2.2 TRANSIENT OVER-VOLTAGES

Transient over-voltages may be either internally generated in the system or externally generated, and may be assessed by an over-voltage factor which is the ratio between the peak over-voltage and the rated peak system-frequency phase voltage of the system. This ratio may also be referred to as an amplitude factor or as per-unit (p.u.).

The internal over-voltages arise from abrupt changes in system conditions due in most cases to switching parts of the system in or out of circuit, and these are discussed in Chapter 2, and in their relation to flashover of system insulation in section 5.4. Fuse operation can also cause over-voltages due to the rapid formation of a large number of arcs in series, and fuses are rejected at over-voltage factors varying from about 6·8 at 3·3 kV to 2·7 at 132 kV, though element design usually limits their peak voltages to factors of about 5·5 and 2·2 respectively at these two system voltages.

Apart from switching, the chief cause of transient over-voltages is lightning and on lower voltage systems the latter causes the higher transients. At around 200 to 300 kV the magnitudes of switching and lightning over-voltages are of the same order, whilst above about 300 kV switching produces the larger surges. The external over-voltages caused by lightning which have a much more rapid rise of voltage than switching surges, are discussed in the next section.

5.2.3 OVER-VOLTAGES DUE TO LIGHTNING

The thunderclouds from which lightning discharges occur are formed by ascending warm, moist air which precipitates water droplets as it cools. Upward air currents acting on the suspended water and ice particles cause charge separation between the upper and lower regions of the cloud. The majority of lightning strokes take place from clouds which have positively-charged upper regions, with the rest negatively-charged except for some localised positive charge in parts of the cloud base. The lightning stroke consists basically of two components, a leader stroke from the cloud which initiates an upward streamer from some irregularity on the earth and completes a conducting channel along which a return stroke then passes. In many cases the stroke is multiple with a number of leader and return strokes. The return stroke, in the course of which the current reaches its peak, may have a current as low as hundreds of amperes, but is more frequently between a few kA and about 100 kA. As much as 180-kA has been reported in a discharge from a positively-charged cloud base. About half of strokes exceed 15 kA, 10% exceed 50 kA and about 1% exceed 130 kA, and about 80% are of negative polarity, i.e. negative charge flows to earth. The current waveform is generally a unidirectional pulse rising to a peak value in about 3 μs and falling to small values in several tens of microseconds.

Lightning can cause two kinds of voltage surge, one induced by indirect stroke when bound charges are dissipated following a lightning discharge to an object near the line conductor, and the other by direct stroke to a line conductor. In the case of an indirect stroke a negative charge in a cloud base causes bound positive charges on the conductors of a nearby transmission line, and discharge of the cloud due to a lightning stroke in the neighbourhood causes the conductor potential to rise to v since high insulation resistance prevents rapid dissipation of the charge. The induced voltage surge has a peak value of $v \simeq Eh$ where E is the mean electric field near the earth's surface under a thunder cloud (which may reach 50 or even as much as 280 kV/m), and h is the height of the phase conductor above earth. Applying [4.17] and [4.18] to this case where the total conductor current $i = 0$ gives $i_f = -i_b$ and $v_f = v_b = v/2$, so that forward- and backward-travelling waves of voltage and current set off in opposite directions along the transmission line from the place nearest to the cloud discharge. The amplitudes of voltages induced indirectly by lightning strokes to a tower, earth wire or nearby ground or object, are however normally much less than those caused by direct strokes to a line conductor, but they are of importance in systems at 33 kV and below and may even be significant above this voltage.

The current in a direct strike to a phase conductor is hardly affected by the voltage which it sets up between conductor and earth, so that a lightning stroke approximates to a constant-current source. If a current I flows in a direct stroke to a phase conductor, forward- and backward-travelling currents i_f and i_b each of $I/2$ flow in the conductor away from the point of strike so that the potential of this phase conductor to earth rises by $IZ_c/2$, where the characteristic impedance Z_c of an E.H.V. line may be about 350 Ω. Since more than half of the lightning currents exceed 10 kA the voltage rise may frequently cause flashover of line insulation except at very high transmission voltages, because typical impulse flashover voltages for overhead lines on steel towers are about 760 to 1100 kV for 132-kV lines, and 1440 to 1950 kV for 275-kV lines. On the other hand the lightning surge is rapidly attenuated partly due to corona as it travels along the line. The number of direct strokes to line conductors can be reduced by an over-running earth wire (or two wires). However, when lightning current discharges from the earth wire or other earthy part of the structure to earth, a high value of tower footing resistance (aided somewhat by tower inductance for very tall towers) may cause the voltage to earth at the top of the tower to be so high that a back-flashover may occur from the tower metalwork across an insulator string to a phase conductor. Reduction of the tower footing resistance below about 10 Ω is not usually worthwhile since below this value the tower inductance predominates.

The incidence of lightning varies greatly from one part of the world to another. One piece of meteorological information which is relevant is the isoceraunic level, which is the number of days in the year on which thunder is heard, and whereas in Britain the average is only about 10, it can exceed 100 in some tropical areas. A more directly relevant factor is the number of lightning strokes to earth per square mile in one year, which is difficult to determine, and can vary from about the isoceraunic level to one eighth of it, with the fraction averaging around one-half for temperate zones to one-third in the tropics. In Great Britain the average number of strokes per square mile per annum is about 4.

5.3 Impulse test levels of equipment

The impulse withstand strength against lightning of system equipment is determined by application of a standard impulse wave of voltage $v = v_0(e^{-\alpha t} - e^{-\beta t})$ where α and β are of the order $1 \cdot 5 \times 10^4$ and 6×10^6 s^{-1} respectively for a 1/50 wave with t measured in seconds. This wave is obtained from an impulse generator where basically capacitors are charged in parallel and discharged in series. The standard 1/50 waveshape in Great Britain has a time from zero to peak value of 1 μs and falls again to half peak value in 50 μs (see Fig. 4.2). The wavefront time t_1 is difficult to measure directly from an oscillogram and t_1 may sometimes be estimated from the times to reach 90% and 30% (or 10%) of peak value by a relationship such as $t_1 = 1 \cdot 67(t_{0.9} - t_{0.3})$, see Fig. 5.1.

When, under impulse test, the apparatus undergoing test flashes over, it 'chops' the impulse voltage wave, and as shown in Fig. 5.2 this may occur on the wavefront if the peak voltage is sufficiently

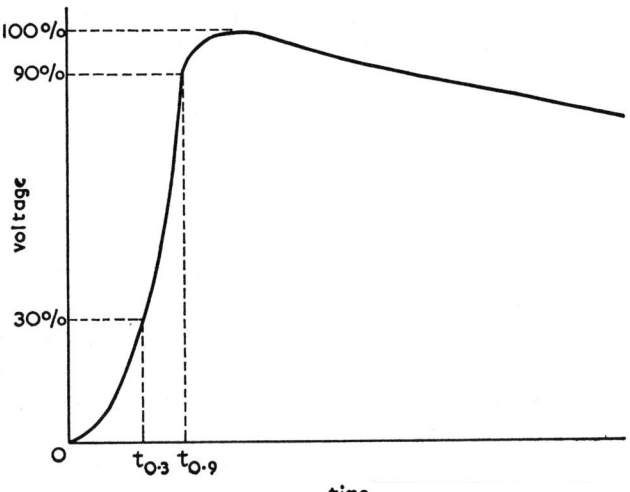

Fig. 5.1. Impulse voltage waveform.

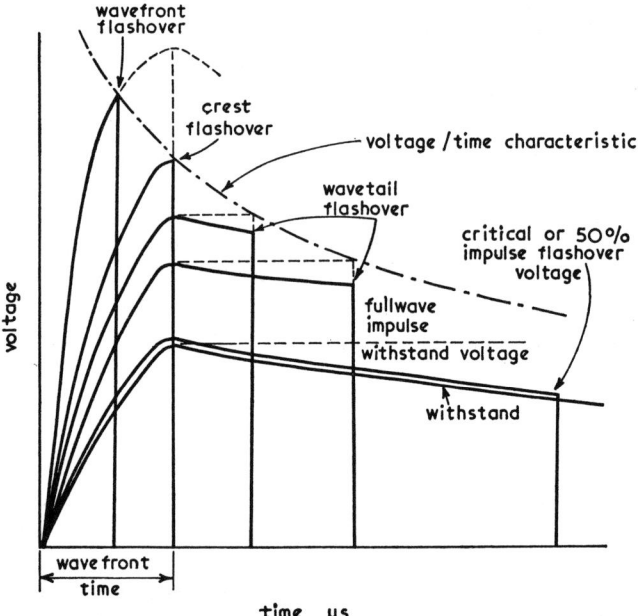

Fig. 5.2. Voltage/time characteristic of equipment under
impulse voltage test.

high, or on the wavetail for lower peak voltages. Fig. 5.2 shows that keeping the wavefront time constant (at say 1 μs), and gradually raising the peak voltage from relatively low values, a peak voltage is reached which is the highest at which the test object does not flashover. This peak or crest value is the full-wave impulse withstand voltage, and, as the voltage is raised gradually above this, values are reached which are called the 10%, 50% and 90% impulse flashover probability voltages, because if these impulses are repeatedly applied (at suitable intervals) the test object flashes over on about 10%, 50% or 90% of the voltage applications. As the peak voltage is raised still further the flashover occurs earlier and earlier on the wavetail until it occurs at the crest, and at even higher voltages the flashover is on the wavefront. A line connecting each point given by the value of crest voltage for a wavetail flashover, or actual flashover voltage for a wavefront flashover, and the corresponding time after zero voltage at which flashover occurs, gives a voltage/time characteristic for that particular apparatus. Power transformers may be subjected to a sudden change in voltage in order to test the inter-turn insulation, by applying an impulse voltage which is chopped to zero by the flashover of an air gap in parallel with the test object some 2 or 3 μs after crest voltage. As Fig. 5.2 shows, apparatus can then withstand a higher crest voltage unless the sudden voltage chop causes flashover or insulation failure, and a 132-kV transformer which has a 550-kV full-wave impulse withstand voltage may have a 630-kV chopped-wave withstand voltage. For much of the system insulation including air gaps, suspension and post insulators, the voltage/time characteristic is higher for negative polarity impulses than it is for positive polarity over some or all of the time scale, though factors such as the addition of arcing fittings to an insulator string can modify and even reverse this. For rod/plane gaps in atmospheric air the ratio of the negative breakdown voltage (i.e. point electrode as cathode) is about 2·2 times that for positive polarity up to gap lengths of several metres, whereas for rod/rod gaps the ratio is not much above unity. The ratio of the 50% impulse flashover voltage to the peak value of the power frequency flashover voltage is rather greater than unity for most insulation.

System equipment has specified impulse test voltages, e.g. B.S. 116 specifies levels of 1/50 impulse voltages for outdoor oil circuit breakers and some of these are quoted in Table 5.1.

These levels apply generally to transformers as is indicated by Table 5.2 (which also gives power-frequency test voltages) and they are in good agreement with American standard basic insulation levels (B.I.L.s), as well as experience of the levels at which surge

TABLE 5.1. (*B.S.* 116). *Impulse Test Voltages for Oil Circuit Breakers*

System rated voltage kV	Non-effectively earthed systems kV	Effectively earthed systems kV	All circuit breakers for systems up to 88 kV
33	—	—	200
132	650	550	—
275	*	1050	—

*All 275-kV systems are considered to be effectively earthed for this specification.

TABLE 5.2 (*B.S.* 171)

System highest voltage	Standard 1 insulation level		Standard 2 insulation level	
	Impulse test voltage kV peak	Power-frequency test voltage kV r.m.s.	Impulse test voltage kV peak	Power-frequency test voltage kV r.m.s.
72·5	350	140	300	115
145	650	275	550	230
245	1050	460	900	395
300	—	—	1050	460
420	—	—	1425	630

Standard 1 insulation level would usually apply to systems with isolated neutral, earthed through an arc suppression coil or non-effectively earthed through a neutral impedance.

Standard 2 insulation level would usually apply to systems with the neutral effectively earthed.

diverters can give protection. Additional factors such as rain, insulator surface pollution and altitude can be dealt with by testing under artificial rain or pollution conditions, specification of minimum creepage distances on insulators, and an increase of about 3·5% in the impulse and power-frequency wet withstand voltages for each 300 m in excess of 1000 m above sea level.

The detailed behaviour of transformers and rotating machines when surge voltages are applied to their terminals, is a study for the designers of such equipment to a greater extent than it is for power system engineers, and in view also of the available space, these phenomena are not discussed in this book.

5.4 Switching surge strength of insulation in E.H.V. systems

The amplitudes of switching surges and the switching surge strength of overhead line substation insulation are extremely important factors in the selection of insulation levels and in insulation co-ordination for E.H.V. systems, particularly those operating at about 300 kV and above. This occurs because

(a) The over-voltages set up in E.H.V. systems by lightning are less likely to cause breakdowns at the higher system voltages, because they are lower relative to the insulation strength than they are at lower system voltages. Furthermore, lightning surges usually originate in transmission lines and are reduced in steepness, and attenuated, due partly to corona, and would be chopped if line insulation flashes over. On the other hand the switching surges originate at points where major equipment is exposed and their amplitudes rise with the increase in system voltage, and can exceed the insulation strength of equipment in E.H.V. systems. When they cause breakdown in system insulation it is nearly always during the rise of voltage to the first peak, except in the case of some solid and liquid dielectrics.

(b) The wavefront times of many of the switching over-voltages (see Chapter 2) which cause surges to travel along transmission lines (see Chapter 4) are of the order of 0·1 ms to 1 ms, and the sparkover voltages of long air gaps and insulators have minimum values for wavefronts between about 0·05 and 0·3 ms, particularly for positive polarity surges.

The observation on switching surge breakdown generally occurring before the first peak for gaseous insulation, makes it possible to simulate switching surges with natural frequencies of the order of some hundreds of Hz, by unidirectional impulse voltages with wavefront times of the order of 1 ms. Application of switching impulse voltages with wavefront times varying between about 10 and 600 μs to long rod/rod and rod/plane air gaps and to suspension and post insulators (see Volume 1, section 4.1), shows minimum sparkover voltages at times of the order of 1 ms. Fig. 5.3 shows the type of curve obtained, in this case for a string of 15 suspension insulators, and it shows the typical marked polarity effect, whereby there is greater reduction in strength for positive surges than for negative surges.

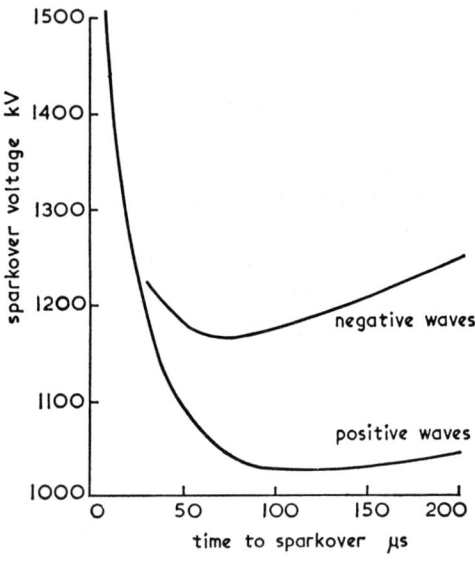

Fig. 5.3. Sparkover voltage as a function of time to spark-over for 15 ball and socket suspension insulators when dry.

With air gaps above about 2 m in length and for time to sparkover between about 50 and 300 μs, the minimum sparkover voltage is generally below the peak power-frequency voltage at which break-down occurs. This is also true for suspension insulator strings longer than 2 m when positive and negative switching-type surges are applied with the insulators dry, whereas when they are wet is is only true for positive polarity. For post insulators it only applies when they are dry and for positive polarity. In general, clearances in E.H.V. systems have to be greater than they would need to be for 1/50 waves in order to give insulation co-ordination with switching surges. If for this reason more units are used in the vertical suspension insulator string of a transmission line than would be required for lightning waves, then larger negative lightning waves may reach substation equipment because they do not cause flashover of the line insulators. It is possible if system transmission voltages rise in time beyond 1000 kV that a limit may be reached perhaps at about 1500 kV due to the reduction in strength of very long gaps at switching surge wavefronts.

The strengths of various parts of transformer insulation under switching surge conditions can be as low as 80% of the strength under 1/50 wave conditions, and a similar reduction can occur in oil-filled cables with oscillating switching impulses which take several cycles to decay to half the initial value. If switching surges could be limited to about 2·3 p.u. they would not necessitate insulation beyond that required to withstand power frequency and 1/50 impulse tests.

Because of the difficulties just discussed, it is likely that as system voltages rise above about 700 kV, new methods of controlling switching over-voltages will be necessary. These may include

(a) Using resistors, reactors or other means of draining away trapped line charges after opening (see sections 2.10 and 2.11).

(b) Connecting into circuit, series resistors of carefully controlled values on closing a circuit breaker, with the resistor subsequently short-circuited by a second switch (see section 2.12).

(c) Ensuring that the three poles of a breaker closing on to a trans-former-terminated line close within about $\frac{1}{8}$ cycle (see section 2.12).

(d) Ensuring that when closing on to a line which is not terminated by a transformer, the time between the closing of the first and second, and second and third poles exceeds twice the line transit time.

(e) Using electronic control of circuit-breaker operation so that the most dangerous conditions give negative polarity over-voltages and not positive ones, since at least for dry insulation the negative sparkover voltage is generally considerably higher and more consistent than that with positive polarity surges.

5.5 Insulation levels for overhead lines

Most overhead lines are carried on steel towers which are bonded to earth so as to make the tower footing resistance as low as possible, and the towers carry one or two over-running earth wires. There are electrodes at either end of the insulator strings so that over-voltages will cause them to flashover, which is preferable to flashover from a live conductor to earth which may damage the conductor. These electrodes should also prevent a power-follow-through arc from damaging the insulator sheds after gap breakdown or flashover across a polluted insulator surface. In some cases at the lower voltages and where a low earth resistance cannot be obtained, wood poles and cross-arms are used and earthed metal fittings including the earth wires are eliminated. The insulation level of the line is then increased by about 15 kV/m of wood even when wet, but sub-station equipment is subjected to more severe surges which may necessitate the installation of surge diverters. All-wood construction has rarely been used in Britain (except at 11 kV) and the pole top metalwork is usually bonded and earthed, which gives a reduced flashover voltage but gives protection against leakage.

At 400 kV and above, the insulation required on overhead lines to withstand switching surges is such that few flashovers take place due to lightning, because in general the clearances required are greater than those required to withstand standard 1/50 waves or lightning surges. At 275 kV and below it is not economical to insulate overhead lines so that they withstand lightning over-voltages of more than moderate severity, but they must withstand all internal over-voltages even under rain, atmospheric pollution and fog conditions. The main aim, therefore, in the design of overhead-line (and outdoor-substation) insulators has been to reduce the detrimental effects of weather and industrial and/or salt pollution on the power-frequency flashover voltage. The over-voltages arising from such conditions as the interruption of capacitance charging current (section 2.10) and chopping of transformer or shunt reactor magnetising currents (section 2.9), very seldom exceed four times peak phase voltage and in most cases the ratio does not exceed 3 p.u. for earthed-neutral systems. For the highest transmission voltages, effective neutral earthing is used and even in Europe where resonant earthing using an arc suppression coil has long been adopted, this method is not employed above 220 kV, while in Britain it has only been used at lower voltages. The usual requirements for overhead line insulators is that when wet they must withstand three to four times the nominal phase voltage, and that they must have a creepage path the length of which depends upon the amount of pollution in that area. In certain places where pollution is negligible compared with lightning, the power frequency requirement can be relaxed somewhat provided that the impulse level is adequate. Although the arcing fittings on an insulator reduce its impulse withstand voltage, it is still in general greater than the impulse test required for the basic insulation level (B.I.L.) of the system because of the power-frequency withstand voltage requirement. For about 1·5 km along a line leading in to a substation there will be earth wire protection (even if it were absent in the rest of the line), and if the substation has rod gaps instead of surge diverters, the insulation of this part of the line will be only about equal to the B.I.L., so as to control the amplitude of surges reaching the station and prevent undue operation of its rod gaps. The rest of the line insulation may have a strength increased above the B.I.L. depending upon a number of factors such as high isoceraunic level and high tower footing resistances. On lines where a trip-out due for example to power-follow-through current after back-flashover may cause a supply interruption, high-speed automatically-reclosing switchgear or a Petersen coil may be used.

5.6 Protection of equipment in high-voltage outdoor substations

The equipment in high-voltage outdoor sub-stations such as transformers, circuit breakers, isolating switches, support and suspension insulators and busbars, may be protected by such means as air gaps, expulsion gaps (protector tubes), surge diverters and aerial earth wires. The risk of lightning strokes to the line near the substation may be lessened by providing a second earth wire for about 1·5 km, or in some cases surges entering along a line may be reduced by terminating the line a distance of 0·8 to 1·5 km from the substation, and connecting it to the station with a cable (see section 4.4.2). Devices known as surge filters or absorbers exist which reduce the amplitude of the surge and/or the steepness of its wavefront, but they are not widely used. If *n* lines terminate at a sub-station then if

$n > 2$ there is a reduction in surge amplitude when a surge V arrives along one line since the voltage rises to $2V/n$ (see section 4.4.2).

Air gaps may be between electrodes of various forms such as rods or rings which are mounted as arcing horns on insulators and bushings, or they may be separate co-ordinating rod gaps on or near transformers. The rod/rod gap flashover voltage (generally quoted as the 50% probability value with 100% and 0% values differing by about $\pm$ 15%) is proportional to gap distance to a power of about 0·9, and if the gap is vertical it may be up to about 30% greater for a negative surge than for a positive one. Co-ordinating gaps and surge diverters can deal with switching over-voltages as well as those due to lightning. If a rod/rod gap is used to protect a clearance which frequently approximates to a rod/plane, then the criterion for satisfactory co-ordination is sometimes taken as having the 90% probability flashover voltage of the former equal to the 10% probability voltage of the latter. Expulsion gaps (or protector tubes) which are for lightning protection, have an insulating cylinder with one open end surrounding the air gap, so that the power-follow-through current after flashover sets up a gas pressure which within a certain current range enables the gap to be self-extinguishing. Though this overcomes one disadvantage of an air gap, that the power current after flashover requires circuit breaker action to interrupt it, the expulsion gap is of limited application. A rod gap has the further disadvantage beside requiring a breaker to clear the power-frequency arc, that it has a relatively long time lag for an impulse voltage of short duration, so that if its length is set to provide protection for apparatus at long times it may not give protection at short times. If on the other hand its gap is set to give protection at short times, its minimum flashover voltage may be too low, and thus can cause undue operation on minor surges, so that the actual setting adopted must be a compromise. Co-ordinating rod gaps may be set to have a 50% impulse flashover voltage of about 80% to 85% of the transformer impulse withstand level. The disadvantage of the air gap power current requiring clearance by a circuit breaker is overcome if the latter automatically recloses within a few tens of cycles, so that in the majority of cases the supply is not effectively interrupted, because the fault path has become de-ionised in this time interval. If on the other hand the fault is persistent rather than transient, the fault current will again flow when the breaker automatically recloses, and after a pre-set number of reclosures the breaker will stay open.

Surge diverters are more expensive than air gaps but their installation to protect particular equipment can sometimes be justified, even in Great Britain where the isoceraunic level is relatively low, where most lines are shielded with earth wires, and where most tower footing resistances are low and duplicate circuits exist. They may for example be installed at a generator transformer or at a transformer at the end of a wood-pole overhead line without earth wires. A surge diverter consists basically of a stack of non-linear resistors in series with a number of spark gaps sealed in an insulated housing with sheds similar to those of a bushing. The spark gaps may be shunted by high-impedance non-linear resistors carrying about 1 mA at normal circuit voltage so as to give a uniform voltage distribution between gaps. The air gaps must not only, not flashover under peak system-frequency phase (line-to-earth) voltage, but if an earth fault occurs on one phase, they must not flashover when the voltage to earth on the other two phases reaches line voltage on a system which is not effectively earthed, or reaches 80% of line voltage on an effectively-earthed system. Thus, for example, a surge diverter of r.m.s. voltage rating 245 kV could be used in a non-effectively earthed system having a line voltage up to 245 kV, but in a solidly-earthed system up to a line voltage of 300 kV.

The impulse sparkover voltage of the diverter gaps is often between two and three times the r.m.s. voltage rating. Upon flashover of the gaps, the high voltage across the main series non-linear resistors causes their resistance to fall to a very low value, so that a high current discharges to earth, with the voltage across the diverter remaining nearly constant during this time of discharge, at a value a little above or below the sparkover voltage (see Fig. 5.4). This almost constant voltage is known as the residual or discharge voltage and may be quoted for a current waveshape such as an 8/20 μs wave of 5 or 10 kA which corresponds approximately to the waveshape of current flowing through the non-linear resistors when a 1/50 μs voltage wave is applied to the diverter. It is not necessarily true, however, that the diverter voltage will not exceed the discharge value for all switching surge conditions. The power-follow-through current arc which follows the lightning current discharge can be interrupted by the air gaps (employing such means as magnetic forces if necessary), and this is facilitated by the large increase in resistance of the non-linear resistor stack which reduces the current when the voltage falls to the relatively low power-frequency system voltage. Switching surges have energies which may be between 10

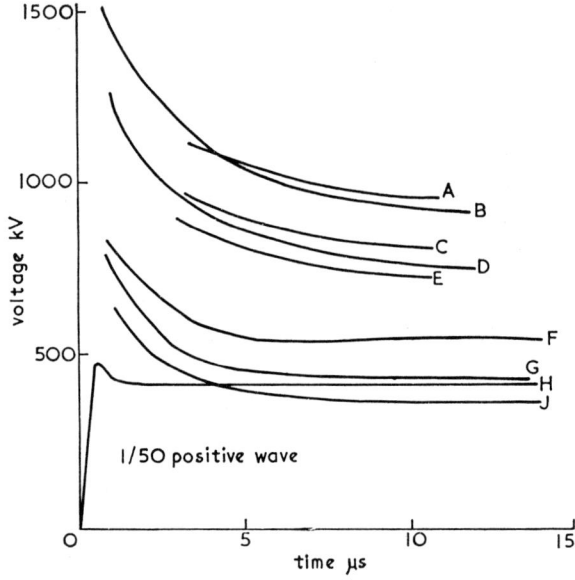

Fig. 5.4. Insulation co-ordination for a 132-kV sub-station.

    A.   tension and suspension insulators; 12 units
    B.   post insulators; 5 units
    C.   tension and suspension insulators; 10 units
    D.   post insulators; 4 units
    E.   tension and suspension insulators; 9 units
    F.   transformer withstand impulse level.
    G.   rod gap; 66 cm
    H.   surge diverter; 116-kV
    J.   rod gap; 56 cm

and 50 times the energy of a typical lightning wave and the diverter must be able to absorb this energy without damage. Diverters may flashover on switching surge at about 80% of the 1/50 flashover voltage.

    When a surge travelling along a line enters a substation where there is a surge diverter which flashes over, a negative voltage surge is reflected back along the line due to the diverter discharge voltage being less than the incoming surge voltage. The reduced surge equal to the discharge voltage continues as a forward wave, and if it reaches a transformer then positive reflection occurs (see Chapter 4). The voltage to earth across the transformer phase thus exceeds the diverter discharge voltage for twice the time taken for a surge to travel between diverter and transformer. Since, as Fig. 5.4 indicates, the increased voltage should not exist across the transformer for even as much as 1 μs to give adequate protection, and a surge on an overhead conductor travels at about 300 m/μs, the three surge diverters (one for each phase) should be placed within a few metres of the transformer which they protect. To allow for voltage drop in the earth connection and scatter due to manufacturing tolerances and waveshape variation, the protective level is often taken as (1·15 × residual voltage) +30 kV.

If a surge diverter operates, due perhaps to a lightning stroke, at a time when the power-frequency voltage is above the normal value it may be unable to clear the power-follow-through current, and also a circuit breaker is more likely to restrike. Capacitive loads are a common cause of a rise in power-frequency voltage and under disturbed conditions this rise depends upon the ratio of the short-circuit MVA to the capacitive MVAr load due particularly to the zero-sequence capacitance of unloaded lines and cables. As the voltage of E.H.V. systems rises this ratio tends to diminish and the power-frequency voltage can at times be increased in consequence, and apart from the above effects this increases switching surges.

Fig. 5.4 illustrates an example of the co-ordination of insulation in a 132-kV outdoor substation based on the voltage/time characteristics of the equipment for waves of positive polarity. Since the system concerned is effectively earthed, reduced insulation of 550-kV impulse withstand level may be used on the transformers instead of 650-kV (Table 5.2) with an appreciable saving in cost. If the 650-kV level is retained instead of 550-kV for the switchgear (Table 5.1) and busbars, the extra cost is relatively small, and the chance of failure of equipment common to a number of circuits is reduced. Co-ordination between the open-gap insulation of switchgear and the insulation to earth is important, because of possible danger to personnel if the gap flashes over due perhaps to a switching surge from another breaker, and this, like other co-ordination aspects, is studied statistically.

Diverters of reduced rating (116- or 121-kV), sometimes called '80% diverters', may be used to protect the transformer provided that they are within about 20 m of it, or rod gaps set at 56 or 66 cm may be used. The choice between these two gaps is based upon engineering experience, taking account of such factors as the extra flashovers on switching surges at the lower gap, and the incidence of lightning in the area, so that in Great Britain the 66-cm gap is usually adopted. It can be seen that this may not protect the transformer against very steep waves, but these are rare when the incoming overhead lines are shielded by earth wires.

The four-unit post insulator of the type having characteristic D meets the 650-kV impulse withstand level required. When the atmospheric pollution is normal a creepage distance to earth of about 1·7 cm/kV (i.e. 224 cm for 132 kV) is adequate, and the four-unit post insulator with a creepage length of 302 cm, of which 216 cm is protected, would be suitable. In an area where pollution is heavy and experience indicates the need for a creepage distance of about 2·5 cm/kV, the five-unit post insulator of curve B could be used. Similarly, a nine-unit string of cap and pin suspension and tension insulators meets the impulse voltage requirements, but ten units with 310 cm total creepage length of which 208 cm are protected may be used for the suspension strings. As many as twelve units may be used for tension strings because they are nearly horizontal, so that their flashover voltage is reduced below that for a vertical suspension string (where the undersides of the insulators remain dry) by rain and its effect on surface pollution. The reduction in flashover voltage due to artificial rain is in general greater for the power-frequency voltage test than for impulse voltage.

5.7 Insulation co-ordination in h.v.d.c. systems

H.V.D.C. systems can, in general, be subjected to similar external and internal over-voltages to a.c. systems, but in addition periodic internal over-voltages arise due to commutation, and transient over-voltages may be caused by failure of commutation or of control circuits or by a short-circuit. Apart from the possibility of appreciable reduction in the cost of convertor equipment by keeping the insulation to a minimum, reduction of over-voltages lessens the chance of mercury-arc convertors restriking or of breakdown of thyristors.

The two main methods of limiting over-voltages are by damping and by diverters or spark gaps. Both internal and external over-voltages may be damped by one of a number of circuits connected in parallel with the convertors. These also serve to give a uniform voltage distribution if a number of convertor units are connected in series. For surge diverters the requirements of low residual voltage, low spark-over voltage, high discharge capacity and self-quenching especially where there is no natural current zero, conflict to some extent. However, the use of magnetic forces to extend the arc and efficient cooling have enabled diverters to be developed for h.v.d.c. systems. Their thermal capacity is not always sufficient for limiting quasi-steady-state over-voltages or transients with a large energy to be dissipated, such as in the discharge of a long cable. To meet these difficulties, air or vacuum spark gaps may be used in special cases in conjunction with diverters, although they are not themselves self-quenching.

The withstand levels of equipment can be based on the protection levels of the surge diverters (i.e. on whichever is the higher of the maximum spark-over voltage and the residual voltage). The switching surge levels, which will be below those for impulse voltages of the 1/50 type, may in some cases be nearly as low as 1·1 times the diverter protection level.

REFERENCES

CLIFF, J. S. 1954. The co-ordination of insulation of high-voltage electrical installations. *Proc. I.E.E.* **101**, Part 1, 39–56.

GOLDE, R. H. 1956. Lightning protection of tropical transmission systems. *Electrical Times*, 13 and 20 September.

HEISE, W. *et al.* 1968. Co-ordination of the insulation in H.V.D.C. systems. *E.T.Z.*, A, **89**, 204–208. (E.R.A. translation/IB2653).

HUGHES, R. C. & ROBERTS, W. J. 1965. Application of flashover characteristics of air gaps to insulation co-ordination. *Proc. I.E.E.*, **112**, No. 1, 198–202.

MORTLOCK, J. R. 1956. *A.C. Switchgear*: Volume 1. Chapman and Hall, London.

NEWMAN, S. E. & PARISH, A. R. 1953. Insulation co-ordination in high-voltage stations. *English Electric Journal*, **13**, No. 3.

OAKESHOTT, D. F. 1968. Lightning and switching surge protection. *Borough Polytechnic Course on System Protection.*

THOMAS, A. M. 1966. The switching surge strength of insulating arrangements for systems operating at voltages above 100 kV. *E.R.A. Report* No. 5080.

THOMAS, A. M. & OAKESHOTT, D. F. 1956. Choice of insulation and surge protection of overhead transmission lines of 33 kV and above. *E.R.A. Report* O/T 14.

Chapter 6

RECTIFICATION, INVERSION AND HIGH-VOLTAGE D.C. SYSTEMS

6.1 Introduction

In this chapter an introduction is given to the principles involved in energy conversion from a.c. to d.c. which is called rectification, and to conversion from d.c. to a.c. which is called inversion. In deriving the relationships for the various direct and alternating voltages and currents, the connection of the valves or semiconductor devices and of the main transformer must be considered. Two particular connections are described here, one of which, the 3-phase bridge, is of importance in high-voltage d.c. schemes and in machine drives, in each of which power flow in either direction is required. In the power system itself convertors which may operate either as rectifiers or invertors, are encountered mainly in high-voltage d.c. schemes and semiconductor devices have only recently been developed sufficiently to meet the requirements of high-voltage high-current conversion with its over-voltage problems (see section 5.7). For this reason some reference is made here to mercury-arc valves without giving any details of their construction. Some references are given at the end of this chapter on applications of convertors to such industrial processes as machine drives, where thyristors and semiconductor diodes have largely superseded mercury-arc valves.

6.2 Rectification

Of the various connections which may be used for convertors, two only will be discussed.

6.2.1 DOUBLE-STAR CONNECTION

This is a connection which has often been employed in rectifiers used in industrial installations. As Fig. 6.1 shows, each of the three phase windings of the main transformer secondary is in two separate parts, so that two star-connected windings are formed with the two star points connected by a centre-tapped interphase reactor. The six phase ends are joined each to one anode of a 6-anode valve or to the

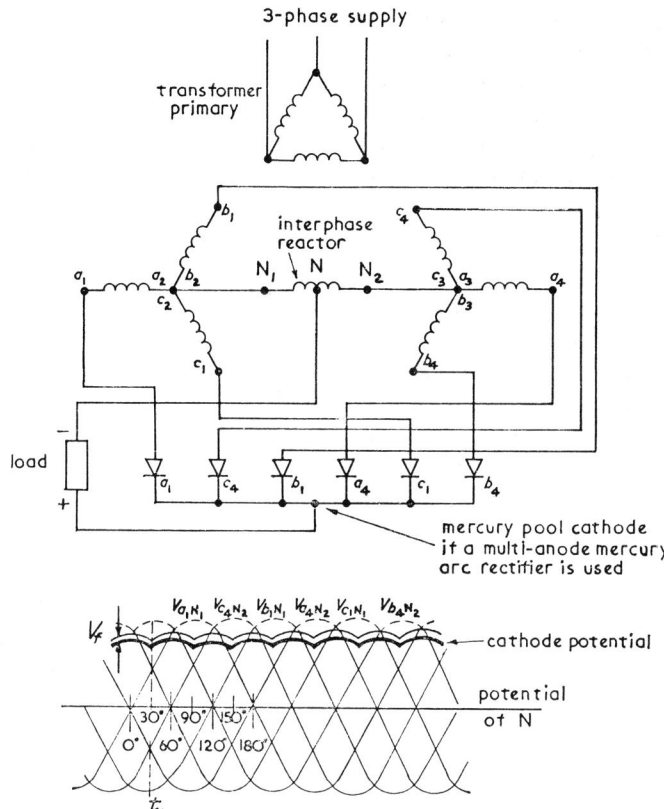

Fig. 6.1. Double-star connection and voltage waveforms.

anodes of six separate valves or diodes as shown. From no-load up to a transition current of about 1 to 2 % of full-load current, the anode potentials during conduction are as shown by the dotted lines. Thus each anode or valve conducts (in the absence of grid control—see section 6.2.3) from the instant when its anode potential (relative to N which is at virtually the same potential as N_1 and N_2) exceeds that of the previously conducting anode, to the instant when its potential falls below that of the next anode. This natural commutation of the arc from one anode or valve to the next when there is no grid control or when the grid delay angle (section 6.2.3) is zero, gives for these very low currents conduction to each anode for 60°. This is 6-phase working, e.g. anode c_4 conducts from 120° to 180°, whereas a 3-anode system gives 120° conduction for each anode or valve. The commutation from a_1 to c_4 to b_1 etc. causes the current in the interphase reactor to change direction at 3 times supply frequency. Above the transition current, sufficient e.m.f. is induced in the reactor to enable the current in a_1 to start earlier and to be prolonged until b_1 conducts and so on. c_4 now conducts from 90° to 210°. From 90° to 150°, c_4 and a_1 conduct in parallel and from 150° to 210°, c_4 and b_1 conduct in parallel. The difference between the two secondary output voltages appears across the interphase reactor and the output d.c. voltage is the mean height of the area between the potential of N and the full curve of the cathode potential. The latter is the mean of the secondary voltages of the two anodes conducting simultaneously, less the forward voltage drop V_f. There are, therefore, two three-phase rectifiers operating in parallel, with each anode carrying half the d.c. output current for 120°. The connection has the advantages of less ripple in the output voltage than if three anodes only were used, and of a higher utilisation factor (section 6.2.4), than for a 6-phase diametral connection (N_1 and N_2 connected together).

6.2.2 3-PHASE BRIDGE CONNECTION

Fig. 6.2 shows that in this connection which is used in high-voltage d.c. convertors, each terminal of a 3-phase transformer secondary is connected to the anode of one valve and the cathode of a second. Current can flow to or from each secondary phase during the central 120° (for natural commutation) of each positive and negative half-cycle. At any instant current is flowing through two valves in series, e.g. during time interval t_1 to t_2 current flows out of phase a through valve 1 and returns from the load through valve 6 to phase b. Following commutation from 6 to 2 at t_2, current flows out of phase a through valve 1 and returns via 2 to c. The output d.c. voltage (neglecting the forward voltage drop in the valves) is the mean height

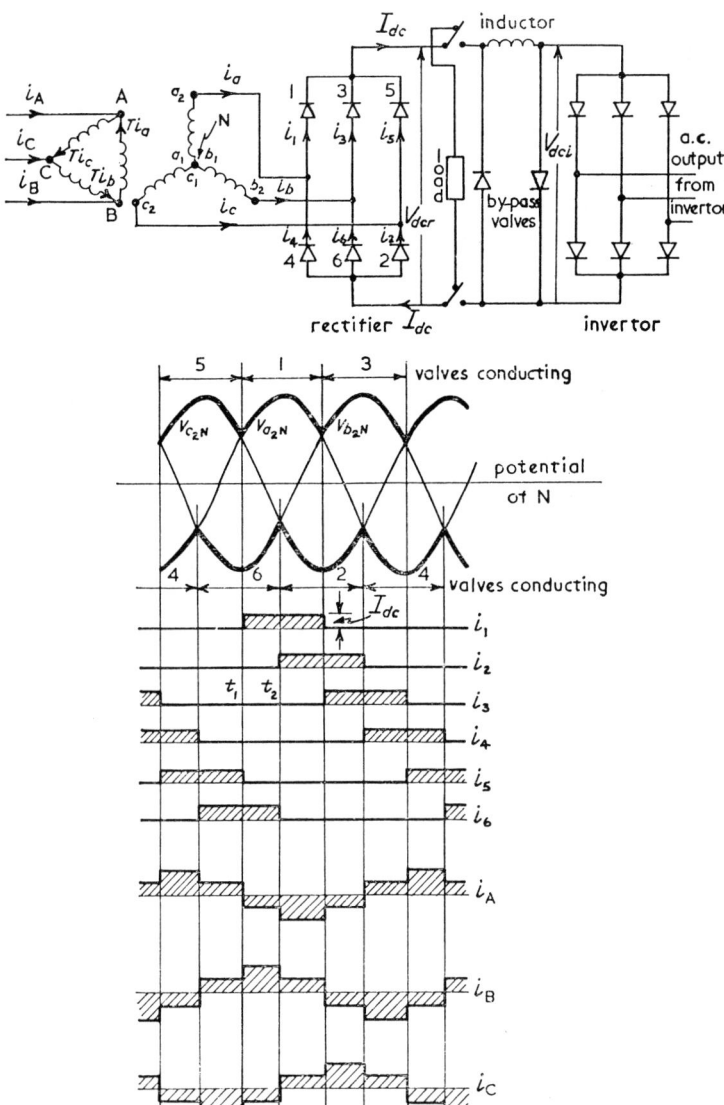

Fig. 6.2. 3-phase bridge connection and waveforms.

of the area between the two heavy curves which are equally spaced about the potential of N.

Fig. 6.2 shows the idealised valve currents i_1 to i_6 which are assumed to be constant at I_{dc} due to inductance. The delay in the rise and fall of current is neglected until it is dealt with in the next section. If T is the ratio of secondary phase turns to primary phase turns on the rectifier transformer, the line currents taken from the a.c. supply are $i_A = T(i_c - i_a)$, $i_B = T(i_a - i_b)$ and $i_C = T(i_b - i_c)$, and their waveforms may be deduced by considering each interval during which two valves conduct. For example, while valves 4 and 5 conduct together $i_c = i_5$ and $i_a = -i_4$ so that $i_A = T(i_5 + i_4) = 2TI_{dc}$ etc. The resulting line current waveforms are shown in Fig. 6.2 with T taken as unity, and they show severe harmonic distortion.

6.2.3 GRID OR GATE CONTROL AND COMMUTATION

It is essential for the control of output voltage of a rectifier and to cause a convertor to invert, that the instant of commutation shall be controlled, instead of the changeover of conduction occurring from one valve to the next at the point of natural commutation when the anode voltage of the first valve falls below that of the second. In the case of a mercury-arc valve with a grid between cathode and anode, conduction cannot occur if the grid is biassed negatively with respect to the cathode. Conduction of valve 2 is thus delayed beyond the point of natural commutation by a grid delay angle α (see Fig. 6.3), until positive voltage is suddenly applied to the grid. The grid voltage has no control over conduction once it has begun because there is then a positive ion space charge around the grid. For this reason the grid firing voltage is a short pulse, and the current can only be stopped by the anode voltage falling to zero or by commutation to the next valve when its anode voltage exceeds that of the first and its grid receives a positive pulse. In the case of a thyristor a milliampere pulse of gate or trigger current of short duration serves to initiate conduction. Due to transformer leakage and a.c. system inductance the fall of current in valve 1 and the rise of current in valve 2 are delayed, and there is a short transient period of overlap occupying $\beta°$, where the two valves conduct in parallel (see Fig. 6.3), and the output voltage waveform is the mean of these two anode voltages. β may be called the overlap or commutation angle.

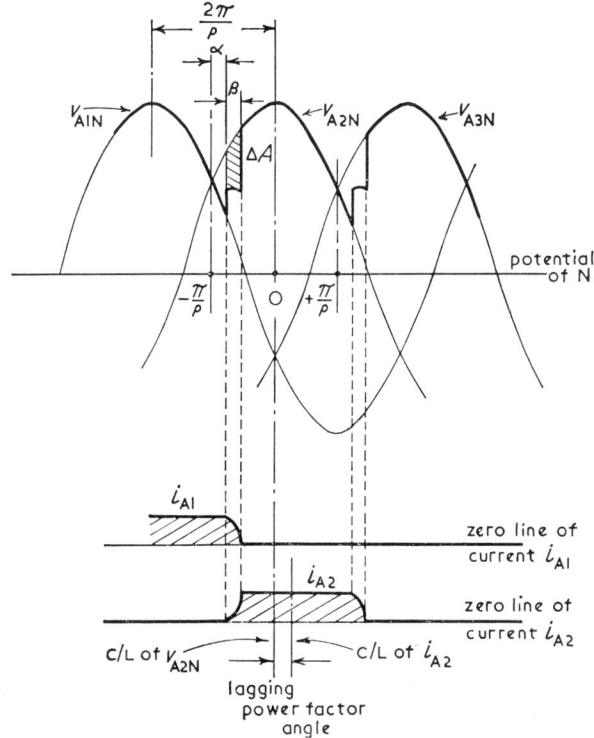

Fig. 6.3. Effect of grid delay and overlap on rectifier waveforms.

6.2.4 VOLTAGE AND CURRENT RATIOS

For a rectifier connection such as the double-star type where there
is no second valve conducting in series, the d.c. output voltage
V_{dcr} at or near no load is the mean height of the area bounded by
the horizontal line representing the secondary star point (or interphase reactor mid-point) potential, and the voltage of the conducting
anode (less the forward voltage drop V_f, which is 15 to 20 V for
mercury-arc, and about 0·5 V for semiconductor rectifiers). If overlap
is neglected (it is zero at no load) then as shown in Fig. 6.3 conduction
to any one anode occurs from $(-\pi/p+\alpha)^\circ$ to $(\pi/p+\alpha)^\circ$ where p

is the number of phases effective in the method of connection (e.g. $p = 3$ for the double-star connection above the transition current and $p = 6$ below that current).

$$V_{dcr} = \frac{p}{2\pi} \int_{(-\pi/p+\alpha)}^{(\pi/p+\alpha)} \sqrt{2}\, V_a \cos\theta\, d\theta - V_f \qquad [6.1]$$

where V_a is the r.m.s. e.m.f. induced in one phase of the transformer
           secondary

   V_f is the forward voltage drop in the rectifier

   $\theta = \omega t = 2\pi f t$

Integration gives

$$V_{dcr} = \frac{\sqrt{2p}}{\pi}\, V_a \sin\frac{\pi}{p} \cos\alpha - V_f \qquad [6.2]$$

[6.2] shows that the maximum no-load voltage $\hat{V}_{dcr}$ of the rectifier which occurs for natural commutation, i.e. for $\alpha = 0$ is given by

$$\hat{V}_{dcr} = \sqrt{2}\frac{p}{\pi}\, V_a \sin\frac{\pi}{p} - V_f \qquad [6.3]$$

and a grid delay angle α causes the voltage to fall by the factor $\cos\alpha$. It should be noted that for the 3-phase bridge connected rectifier where two valves conduct in series it may be seen from Fig. 6.2 that the d.c. voltage is given by doubling the value of [6.2] with $p = 3$.

When overlap of angle β is taken into account, inspection of Fig. 6.3 enables the output voltage to be written as

$$V_{dcr} = \sqrt{2}V_a \frac{p}{2\pi} [\tfrac{1}{2}\{ \int_{(-\pi/p+\alpha)}^{(-\pi/p+\alpha+\beta)} (\cos\theta + \cos(\theta+2\pi/p))\, d\theta\}$$

$$+ \int_{(-\pi/p+\alpha+\beta)}^{(\pi/p+\alpha)} \cos\theta\, d\theta] - V_f \qquad [6.4]$$

because from $(-\pi/p+\alpha)$ to $(-\pi/p+\alpha+\beta)$ the output voltage is the mean of the voltages of the two anodes which conduct together. Integration of [6.4] and expansion of its terms gives

$$V_{dcr} = \frac{\sqrt{2}V_a p}{2\pi} \sin\frac{\pi}{p} [\cos\alpha + \cos(\alpha+\beta)] - V_f \qquad [6.5]$$

and from [6.3] it is possible, if the forward voltage drop is negligible, to write

$$V_{dcr} = \frac{\hat{V}_{dcr}}{2} \left[\cos \alpha + \cos (\alpha + \beta) \right] \qquad [6.6]$$

In order to show the quantitative effect of L, the leakage inductance of each phase of the transformer together with supply circuit inductance, which is the cause of the overlap, [6.6] may be derived in an alternative way as follows. When two anodes or valves conduct simultaneously there is a short-circuit across two phases of the transformer secondary winding which causes an instantaneous current i_s to flow. If $t = 0$ at the point of natural commutation then the voltage between the two phase ends rises sinusoidally from zero to the peak of the line voltage, which is $\sqrt{3}(\sqrt{2}V_a)$ for a 3-phase winding but which in general is $2\sqrt{2}V_a \sin \pi/p$ for a p-phase winding. If the resistance of the transformer windings is negligible compared with the leakage reactance $X = \omega L$ then

$$2L \, di_s/dt = 2\sqrt{2}V_a \sin (\pi/p) \sin \omega t$$

$$i_s = - \frac{\sqrt{2}V_a \sin (\pi/p) \cos \omega t}{\omega L} + K$$

Since $\omega t = \alpha$ where $i_s = 0$; $K = \dfrac{\sqrt{2}V_a \sin (\pi/p) \cos \alpha}{\omega L}$

and when $\omega t = (\alpha + \beta)$, $i_s = I_{dc}$

$$I_{dc} = \frac{\sqrt{2}V_a \sin \pi/p}{X} \left[\cos \alpha - \cos (\alpha + \beta) \right]$$

$$[6.7]$$

The voltage drop ΔV due to overlap is the mean height over an angle of $2\pi/p$ of the area ΔA shaded in Fig. 6.3, i.e. $\Delta V = \Delta A p/2\pi$ where

$$\Delta A = \tfrac{1}{2} \int_{\alpha}^{(\alpha+\beta)} 2\sqrt{2}V_a \sin (\pi/p) \sin \theta \, d\theta = \sqrt{2}V_a \sin \pi/p$$
$$\times \left[\cos \alpha - \cos (\alpha + \beta) \right]$$

$$\Delta V = \frac{\sqrt{2}V_a p \sin \pi/p}{2\pi} \left[\cos \alpha - \cos (\alpha + \beta) \right] \qquad [6.8]$$

Neglecting forward voltage drop the mean output voltage for a grid delay angle α and overlap angle β is

$$V_{dcr} = \hat{V}_{dcr} \cos \alpha - \Delta V \qquad [6.9]$$

and from [6.2] and [6.3]

$$V_{dcr} = \frac{\sqrt{2} p V_a \sin \pi/p}{2\pi} [\cos \alpha + \cos (\alpha + \beta)] \qquad [6.10]$$

and [6.10] is identical to [6.6].

[6.10] and [6.3] show that the effect of overlap on the maximum voltage $\hat{V}_{dcr}$ which occurs when $\alpha = 0$, is to reduce it by the factor $\frac{1}{2}(1 + \cos \beta)$, but that when α and β are both finite their effect differs somewhat from the individual factors of $\cos \alpha$ and $\frac{1}{2}(1 + \cos \beta)$. From [6.7] and [6.8] the voltage drop due to overlap is

$$\Delta V = I_{dc} p X / 2\pi \qquad [6.11]$$

For the 3-phase bridge connection $p = 3$ in [6.11] and ΔV must be doubled because the area ΔA is lost in both upper and lower halves of Fig. 6.2, so that for this connection

$$\Delta V = 3 I_{dc} X / \pi \qquad [6.12]$$

[6.11] shows that overlap due to circuit inductive reactance produces a drop in output voltage proportional to output current and to the reactance, although no power can be dissipated in the reactance. The reduction in d.c. output power arising from grid delay angle α and overlap angle β is not due to any change in r.m.s. line currents and voltages on the primary side of the transformer, but to a lagging power factor

$$\cos \phi \simeq \frac{1}{2}[\cos \alpha + \cos (\alpha + \beta)] \qquad [6.13]$$

The lagging power factor angle of a rectifier is illustrated qualitatively in Fig. 6.3 since an anode current centre line lags behind the centre line of its voltage.

If the whole load current flows to any one anode for a time corresponding to $\theta = 2\pi/p$ radians, then neglecting overlap it may be assumed to be constant at I_{dc} due to inductance, whether or not there is grid control. The mean anode current I_m and r.m.s. anode current I_a are given by

$$I_m = \frac{1}{2\pi} \int_{-\pi/p}^{\pi/p} I_{dc} \, d\theta = \frac{I_{dc}}{p} \qquad [6.14]$$

$$I_a = \sqrt{\left(\frac{1}{2\pi} \int_{-\pi/p}^{\pi/p} I_{dc}^2 \, d\theta\right)} = \frac{I_{dc}}{\sqrt{p}} \qquad [6.15]$$

For the double-star connection each valve carries only $I_{dc}/2$ in normal working while $p = 3$ so that the r.m.s. anode current given by [6.15] becomes $I_a = I_{dc}/2\sqrt{3}$, and the volt-ampere rating of the transformer secondary using the value of the r.m.s. voltage between an anode and star point V_a given by [6.3] is

$$6V_a I_a = 6(\hat{V}_{dcr}/1\cdot17)(I_{dc}/2\sqrt{3}) = 1\cdot481\hat{V}_{dcr} I_{dc}$$

If T is defined for this connection as the ratio of the total turns on a secondary phase winding (both halves) to those on a primary phase, the r.m.s. primary phase voltage is $2V_a/T$. The current flow in one phase of the secondary, say phase a in Fig. 6.1, consists of current $I_{dc}/2$ from a_2 to a_1 for 120°, zero current for 60°, $I_{dc}/2$ from a_3 to a_4 for 120° and zero current again for 60°. This gives an r.m.s. current of $I_{dc}/\sqrt{6} = \sqrt{2}I_a$ which is flowing effectively in only half the turns of phase a. (For it to flow effectively in the whole turns of the phase, each half of the winding would have to carry current in both directions with each direction of flow occupying one-third of every cycle.) The r.m.s. primary current is therefore $I_a T/\sqrt{2}$ and the primary volt-ampere rating is

$$3(2V_a/T)(I_a T/\sqrt{2}) = 3\sqrt{2}V_a I_a = 1\cdot047\hat{V}_{dcr} I_{dc}$$

The ratio of d.c. output power to a.c. volt-ampere input which may be deduced in the way just illustrated is sometimes referred to as the utilisation factor, and clearly it is advantageous for this to be high. The peak inverse voltage across any valve which is an important factor in rectification at high voltage, may be obtained from [6.3] and the voltage waveforms; e.g. inspection of Fig. 6.1 shows that at the instant t_1, valve b_1 which is not conducting has its maximum voltage across it of $\sqrt{2}V_a + \frac{1}{2}(\sqrt{2}V_a + 0\cdot5\sqrt{2}V_a)$ which from [6.3] for $p = 3$ gives a peak inverse voltage of $2\cdot09\hat{V}_{dcr}$. Results of calculations of quantities such as these are summarised in Table 6.1 for the two connections considered.

Table 6.1 shows that although the valves of a 3-phase bridge rectifier have to carry twice the current of those of a double-star rectifier, they only have to withstand half the peak inverse voltage and voltage stresses are very critical in high-voltage d.c. convertor valves. Valves with a single anode or several valves connected in parallel to increase the current rating must be used for the bridge connection, but this is not a disadvantage since for the high powers

TABLE 6.1. *Comparison between two connections*

	6-phase double-star connection	3-phase bridge connection
Transformer secondary r.m.s. phase voltage	$0 \cdot 855 \hat{V}_{dcr}$	$0 \cdot 427 \hat{V}_{dcr}$
Peak inverse voltage	$2 \cdot 09 \hat{V}_{dcr}$	$1 \cdot 045 \hat{V}_{dcr}$
Mean anode current	$0 \cdot 167 I_{dc}$	$0 \cdot 33 I_{dc}$
Transformer secondary volt-ampere rating	$1 \cdot 481 \hat{V}_{dcr} I_{dc}$	$1 \cdot 047 \hat{V}_{dcr} I_{dc}$
Transformer primary volt-ampere rating	$1 \cdot 047 \hat{V}_{dcr} I_{dc}$	$1 \cdot 047 \hat{V}_{dcr} I_{dc}$
Peak anode current	$0 \cdot 5 I_{dc}$	I_{dc}

involved in h.v.d.c. schemes, convertors with the anodes of all three (or more) phases in one tank would not in any case be used. The bridge connection has the further advantage that arc-backs (back-fires) can easily be handled by making all grids negative so as to block the valves. An arc-back or backfire is a loss of the undirectional conducting property of a valve when the valve conducts in the wrong direction while the anode is negative with respect to the cathode, due possibly to electron emission from impurities on the anode surface. In the double-star connection, a negative bias on the grids does not prevent alternating current flowing in the faulty anode to the load and returning to the affected phase via the star-point, and the a.c. breaker would have to be opened. In the bridge connection, however, since two valves are always in series to supply the load, the faulty valve cannot continue to carry current if the grids of the other healthy valves are blocked, and the chance of simultaneous arc-backs on two valves is remote. The direct current is temporarily diverted through a by-pass valve (see Fig. 6.2) when a bridge is blocked. In the case of the cross-channel h.v.d.c. interconnection a valve group is de-blocked within no more than 1 second after an arc-back (the interval is less in some cases) and in most cases the faulty valve is then normal again.

The brief comparison here has only been made between two connections, but no other connection is as suitable for h.v.d.c. schemes as the 3-phase bridge circuit.

6.3 Inversion

When a convertor rectifies, i.e. it converts energy from an a.c. system into energy flowing into a d.c. system or load, the current flow, which must be undirectional inside a rectifying valve or thyristor, is from anode to cathode, so that the latter is the positive terminal of the d.c. output. When it is required to operate the convertor as an invertor so that energy is taken from a d.c. system and fed into an existing or self-created a.c. system, the current in the convertor valves or semiconductor devices cannot be reversed. Where the a.c. system is created by the invertor, as for example in variable-speed a.c. motor drives provided by variable-frequency thyristor invertors, the thyristors act as switches which connect d.c. of alternate polarity to the machine at a frequency controlled by the frequency of the gate pulses. On the other hand, where energy is to be supplied to an existing fixed-frequency system with sinusoidal 3-phase voltage and currents, the flow of current must be delayed by grid or gate control until a time when it is driven by the d.c. source in opposition to the voltage produced in that phase of the transformer winding by the existing a.c. system. In the case of an h.v.d.pc.ower application it is necessary to have an a.c. system or synchronous compensator connected to the invertor so as to establish the phase rotation, frequency and voltage, and to cause the current to commutate, and also to supply reactive power (see below). A valve cannot conduct when its anode is negative with respect to its cathode, but only during the time when the direct voltage applied to the invertor input exceeds the alternating voltage induced in the transformer winding from the a.c. system, so that the anode is positive with respect to the cathode by the amount of the forward voltage drop. The negative pole of the d.c. supply must therefore be connected to the valve cathodes as shown in Fig. 6.2, whereas during rectification the cathode is the positive pole of the load circuit.

For inversion the pulse on any grid must be delayed until the alternating e.m.f. makes the corresponding anode negative with respect to the transformer star-point N (this potential still exists even if physical access to it is not provided because of a delta-connected winding), but with this e.m.f. less than the applied d.c. voltage. For example, conduction can be started in valve 1 at time t_1 in Fig. 6.4 which for simplicity illustrates conditions for inversion without a second valve conducting in series, rather than for the bridge-connected invertor of Fig. 6.2. During conduction the potential V_{CN} of the cathode of the conducting valve with respect to the secondary star-point will be below that of its anode (V_{A1N}) by the forward voltage drop V_f. Conduction would cease at t_2 when the d.c. supply voltage no longer exceeds the sum of V_f and the alternating e.m.f. induced in the transformer winding, if it were not for the inclusion of sufficient inductance to the circuit (see Fig. 6.2). Since the current I_{A1} is almost constant from t_1 to t_2 the energy stored in the electromagnetic field of the inductor(s) is approximately given by the first shaded area in Fig. 6.4, since the instantaneous voltage across the inductance is the vertical ordinate between the V_{CN} curve and the horizontal line which represents the potential of the negative d.c. terminal. Conduction can continue beyond t_2 until all this energy is recovered, which would be when the second shaded area becomes equal to the first. If the d.c. supply voltage were too low this energy would have been recovered before the positive pulse of voltage appeared on grid 2, and the current would then fall to zero until this pulse arrived. However, continuous current rather than discontinuous current is normally required. Commutation to valve 2 must in any case occur before t_4 because after that time the voltage of $A2$ becomes negative with respect to that of $A1$, and commutation is no longer possible. Commutation must begin earlier than the angle corresponding to t_4 by a minimum margin of β (the overlap or commutation angle) plus δ_0 which is the angle (or time) taken for valve 1 to become de-ionised, so that it cannot again take over current from valve 2 when it becomes more positive after t_4. After δ_0 when

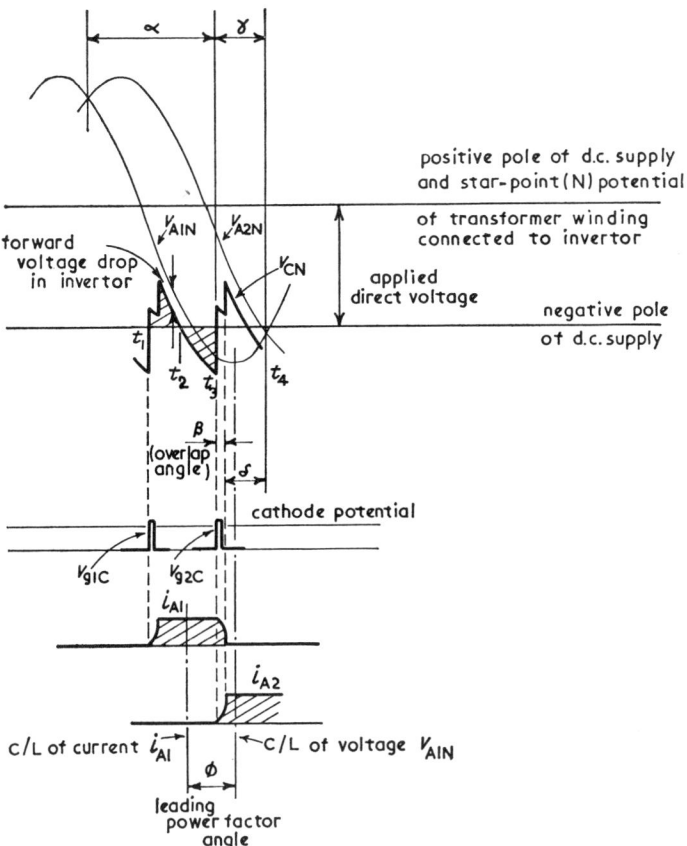

Fig. 6.4. Inversion waveforms:

V_{A1N} is alternating e.m.f. (with respect to N) induced in transformer winding of 1st anode.

V_{A2N} is alternating e.m.f. (with respect to N) induced in transformer winding of 2nd anode.

V_{g1c} is potential of grid of 1st anode with respect to cathode.

V_{g2c} is potential of grid of 2nd anode with respect to cathode.

V_{CN} is potential of cathode with respect to N during conduction.

deionisation is complete $A1$ cannot take over current from $A2$ because it is blocked by the negative bias on its grid. If the margin for deionisation is raised from the minimum of δ_0 to δ, then commutation will begin an angle $(\beta + \delta) = \gamma$ before t_4.

It may be noted from Fig. 6.4 that the invertor is being fired with a grid delay angle α measured from the point of natural commutation as a rectifier of $\alpha = (\pi - \gamma)$ radians. If the overlap angle were neglected, then substitution of $(\pi - \gamma)$ for α in [6.2] or deduction from Fig. 6.4 shows that the applied d.c. voltage for inversion V_{dci} is

$$V_{dci} = \frac{\sqrt{2}pV_a}{\pi} \sin \frac{\pi}{p} \cos \gamma + V_f \qquad [6.16]$$

The d.c. voltage is in fact opposite in polarity to the output voltage as a rectifier as already discussed, and the sign of the forward voltage drop relative to V_{dc} has changed. As in the case of rectification the voltage drop ΔV due to inductive reactance causing an overlap angle β is given by [6.8] and substitution into this equation of $\alpha = (\pi - \gamma)$ gives

$$\Delta V = \frac{\sqrt{2}pV_a}{2\pi} \sin \pi/p[\cos (\gamma - \beta) - \cos \gamma]$$

and as $(\gamma - \beta) = \delta$

$$\Delta V = \frac{\sqrt{2}pV_a}{2\pi} \sin \pi/p[\cos \delta - \cos \gamma] \qquad [6.17]$$

Since for an inverter the applied d.c. voltage has to be increased by the amount of this voltage drop ΔV (as may be seen from Fig. 6.4), V_{dci} for finite commutation angle is given by the sum of [6.16] and [6.17] so that

$$V_{dci} = \hat{V}_{dcr} \cos \gamma + \Delta V + V_f \qquad [6.18]$$

$$V_{dci} = \frac{\sqrt{2}pV_a}{2\pi} \sin \pi/p[\cos \delta + \cos \gamma] + V_f \qquad [6.19]$$

Similar substitution into [6.7] gives

$$I_{dc} = \frac{\sqrt{2}V_a \sin \pi/p}{X} [\cos \delta - \cos \gamma] \qquad [6.20]$$

From [6.17] and [6.20]

$$\Delta V = I_{dc}pX/2\pi \qquad [6.21]$$

which is the same value as for rectification ([6.11]). For the 3-phase bridge connection [6.18] and [6.21] with the latter incorporating the factor of 2 which is required for operation with two valves in series, gives

$$V_{dci} = \hat{V}_{dcr} \cos \gamma + \frac{3I_{dc}X}{\pi} + V_f \qquad [6.22]$$

Alternatively the d.c. voltage may be written from Fig. 6.4 as

$$V_{dci} = \hat{V}_{dcr} \cos \delta - \Delta V + V_f \qquad [6.23]$$

and from [6.21]

$$V_{dci} = \hat{V}_{dcr} \cos \delta - \frac{I_{dc}pX}{2\pi} + V_f \qquad [6.24]$$

Substitution of [6.20] in [6.24] gives [6.19].

If there were no inductance and if deionisation occurred infinitely quickly then β and δ could both be zero and γ would be zero, in which case [6.19] shows that the applied d.c. voltage would be $\hat{V}_{dcr} + 2V_f$, and neglecting the forward voltage drop it would be $\hat{V}_{dcr}$ (as defined by [6.3] with $V_f = 0$). In general but neglecting V_f [6.19] may be written

$$V_{dci} = \frac{\hat{V}_{dcr}}{2} (\cos \delta + \cos \gamma) \qquad [6.25]$$

It can be seen in Fig. 6.4 that the centre line of the anode current is ahead of the corresponding anode voltage, so that an invertor supplies current at a leading power factor to an a.c. system. This is another way of saying that an invertor like a rectifier consumes (lagging) reactive power from the a.c. system to which it is connected, since a source of leading reactive power is equivalent to a sink of lagging reactive power. This power factor is approximately the factor by which the d.c. input voltage has been increased (it was the factor of decrease for the rectifier) so that from [6.25]

$$\cos \phi \simeq \tfrac{1}{2}(\cos \delta + \cos \gamma) \qquad [6.26]$$

Thus although no reactive power can be transmitted along a d.c. interconnector in an h.v.d.c. scheme, both the rectifier at the transmitting end and the invertor at the receiving end consume (lagging) reactive power, partly because of the transformer leakage and a.c. system reactance affecting commutation and partly due to grid control, and in both rectifier and invertor the reactive power consumed is frequently about half the (active) power transmitted. Since the (lagging) reactive power is supplied to the invertor by the a.c. system into which it is itself supplying (active) power, then this is equivalent to the invertor supplying current to the a.c. system at leading power factor as stated above. Most systems do not have sufficient reactive power available because their loads in aggregate also consume (lagging) reactive power. The reactive power is therefore usually supplied to the invertor by synchronous compensators (capacitors) or by static shunt capacitors. The synchronous compensator can also be used to stabilise the frequency and supply additional reactive power to the system loads, but if these extra requirements do not exist then its greater capital cost and losses may cause static shunt capacitors to be used, partly because of their advantage of being combined with the harmonic filters, as in the cross-Channel link. In the case of supplying a.c. to an area such as an island without its own a.c. system, a synchronous capacitor is used to give an alternating voltage at the correct frequency so that the valves may commutate, as well as to supply reactive power to the invertor and to the loads.

6.4 High-voltage d.c. systems

In the context of this book it is only possible to outline certain of the most important aspects of the operation and application of h.v.d.c. links as parts of power systems. Although some introduction is given in section 6.4.1, it is not possible for example to discuss in any detail the compounding of rectifier and of convertor, which is the name given to the setting of their characteristics, and includes automatic control of their firing angles so as to give the required voltage, current and power flow with minimum reactive power consumption, as well as adequate protection against faults. A few references given at the end of the chapter, together with the journal 'Direct Current', would enable a student to begin to study the subject in some depth.

6.4.1 OPERATION

A high-voltage d.c. power link consists basically of an a.c./d.c./a.c. interconnector in which power is taken from an a.c. system via a 3-phase bridge rectifier, fed through a d.c. link which may be a cable, a long transmission line or a link of virtually zero length, at the end of which a 3-phase bridge invertor feeds power into a second a.c. system. The flow of power may be undirectional as in the first commercial scheme where 20 MW is supplied from the mainland of Sweden to the island of Gotland, or the flow can be reversible as in the cross-Channel link, since a convertor can become a rectifier or an invertor by grid control. Since the current direction in the convertors

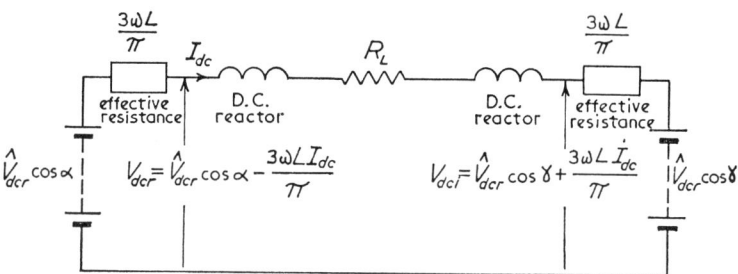

Fig. 6.5. Basic equivalent circuit of a rectifier/inverter link.

cannot be changed, power reversal is accomplished by reversing the direct voltage, although the cable polarity may be kept the same by operating polarity-reversing isolators.

The simple equivalent circuit of Fig. 6.5 represents the d.c. circuit (V_f is negligible in high-voltage rectifiers). The output voltage of the rectifier V_{dcr} is obtained from [6.9] and [6.22] and the input voltage to the invertor V_{dci} from [6.22]. The current I_{dc} flowing from rectifier to invertor is

$$I_{dc} = \frac{V_{dcr} - V_{dci}}{R_L} \qquad [6.27]$$

and the link power P is conventionally referred to as that at the sending end

$$P = V_{dcr} I_{dc} \qquad [6.28]$$

It is most efficient to work at the maximum design value of V_{dci} and to vary P by using grid control of the rectifier to change its output voltage V_{dcr} and thus control I_{dc}.

From [6.3], [6.9], [6.11] and Table 6.1 the r.m.s. a.c. line voltage $V_{ar} = \sqrt{3}V_a$ on the star-connected secondary of the transformer supplying the rectifier (Fig. 6.2), its output voltage V_{dcr} and its actual output current I_{dc} (with a rated value $\hat{I}_{dc}$) are related by

$$V_{dcr} = 1\cdot35\ V_{ar}(\cos\alpha - I_{dc}X/2\hat{I}_{dc}) \qquad [6.29]$$

For the inverter the corresponding equation involving the line voltage V_{ai} on a star-connected transformer winding connected to the invertor may be written from [6.24] as

$$V_{dci} = 1\cdot35\ V_{ai}(\cos\delta - I_{dc}X/2\hat{I}_{dc}) \qquad [6.30]$$

[6.29] and [6.30] show that the a.c. system voltages impose limits on V_{dcr} and V_{dci}, but these can be varied somewhat by tap changing (section 6.7, Volume 1). The current is sensitive to small changes in V_{dcr} and V_{dci} as shown by [6.27] and is kept to the value required for any load condition by closed loop control. V_{dci} may be kept as near as possible to its maximum value $\hat{V}_{dcr}$ (see [6.24]) keeping δ to, or very near, its minimum safe value $\delta_0(\simeq 15°)$ so as to minimise the reactive power consumption of the inverter, which gives a characteristic AB for V_{dci} in Fig. 6.6 which is drooping due to the voltage drop caused by inductance ([6.24]). Similarly the rectifier characteristic for zero grid delay angle α has a negative slope (portion CD in Fig. 6.6), but the control circuits vary α so as to give the characteristic DEF and prevent the current exceeding the desired current limit I'_{dc}. The intersection E of these characteristics is the normal operating condition for this current setting. It is possible that changes in V_{ar} and V_{ai} might tend to cause the rectifier characteristic to fall transiently below the inverter characteristic, during the period of some seconds while the tap-changing gear is operating to restore conditions. Since this would stop the transmission of any power it is prevented by the invertor control increasing the angle γ which reduces V_{dci} ([6.25]) and by giving the characteristic AHG enables operation to continue at H with a slightly reduced current until it is restored by control and tap change. The margin of current between such points as H and E must be sufficient to avoid both rectifier and invertor regulation controls operating simultaneously, and when the current through the d.c. link is to be set to a new value it is necessary to adjust both controls together to maintain the margin. The margin of voltage DE is a compromise between avoiding frequent invertor control with its effect on current and minimising reactive compensation by keeping α as low as possible (see [6.13] for the effect of α upon power factor).

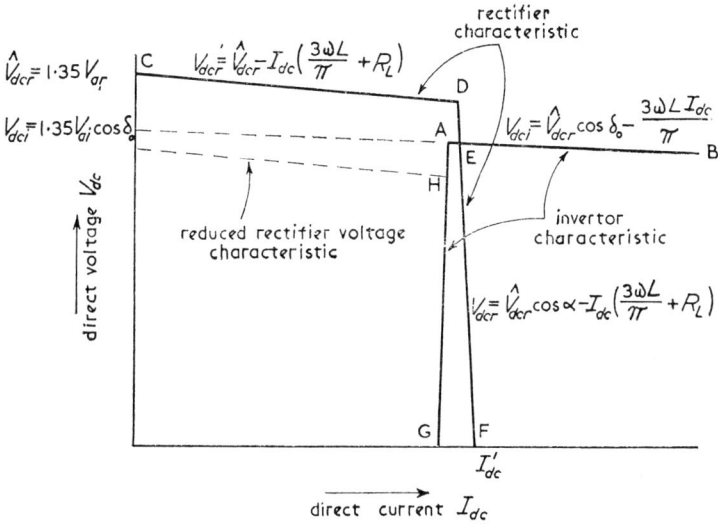

Fig. 6.6. Characteristics of bridge-connected rectifier and invertor with upper and lower current limits.

6.4.2 APPLICATIONS

There are a number of ways in which a high-voltage d.c. system may be technically advantageous in particular circumstances in which it is necessary to extend a power supply system, but comparison with alternative solutions does not always show it to be the most economic scheme. Such economic comparisons vary from one location and conditions to another, and in any case are outside the scope of this book, and most simple comparisons are invalidated by the assumptions made, so that only general discussion will be given.

Three of the chief advantages of d.c. systems being planned as a part of extensions to an existing a.c. system are:

(a) d.c. overhead lines and cables cost less than a.c. lines and cables. High voltage d.c. cables can be longer than corresponding a.c. cables because the steady-state charging current is avoided;

(b) it is possible to bring more power into an a.c. system via a d.c. link without raising the fault level and circuit breaker ratings (fault levels are discussed in Chapter 8 of Volume 1);

(c) a d.c. link is asynchronous, i.e. the two a.c. systems at either end do not have to be controlled in phase or even be at exactly the same frequency as they do for an a.c. link, and the power transmitted can be readily controlled. There is no risk of a fault in one system causing loss of transient stability in the other system.

Also d.c. overhead lines themselves are inherently more economical than a.c. lines if the earthed d.c. neutral conductor is absent (or very lightly insulated and loaded) so that there are two d.c. conductors, one positive and the other negative with respect to earth, and if the towers and insulators have been designed for d.c. working. A d.c. leakage path across an insulator surface tends to be longer than that required for a peak a.c. line-to-earth (phase) voltage equal to the d.c. voltage, because of electrostatic precipitation of atmospheric dirt. This limits the d.c. voltage and though there is a slightly higher current rating for a given conductor with d.c. than with a.c., the increase in load transfer, or reduction in line cost for the same load with d.c., only offsets the increased cost of terminal equipment (convertors, harmonic filters, etc.) for point-to-point transmission if the line is longer than distances which may be 650 km or more depending upon a number of factors.

In the case of underground cables the economic cross-over length is very much less than for lines and may be under 40 km. This is mainly because (as has been discussed in Chapter 5 of Volume 1) the d.c. dielectric strength of E.H.V. cable insulation is about three times its a.c. strength, and partly because the d.c. current rating of a cable exceeds the a.c. current rating due to the absence of charging current, dielectric and sheath losses and lower conductor resistance. Most d.c. cables require only very narrow wayleaves compared with a.c. cables of comparable rating and one effect of d.c. operation would be to reduce the cost ratio between overhead lines and underground cables.

The second advantage, referred to as (b) above, arises from the facts that a d.c. link cannot transmit reactive power and that convertors have very little stored energy since they have no inertia, so that the d.c. link makes only a negligible contribution to the fault current of the a.c. system on either side of it. In order to meet the load growth in Great Britain (which in recent years diminished to a 4% annual growth in maximum demand due to reduced industrial expansion and gas and oil finds in the North Sea, and has since been followed by a temporary decline in total demand), it might eventually become necessary to supersede with a new higher voltage system (of the order of 1 MV) the present primary transmission system (or supergrid, as it is sometimes called) with its 400-kV and 275-kV lines, to which most of the recent very large generating sets are connected. This introduction of a completely new primary system should be delayed as long as possible (or even if possible avoided altogether), for reasons which concern such factors as economy, supply, security and availability of land in the right place for substations and lines. There are two ways in which the system could be reinforced using d.c. links in order to delay or avoid constructing a new a.c. primary system. The first of these would be to divide the present single power system in Great Britain into a number of systems which would be separate except for special links between them, through which very little fault current could flow from one system to another but through which load transfers could take place. The two alternative special links are d.c. links and resonant links (see Chapter 3) and investigations into their relative merits (which may show resonant links to be the more economical in general) and into the whole question of subdivision of the system are in progress.

The second way in which d.c. links can be used to reinforce the system in a manner which may delay the need for a new primary system, is to connect rectifiers to a large generating station or stations, and to supply power over d.c. line or cable to an invertor station connected into the secondary transmission system (or into a bulk supply point in the distribution system), and situated in an area with a high loading. In this way the generating stations which would be difficult to locate near the loads can be situated where fuel and water are readily available, while space for the relatively small invertor station can more easily be found near the load area. In the event of failure of the d.c. link, power transfer can be made by the a.c. connection between the generating stations and the primary transmission system. The Kingsnorth scheme will provide valuable data from which to assess the economical and operating merits of similar links which might be investigated in the future. In this scheme a 500-MW alternator and the 400-kV primary system are connected to a convertor station at Kingsnorth, from where up to 320 MW can be transmitted (and in the opposite direction also if necessary) along a three-wire d.c. cable system at $\pm$ 266 kV to Beddington and a further 320 MW to Willesden, with any out-of-balance current between these two load centres flowing in the neutral cable. In this type of application of high-voltage d.c., the third advantage (c) above plays an important part since it is vital to have flexible control over the power transfer, which would not be so easily obtained with a synchronous connection where the load sharing between parallel circuits depends upon the individual circuit impedances.

This third advantage (c) was particularly important in the case of the cross-Channel 160-MW, $\pm$ 100-kV link, because it would have been necessary to install very expensive frequency-control equipment in Great Britain to obtain with an a.c. link a steady transfer of power, whereas a d.c. link makes this possible even though there are differences in the frequencies in the two countries.

The development of semiconductor rectifiers has now reached the stage where the Inga-Shaba scheme in Zaire, for example, is now nearing completion. This will transmit 560 MW at $\pm$500 kV over 1700 km of overhead line, with each unit containing hundreds of air-cooled thyristors in series with potential grading to equalise the voltages.

REFERENCES

ADAMSON, C. & HINGORANI, N. G. 1960. *High Voltage Direct Current Power Transmission.* Garraway, London.

BRADLEY, D. A., CLARKE, C. D., DAVIS, R. M. & JONES, D. A. 1964. Adjustable-frequency invertors and their application to variable-speed drives. *Proc. I.E.E.*, **111**, 1833–1846.

CAILLEZ, H., CASSON, W., LAURENT, P. & SCHOFIELD, H. R. 1963. Design and construction of the cross-Channel d.c. inter-connector. *Proc. I.E.E.*, **110**, No. 3, 603–618.

CASSON, W., LAST, F. H. & HUDDART, K. W. 1966. The introduction of H.V.D.C. transmission into a predominantly A.C. network. *Electrical Review*, **178**, No. 8, 290–295.

ENGSTROM, L. 1975. Refining copper with HVDC. *I.E.E.E. Spectrum*, 40–55.

GAVRILOVIC, A. 1966. Basic facts of A.C./D.C. Conversion. *Electrical Review*, **178**, No. 8, 296–300.

KING, K. G. 1963. The application of silicon controlled rectifiers to the control of electrical machines. *Proc. I.E.E.*, **110**, No. 1, 197–204.
LANE, F. J. 1969. Engineering unlimited. *Proc. I.E.E.* **116**, No. 1, 89–95.
READ, J. C. 1963. Rectifiers and rectifier applications. *Proc. I.E.E.*, **110**, No. 4, 714–738.

Examples

1. Calculate from first principles, the no-load and full-load mean output voltages for a 6-anode rectifier, the angle of overlap on full-load being 30°. The peak voltage of each secondary phase is 300 V. Assume a constant arc drop of 20 V.

(University of London) (267 V, 248 V)

2. In a 3-anode rectifier the secondary leakage reactance of each phase is 0·38 Ω, the secondary r.m.s. voltage is 390 V and the perfectly-smoothed output current is 220 A. Neglecting all losses in the transformer and rectifier, and deriving any necessary expressions, calculate the angle of overlap between anode currents.

Calculate also the output voltage.

(University of London) (34·4°, 416 V)

3. Each core limb of a 3-phase transformer carries a primary winding of 100 turns and a centre-tapped secondary winding with a total of 160 turns. The three primary windings are connected in delta and are supplied from the 415-V (line voltage), 50-Hz, 3-phase supply. The centre taps of the three secondary windings are connected together, while the six ends of the windings are connected to the six anodes of a mercury-arc rectifier.

The transformer iron loss is 40 W, the primary resistance per phase is 0·16 Ω, and the secondary resistance measured between one anode and the centre tap is 0·05 Ω. If the rectifier is operating with a constant output current of 30 A, an arc voltage of 20 V and a grid delay angle of 60°, calculate from first principles (with overlap neglected), (*a*) the mean output voltage; (*b*) the r.m.s. anode current, and (*c*) the overall efficiency.

(University of Leeds) (204·1 V, 12·25 A, 88·74%)

4. The d.c. supply to a mercury-arc invertor is 200 A at 480 V. The six anodes are connected to a simple six-point star of transformer

primary windings, in each of which a sinusoidal e.m.f. is induced, and thence through the transformer secondary to a balanced 3-phase system.

Assuming the direct current to be steady, and arc commutation to occur 15° later than the successive a.c. voltage zeros, sketch the wave forms of voltage and current for one anode. Determine the r.m.s. transformer e.m.f. in an anode winding and the utilisation factor on the six-phase (primary) side of the transformer. Neglect all internal voltage drops.

(University of London) (503 V, 0·39)

5. Each phase of the 6-phase secondary winding of a transformer connected to a mercury-arc convertor has an r.m.s. voltage of 300 V. The convertor can operate (a) as a rectifier, and (b) as an invertor, in each case with a constant arc-drop of 20 V and a load current small enough for overlap and impedance voltage drops to be neglected.

As a rectifier (a) the direct output voltage is 310 V (average value). As an invertor (b) supplied from a 400-V d.c. source with negligible circuit inductance, conduction is discontinuous and each anode conducts for $\frac{1}{8}$ period (45°). Calculate from first principles the difference in grid delay-angle between conditions (a) and (b).

(University of London) (103° 30′)

6. A three-anode mercury-arc convertor is supplied from a 415-V, 3-phase supply through a delta/star step-up transformer with a phase turns ratio of 1 : 2. The convertor is operating as an invertor with a grid delay angle (measured from the point of natural commutation as a rectifier) of 170°, and an arc voltage drop of 20 volts. Between the end of current flow in one anode and the beginning of current flow in the next anode, there is a period of non-conduction corresponding to an angle of 80°. The circuit inductance is negligible. Calculate the voltage of the d.c. source. State any assumptions which are made in obtaining a solution.

(University of Leeds) (1037 V)

Chapter 7

3-PHASE NETWORK MATRICES AND CO-ORDINATE SYSTEM TRANSFORMATIONS

7.1 Introduction

The previous chapters in this book and in Volume 1 have dealt with such individual aspects of power systems as a particular class of equipment, the principles in the calculation of fault currents in a system or the protection of parts of the system against over-currents or over-voltages and the avoidance of loss of synchronism. This chapter and the following one, on the other hand, deal with the principles involved in methods of simplifying and reducing to a systematic routine process using a digital computer, the various simultaneous linear equations involved in a large part of a power system with all the diverse equipment playing its part simultaneously. These equations may be those for normal load flow in the system (see section 8.4), or those for unbalanced faults (see Chapter 8 of Volume 1, and section 7.6) or those equations relating to the transient stability of the system (see Chapter 3).

The systematic computation of currents and voltages in electrical circuits when employing a digital computer, may most conveniently be achieved by representing the simultaneous linear equations in matrix form with these variables expressed in one of several co-ordinate systems. The most common systems encountered in power system calculations are mesh current equations (section 8.2), nodal voltage equations (section 8.3) and where unbalance occurs in 3-phase circuits the currents and voltages may be represented by their symmetrical components (Chapter 8 of Volume 1). An alternative system to the latter which is used occasionally for special purposes in unbalanced 3-phase circuits is the system of α, β and zero components (Clarke, 1943, and section 7.8). When the small differences between the polar and interpolar axes of a cylindrical-rotor synchronous machine or the larger differences in a salient-pole machine cannot be neglected, then the three-phase currents and voltages may be transformed into direct and quadrature-axis components (Volume 1, Chapter 7, and section 7.7).

All of these are examples of the use of substitute variables in place of actual phase currents and voltages in order to simplify and systematise the calculations. The manner in which individual branches or separate circuits are interconnected to form a circuit in which currents are defined around the meshes, or in which the potentials of all nodes are defined with respect to one reference node, can be specified by a transformation matrix known as a connection matrix (see sections 8.2 and 8.3 respectively). A set of unbalanced 3-phase currents or voltages can be transformed into its corresponding set of symmetrical components by a transformation matrix (see section 7.6), so that in effect a substitute circuit is provided.

In order to evaluate the currents and voltages in a large network such as a 3-phase power system, a much simpler circuit known as a primitive network is first considered, for which the equations can readily be stated. The currents and voltages in this primitive or related network, which will be denoted here by a prime, e.g. $\mathbf{I}'$ and $\mathbf{V}'$, are then related to the actual system currents and voltages, $\mathbf{I}$ and $\mathbf{V}$, by a connection or transformation matrix. Thus for each of the coordinate systems mentioned above there is a primitive network: for mesh currents it is that of the network branches separated from one another and short-circuited (see Fig. 8.1(c)); for nodal voltages it is that of the network branches separated from each other and open-circuited, each with its own nodal current impressed (see Fig. 8.3(b); for symmetrical components it is the positive-, negative- and zero-sequence networks relating the sequence currents and voltages of the reference phase (see Chapter 8 of Volume 1); for rotating electrical machines the primitive network or machine is one in which there are coils stationary in space on two perpendicular axes, the direct and quadrature axes, with rotational and transformer e.m.f.s induced in these coils.

In this chapter the relationships involved in these transformations are derived before going on in Chapter 8 to apply nodal voltages (and to a lesser extent mesh currents) to the systematic solution by computer of the load flow equations in a power system. The most common restraint relating the currents and voltages in the power system to those in the primitive system is that of the same power in each, which is known as power invariance. This is the most logical condition and might at first be thought to be the only one which would be employed, but it will be shown in sections 7.5 and 7.6 that the convenience of having the same transformation matrix for both voltages and currents, and of a multiplying factor of unity, has led to the symmetrical component transformation being applied most commonly without power invariance.

This chapter begins by applying power invariance to the mesh current and nodal voltage transformations. Using the results obtained for the mesh current case, a 3-phase circuit with unbalanced mutual impedances between the phases and unbalanced self-impedances in the phases is considered. When each of these sets of impedances is balanced, an equivalent single-phase representation is obtained and this justifies the single-phase treatment of balanced 3-phase circuits without explicitly including the mutual impedance except as part of an effective self-impedance, as was done in Volume 1. After giving in section 7.4 the general case of the voltage transformation being unrelated to the current transformation by a condition such as power invariance, the next section deals with the symmetrical component transformation matrix which is the same for both voltages and currents, and it is shown that unless a particular value of multiplying factor is used (which generally is not the case), then the power is not invariant between the power system and the symmetrical component sequence networks. The use of matrix methods shows, more completely than was possible in section 8.3.2 of Volume 1, the conditions required for there to be no mutual coupling between the sequence networks. Finally, in this chapter, the relationships are given for transformations to fixed rotor axes in a synchronous machine, and for the alpha, beta and zero components sometimes known as Clarke's components.

7.2 Transformations with invariance of power

The column vectors (see Appendix A.1.1) of the currents and voltages of the primitive network are denoted by $[I']$ and $[V']$, while those of the actual network to be studied are $[I]$ and $[V]$. The currents of the primitive network may be related to those of the actual system by means of a connection matrix $[C]$,

$$[I'] = [C][I] \qquad [7.1]$$

In the primitive system the voltages will be related to the currents by a square impedance matrix $[Z']$,

$$[V'] = [Z'][I'] \qquad [7.2]$$

Similarly the voltages and currents in the actual system are related by a square impedance matrix $[Z]$,

$$[V] = [Z][I] \qquad [7.3]$$

[7.3] does not give any information as to the co-ordinate system employed. It could, for example, be the general form of [8.2], in which case the mesh current system is being used and $[Z]$ is a mesh impedance matrix; or it could represent [8.32] in which case the equations are written in nodal voltage form and $[Z]$ is a nodal impedance matrix.

The restraint which is most frequently employed in relating the currents and voltages of the actual system to those of the primitive system is that the complex power in each system is the same. In any branch carrying a current $\mathbf{I}$ with a potential difference $\mathbf{V}$ across it, the complex power $\mathbf{S}$ is given by (see Appendix 4 of Volume 1)

$$\mathbf{S} = P + jQ = \mathbf{VI^*}$$

Since $[V]$, $[I]$, $[V']$ and $[I']$ are column vectors, one of each set must be transposed to form a row vector before multiplication. Thus the condition of power invariance gives

$$[V']_t[I'^*] = [V]_t[I^*] \qquad [7.4]$$

Substituting in [7.4] for $[I'^*]$ obtained from [7.1] gives

$$[V']_t[C^*][I^*] = [V]_t[I^*]$$

$$([V']_t[C^*] - [V]_t)[I^*] = 0$$

Since $[I^*] \neq 0$, then either $([V']_t[C^*] - [V]_t) = 0$ or its product with $[I^*] = 0$. The latter cannot be true for a linearly independent set of currents $[I^*]$ (Bandyopadhyay, 1969). Hence

$$[V']_t[C^*] = [V]_t$$

This equation does not follow directly from that following [7.4] because $[I^*]$ is a singular matrix so that its inverse does not exist and hence 'cancellation' of [I*] from both sides of the equation is not permissible.

Transposing (see Appendix A.1.5.1) gives

$$[V] = [C^*]_t[V']$$ [7.5]

[7.5] expresses the constraint placed by the condition of complex power invariance, upon the relationship between the voltages of the actual and primitive systems when the currents are related by [7.1]. [8.5(a)] is an example of applying [7.5] and in this particular case it is applied to mesh current analysis.

The relationship between the impedance matrix of the actual network using a particular co-ordinate system, and the impedance matrix of the primitive network may now be found by combining [7.3] and [7.5] to give

$$[V] = [Z][I] = [C^*]_t[V']$$

Substitution of [7.2] and [7.1] then gives

$$[Z][I] = [C^*]_t[Z'][C][I]$$

so that

$$[Z] = [C^*]_t[Z'][C]$$ [7.6]

7.2.1 APPLICATION TO MESH CURRENTS

In section 2 of the next chapter the voltages and currents of the separated branches of the primitive system are denoted by [e] and [i], and these correspond to [V'] and [I'] as used here, while the mesh driving voltages of the actual system [E] correspond to [V]. The equations which correspond are thus [7.1] and [8.3], [7.5] and [8.5(a)], [7.6] and [8.10]. The matrix [Z] of the last two of these equations is known as the mesh impedance matrix of the network.

7.2.2 APPLICATION TO NODAL VOLTAGES

If the relationships between the sets of currents and voltages are expressed in terms of admittances, as is done in nodal voltage equations, then instead of [7.2] and [7.3] there will appear

$$[I'] = [Y'][V']$$ [7.7]

$$[I] = [Y][V]$$ [7.8]

If the constraint of power invariance is again applied by using [7.4], with the voltages now related by

$$[V'] = [C][V] \qquad [7.9]$$

in place of the current relationship of [7.1], then substitution as before gives

$$[I] = [C^*]_t[I'] \qquad [7.10]$$

and

$$[Y] = [C^*]_t[Y'][C] \qquad [7.11]$$

These last two equations correspond to [8.17] and [8.21] of section 8.3 of the next chapter. [Y] is known as the nodal admittance matrix of the network.

7.3 Mesh current connection matrices applied to a 3-phase network

In addition to the self-impedances shown in the 3-phase network of Fig. 7.1, mutual impedances

$$z_{ab} \quad z_{bc} \quad z_{ca} \quad z_{an} \quad z_{bn} \quad \text{and} \quad z_{cn}$$
$$z_{ba} \quad z_{cb} \quad z_{ac} \quad z_{na} \quad z_{nb} \quad \text{and} \quad z_{nc}$$

exist in the equipment in the system. In the general case these mutual impedances may both be unbalanced so that $z_{ab} \neq z_{bc}$ etc., and $z_{an} \neq z_{bn}$ etc., and non-reciprocal so that $z_{ab} \neq z_{ba}$ etc. The simple equivalent circuit of Fig. 7.1 has, however, only limited representation as it applies to a system containing static equipment only, or approximately to the first one or two cycles after a fault occurs in a system containing synchronous machines (see section 7.6.2). This is because it does not include any time-varying interaction with the field circuits of the synchronous machines in the system, so that any circuit containing a set of machine impedances could only apply to a single sequence for one time regime; e.g. it could contain the values of steady-state positive-sequence impedances for all the equipment. For initial positive-sequence conditions the machine impedances could be changed to subtransient or transient values, and they would have to be changed again for negative-sequence conditions (see Chapter 7, Volume 1).

It is assumed that there are no harmonics present in the circuit currents and voltages. The four r.m.s. branch currents are given by I_a, I_b, I_c and I_n, and these branch currents are related to the r.m.s. mesh currents I_{ae}, I_{be} and I_{ce} by a mesh connection matrix [C], where

$$[C] = \begin{bmatrix} 1 & 0 & 0 \\ 0 & 1 & 0 \\ 0 & 0 & 1 \\ -1 & -1 & -1 \end{bmatrix}$$

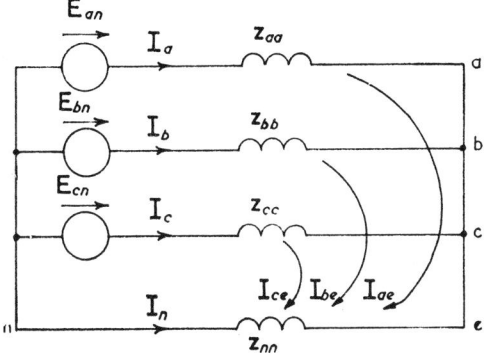

Fig. 7.1. Mesh currents in a 3-phase system.

From [7.6] the mesh impedance matrix is given by

$$[Z] = [C^*]_t[z][C]$$

where the branch impedance matrix [z] is

$$[z] = \begin{bmatrix} z_{aa} & z_{ab} & z_{ac} & z_{an} \\ z_{ba} & z_{bb} & z_{bc} & z_{bn} \\ z_{ca} & z_{cb} & z_{cc} & z_{cn} \\ z_{na} & z_{nb} & z_{nc} & z_{nn} \end{bmatrix}$$

[7.12]

$[\mathbf{Z}] =$

$$
\begin{bmatrix}
(z_{aa}-z_{na}-z_{an}+z_{nn}) & (z_{ab}-z_{an}-z_{nb}+z_{nn}) & (z_{ac}-z_{an}-z_{nc}+z_{nn}) \\
(z_{ba}-z_{bn}-z_{na}+z_{nn}) & (z_{bb}-z_{bn}-z_{nb}+z_{nn}) & (z_{bc}-z_{bn}-z_{nc}+z_{nn}) \\
(z_{ca}-z_{cn}-z_{na}+z_{nn}) & (z_{cb}-z_{cn}-z_{nb}+z_{nn}) & (z_{cc}-z_{cn}-z_{nc}+z_{nn})
\end{bmatrix}
$$

$$[7.13a]$$

$$
[\mathbf{Z}] =
\begin{bmatrix}
Z_{aa} & Z_{ab} & Z_{ac} \\
Z_{ba} & Z_{bb} & Z_{bc} \\
Z_{ca} & Z_{cb} & Z_{cc}
\end{bmatrix}
\qquad [7.13b]
$$

In the next subsection [7.13] is used to examine the effect of introducing balance into the system.

7.3.1 SINGLE-PHASE REPRESENTATION OF A BALANCED 3-PHASE NETWORK

It is generally assumed in power-system analysis that the self- and mutual impedances of machines and transformers are balanced for all three phases. If, for a long transmission line, the inequality in phase-to-phase and phase-to-earth spacings were sufficient to cause significant unbalance, then the conductors could be transposed, so that again the self-and mutual impedances would be balanced (see Chapters 2 and 3 of Volume 1).

Thus for the system shown in Fig. 7.1 which represents the network up to the terminals a, b, c and e where an unbalanced fault will be considered in later sections to occur,

$$z_{aa} = z_{bb} = z_{cc}; \; z_{ab} = z_{bc} = z_{ca}; \; z_{ba} = z_{cb} = z_{ac};$$

$$z_{an} = z_{bn} = z_{cn} = z_{na} = z_{nb} = z_{nc}$$

For these conditions [7.13] may be written as

$$
[\mathbf{Z}] =
\begin{bmatrix}
Z_p & Z'_m & Z''_m \\
Z''_m & Z_p & Z'_m \\
Z'_m & Z''_m & Z_p
\end{bmatrix}
\qquad [7.14]
$$

where

$$Z_p = z_{aa} - 2z_{an} + z_{nn}$$

$$Z'_m = z_{ab} - 2z_{an} + z_{nn}$$

$$Z''_m = z_{ba} - 2z_{an} + z_{nn}$$

If the voltages and currents in the system of Fig. 7.1 are balanced so that they are of positive-sequence only then using the h operator discussed in Appendix 3 of Volume 1 together with [7.14] gives the relationship

$$
E_{an1}
\begin{bmatrix}
1 \\
h^2 \\
h
\end{bmatrix}
=
\begin{bmatrix}
Z_p & Z'_m & Z''_m \\
Z''_m & Z_p & Z'_m \\
Z'_m & Z''_m & Z_p
\end{bmatrix}
I_{ae1}
\begin{bmatrix}
1 \\
h^2 \\
h
\end{bmatrix}
$$

and $E_{an1} = E_{an}$ because $E_{an2} = E_{an0} = 0$, and similarly $I_{ae1} = I_a$

$$E_{an} = I_a(Z_p + h^2 Z'_m + h Z''_m)$$

Substituting values of impedance from [7.14] into this result gives

$$E_{an} = I_a(z_{aa} + h^2 z_{ab} + h z_{ba}) \qquad [7.15]$$

In most cases reciprocal mutual inductances (i.e. $M_{ab} = M_{ba}$ etc.) are assumed without question, as this is always true for passive networks and energy considerations can show this to be true for a machine under ideal non-saturating conditions. In a real iron-cored machine, reciprocity does not necessarily hold, but experiments have shown that where the saturation conditions are carefully reproduced, the difference between the two mutual inductances is very small, even when measured between a distributed winding and a lumped winding (Jones, 1958). It may, therefore, be generally assumed that $z_{ab} = z_{ba}$ etc. so that

$$E_{an} = I_a(z_{aa} - z_{ab}) \qquad [7.16]$$

Thus in a balanced 3-phase circuit the effective self-impedance of each phase is the series phase impedance minus the mutual impedance between phases.

7.4 General transformation

It has been shown in section 7.2 that if the currents of the real and primitive systems are related by the connection matrix [C] in the mesh current method, the voltages are related by $[C^*]_t$ when the power is the same in the two systems. If this condition is not applied then in general the voltages may be related by [B] which may or may not be related to [C].

If [7.5] and [7.10] are pre-multiplied by $[C^*]_t^{-1}$ then they become

$$\text{for mesh currents } [V'] = [C^*]_t^{-1}[V] \qquad [7.17]$$

$$\text{for nodal voltages } [I'] = [C^*]_t^{-1}[I] \qquad [7.18]$$

Comparing these equations with [7.1] and [7.9] respectively shows that it is only possible to write [7.17] and [7.18] when [C] is a square matrix.

The relationships developed in section 7.2 represent only the special case of power invariance between the two systems, whereas the more general form of [7.5] for the case where the currents are related by [7.1] is

$$[V] = [B][V'] \qquad [7.19]$$

Similarly, for the case where the nodal voltages are related by [7.9] the current relation could in general be written as

$$[I] = [D][I'] \qquad [7.20]$$

instead of the special case of [7.10]. The general form of [7.6] then becomes

$$[Z] = [B][Z'][C] \qquad [7.21]$$

Similarly for the nodal admittance matrix the general form of [7.11] becomes

$$[Y] = [D][Y'][C] \qquad [7.22]$$

In practice [B] or [D] will be related to [C] by some condition such as power invariance, and in the next section a further restraint beyond this one is imposed.

7.5 Equal transformation matrices for currents and voltages and orthogonal transformations

Another special case beside that of power invariance is where the same transformation matrix [T] is used for both currents and voltages.

$$[I'] = [T][I] \qquad [7.23]$$

$$[V'] = [T][V] \qquad [7.24]$$

If in addition to this restriction of the same matrix for both currents and voltages, the condition of power invariance is also imposed, then provided that [T] is a square matrix (i.e. the transformation is between two systems having the same number of dimensions), comparison of [7.23] and [7.24] with [7.1] and [7.17] shows that

$$[T] = [T^*]_t^{-1} \qquad [7.25]$$

or

$$[T]^{-1} = [T^*]_t \qquad [7.26]$$

which indicates that [T] is an orthogonal matrix (see Appendix A.1.9.1). Pre-multiplying both sides of [7.26] by [T] gives

$$[T][T^*]_t = [U] \qquad [7.27]$$

where [U] is the unit matrix. Since the determinant $|T|$ of [T] is unaltered by interchanging rows and columns, then

$$|T| \times |T^*| = 1 \qquad [7.28]$$

In some cases, such as a symmetrical component transformation matrix [T_s], the condition of the same transformation matrix for both currents and voltages is satisfied but the condition of power invariance is not. The latter condition would only be met if the transformation matrix [T_s] were multiplied by a factor k, i.e. if [T] = k[T_s]. [7.26] then becomes

$$(1/k)[T_s]^{-1} = k[T_s^*]_t$$

so that

$$k^2[T_s][T_s^*]_t = [U]$$

and if [T_s] is a square matrix of n rows and n columns [7.28] is replaced by

$$k^{2n}|T_s| \times |T_s^*| = 1$$

from which the factor k is given as

$$k = \frac{1}{2n\sqrt{(|T_s| \times |T_s^*|)}} \qquad [7.29]$$

This factor k is evaluated in the next section, but it is shown that the method of symmetrical components can be, and usually is, applied with $k = 1$, which does not satisfy the condition of power invariance.

7.6 Symmetrical component transformation matrix

In this section, after dealing with the value of k in [7.29], the relationship between the symmetrical component and actual system mesh impedance matrices is obtained. In the following two subsections this is applied to an unbalanced and then to a balanced 3-phase circuit in order to show the condition for no mutual coupling between the three sequence networks. In section 7.6.3 the simplification which results if an unbalanced fault is symmetrical with respect to the phase which is taken as reference is illustrated, and section 7.6.4 indicates a way in which to deal with a system with energy in-feed on both sides of an unbalanced fault.

As described in Chapter 8 of Volume 1, unbalanced 3-phase voltages and currents in a linear circuit may be represented by the addition of three sets of balanced components. With phase a taken as the reference phase, the r.m.s. phase e.m.f.s which are the driving voltages in the three meshes of the circuit of Fig. 7.1 are given by

$$\begin{bmatrix} \mathbf{E}_{an} \\ \mathbf{E}_{bn} \\ \mathbf{E}_{cn} \end{bmatrix} = \begin{bmatrix} 1 & 1 & 1 \\ h^2 & h & 1 \\ h & h^2 & 1 \end{bmatrix} \begin{bmatrix} \mathbf{E}_{an1} \\ \mathbf{E}_{an2} \\ \mathbf{E}_{an0} \end{bmatrix} \qquad [7.30]$$

and similarly the mesh currents may be written as

$$\begin{bmatrix} \mathbf{I}_{ae} \\ \mathbf{I}_{be} \\ \mathbf{I}_{ce} \end{bmatrix} = \begin{bmatrix} 1 & 1 & 1 \\ h^2 & h & 1 \\ h & h^2 & 1 \end{bmatrix} \begin{bmatrix} \mathbf{I}_{ae1} \\ \mathbf{I}_{ae2} \\ \mathbf{I}_{ae0} \end{bmatrix} \qquad [7.31]$$

These equations may be written as

$$[E] = [T_s][E_s] \qquad [7.32]$$

and

$$[I] = [T_s][I_s] \qquad [7.33]$$

where $[T_s]$ may be called a symmetrical component transformation matrix.

From [7.29] it can be seen that if the constraint were to be applied that this power shall be invariant between the actual system voltages and currents and their symmetrical components, then $k[T_s]$ would need to be used in place of $[T_s]$ where

$$k = \frac{1}{\sqrt[6]{(|T_s| \times |T_s^*|)}} = \frac{1}{\sqrt{3}} \qquad [7.34]$$

(The student may verify, if he wishes, as an exercise in manipulation, that the value of k is indeed $1/\sqrt{3}$.)

It is not in fact necessary or customary to multiply $[T_s]$ by $1/\sqrt{3}$ so as to keep the power invariant in the two systems, since the sequence voltages and currents are evaluated as shown in Chapter 8 of Volume 1, and are then synthesised to give the actual system voltages and currents by means of [7.30] and [7.31], so that the transformation is applied and then reversed before power is calculated.

Pre-multiplying [7.32] and [7.33] by $[T_s]^{-1}$ gives

$$[E_s] = [T_s]^{-1}[E] \qquad [7.35]$$

and

$$[I_s] = [T_s]^{-1}[I] \qquad [7.36]$$

where

$$[T_s]^{-1} = \tfrac{1}{3} \begin{bmatrix} 1 & h & h^2 \\ 1 & h^2 & h \\ 1 & 1 & 1 \end{bmatrix} \qquad [7.37]$$

The mesh e.m.f.s. and currents are related by the mesh impedance matrix $[Z]$

$$[E] = [Z][I] \qquad [7.38]$$

and the symmetrical components of e.m.f. and current by

$$[E_s] = [Z_s][I_s] \qquad [7.39]$$

where $[\mathbf{Z}_s]$ is the symmetrical component mesh impedance matrix. [7.32] and [7.36] correspond to [7.19] and [7.1] so that from [7.21] the actual system mesh impedance matrix $[\mathbf{Z}]$ (see [7.13]) is related to $[\mathbf{Z}_s]$ by

$$[\mathbf{Z}] = [\mathbf{T}_s][\mathbf{Z}_s][\mathbf{T}_s]^{-1}$$

and this may be rewritten as

$$[\mathbf{Z}_s] = [\mathbf{T}_s]^{-1}[\mathbf{Z}][\mathbf{T}_s] \qquad [7.40]$$

This is exactly the same result as would be obtained if power invariance were applied because k would appear on both sides of [7.39] and is thus cancelled.

7.6.1 GENERAL UNBALANCED 3-PHASE CIRCUIT

For the circuit of Fig. 7.1 the mesh impedance matrix is given by [7.13]. From [7.13], [7.30], [7.32], [7.37] and [7.40] the symmetrical component mesh impedance matrix is

$$[\mathbf{Z}_s] = \begin{bmatrix} \mathbf{Z}_{11} & \mathbf{Z}_{12} & \mathbf{Z}_{10} \\ \mathbf{Z}_{21} & \mathbf{Z}_{22} & \mathbf{Z}_{20} \\ \mathbf{Z}_{01} & \mathbf{Z}_{02} & \mathbf{Z}_{00} \end{bmatrix} \qquad [7.41]$$

where

$$\mathbf{Z}_{11} = \tfrac{1}{3}((\mathbf{Z}_{aa}+\mathbf{Z}_{bb}+\mathbf{Z}_{cc})+\mathrm{h}^2(\mathbf{Z}_{ab}+\mathbf{Z}_{bc}+\mathbf{Z}_{ca})+\mathrm{h}(\mathbf{Z}_{ba}+\mathbf{Z}_{ac}+\mathbf{Z}_{cb}))$$

$$\mathbf{Z}_{12} = \tfrac{1}{3}((\mathbf{Z}_{aa}+\mathrm{h}^2\mathbf{Z}_{bb}+\mathrm{h}\mathbf{Z}_{cc})+\mathrm{h}(\mathbf{Z}_{ab}+\mathbf{Z}_{ba})+(\mathbf{Z}_{bc}+\mathbf{Z}_{cb}) \\ +\mathrm{h}^2(\mathbf{Z}_{ac}+\mathbf{Z}_{ca})$$

$$\mathbf{Z}_{10} = \tfrac{1}{3}((\mathbf{Z}_{aa}+\mathrm{h}\mathbf{Z}_{bb}+\mathrm{h}^2\mathbf{Z}_{cc})+(\mathbf{Z}_{ab}+\mathbf{Z}_{ac})+\mathrm{h}(\mathbf{Z}_{ba}+\mathbf{Z}_{bc}) \\ +\mathrm{h}^2(\mathbf{Z}_{cb}+\mathbf{Z}_{ca}))$$

$$\mathbf{Z}_{21} = \tfrac{1}{3}((\mathbf{Z}_{aa}+\mathrm{h}\mathbf{Z}_{bb}+\mathrm{h}^2\mathbf{Z}_{cc})+\mathrm{h}^2(\mathbf{Z}_{ab}+\mathbf{Z}_{ba})+(\mathbf{Z}_{bc}+\mathbf{Z}_{cb}) \\ +\mathrm{h}(\mathbf{Z}_{ac}+\mathbf{Z}_{ca}))$$

$$\mathbf{Z}_{22} = \tfrac{1}{3}((\mathbf{Z}_{aa}+\mathbf{Z}_{bb}+\mathbf{Z}_{cc})+\mathrm{h}(\mathbf{Z}_{ab}+\mathbf{Z}_{bc}+\mathbf{Z}_{ca})+\mathrm{h}^2(\mathbf{Z}_{ba}+\mathbf{Z}_{ac}+\mathbf{Z}_{cb}))$$

$$\mathbf{Z}_{20} = \tfrac{1}{3}((\mathbf{Z}_{aa}+\mathrm{h}^2\mathbf{Z}_{bb}+\mathrm{h}\mathbf{Z}_{cc})+(\mathbf{Z}_{ab}+\mathbf{Z}_{ac})+\mathrm{h}^2(\mathbf{Z}_{ba}+\mathbf{Z}_{bc}) \\ +\mathrm{h}(\mathbf{Z}_{ca}+\mathbf{Z}_{cb}))$$

$$Z_{01} = \tfrac{1}{3}((Z_{aa}+h^2 Z_{bb}+h Z_{cc})+(Z_{ba}+Z_{ca})+h^2(Z_{ab}+Z_{cb})$$
$$+h(Z_{ac}+Z_{bc}))$$

$$Z_{02} = \tfrac{1}{3}((Z_{aa}+h Z_{bb}+h^2 Z_{cc})+(Z_{ba}+Z_{ca})+h(Z_{ab}+Z_{cb})$$
$$+h^2(Z_{bc}+Z_{ac}))$$

$$Z_{00} = \tfrac{1}{3}((Z_{aa}+Z_{bb}+Z_{cc})+(Z_{ab}+Z_{ba})+(Z_{bc}+Z_{cb})+(Z_{ca}+Z_{ac}))$$

The fact that the non-diagonal terms in $[Z_s]$ are not zero indicates that there is mutual coupling between the three sequence networks due to the unbalance in the 3-phase system. It is not, therefore, possible to apply to a 3-phase circuit, which is unbalanced before the point of fault, the method of symmetrical components which is described in Chapter 8 of Volume 1, since the method assumes that currents of any one sequence do not set up voltages of any other sequence.

7.6.2 COMPLETELY BALANCED 3-PHASE CIRCUIT UP TO THE POINT OF FAULT

If the 3-phase circuit is completely balanced up to the point of fault then $z_{aa} = z_{bb} = z_{cc}$, $z_{ab} = z_{bc} = z_{ca} = z_{ba} = z_{cb} = z_{ac}$ and $z_{an} = z_{na} = z_{bn} = z_{nb} = z_{cn} = z_{nc}$ and

$$[Z] = \begin{bmatrix} (z_{aa}-2z_{an}+z_{nn}) & (z_{ab}-2z_{an}+z_{nn}) & (z_{ba}-2z_{an}+z_{nn}) \\ (z_{ba}-2z_{an}+z_{nn}) & (z_{aa}-2z_{an}+z_{nn}) & (z_{ab}-2z_{an}+z_{nn}) \\ (z_{ab}-2z_{an}+z_{nn}) & (z_{ba}-2z_{an}+z_{nn}) & (z_{aa}-2z_{an}+z_{nn}) \end{bmatrix}$$

$$[7.42]$$

so that

$$Z_{aa} = Z_{bb} = Z_{cc}, \; Z_{ab} = Z_{bc} = Z_{ca} = Z_{ba} = Z_{cb} = Z_{ac}$$

This shows that the same relationships exist between the mesh impedances as between the branch impedances.

Substituting these conditions in [7.41], it can be seen that there are no mutual impedances between the three sequence networks because

$$[Z_s] = \begin{bmatrix} Z_{11} & 0 & 0 \\ 0 & Z_{22} & 0 \\ 0 & 0 & Z_{00} \end{bmatrix} \qquad [7.43]$$

where
$$Z_{11} = Z_{22} = Z_{aa} - Z_{ab} = z_{aa} - z_{ab}$$
$$Z_{00} = Z_{aa} + 2Z_{ab}$$
$$[7.44]$$

It has been shown, therefore, that the normal application of the method of symmetrical components, with no mutual coupling between the three sequence networks, is restricted to a 3-phase system which up to the point of fault has completely balanced impedances in all three phases, and completely balanced mutual impedances.

For these conditions of [7.43], where current of any one sequence cannot produce voltage drops in any other sequence network, each piece of equipment in the system has a self-impedance to each sequence:

Z_{11} is written as Z_1 and is known as the positive-sequence impedance

Z_{22} is written as Z_2 and is known as the negative-sequence impedance

Z_{00} is written as Z_0 and is known as the zero-sequence impedance.

The reason for equality of Z_{11} and Z_{22} in [7.44] is that, as noted in section 7.3, the circuit of Fig. 7.1 does not include any interaction with synchronous machine field windings. [7.44] applies, therefore, to a circuit with static equipment only, though it is approximately true for the initial conditions after an unbalanced fault occurs in a system containing synchronous machines. This is because the values of machine positive- and negative-sequence impedances are then similar (see Chapter 7 of Volume 1), and they are small enough for the rest of the system impedances to tend to swamp the slight differences between them. When the system is transformed into three sequence networks there is then no difficulty in inequalities between Z_1 and Z_2.

At this stage only the power system up to the point of fault but not including an unbalanced shunt fault has been considered, and the latter may now be applied at the terminals a, b, c and e in Fig. 7.1. The potentials of these terminals with respect to earth are

$$\begin{bmatrix} V_{ae} \\ V_{be} \\ V_{ce} \end{bmatrix} = \begin{bmatrix} E_{an} \\ E_{bn} \\ E_{cn} \end{bmatrix} - \begin{bmatrix} Z_{aa} & Z_{ab} & Z_{ac} \\ Z_{ba} & Z_{bb} & Z_{bc} \\ Z_{ca} & Z_{cb} & Z_{cc} \end{bmatrix} \begin{bmatrix} I_{ae} \\ I_{be} \\ I_{ce} \end{bmatrix} \qquad [7.45]$$

where the mesh impedances in [7.45] are defined by [7.13] in terms of branch self and mutual impedances which now include any impedance at the point of unbalanced fault. [7.45] may be written as

$$[V] = [E] - [Z][I] \qquad [7.46]$$

Transforming [7.46] into its symmetrical components

$$[V_s] = [E_s] - [Z_s][I_s] \qquad [7.47]$$

For the power system which is completely balanced up to the point of fault, the sequence components of the reference phase potential with respect to earth are given from [7.43] and [7.47] as

$$\begin{bmatrix} V_{ae1} \\ V_{ae2} \\ V_{ae0} \end{bmatrix} = \begin{bmatrix} E_{an1} \\ 0 \\ 0 \end{bmatrix} - \begin{bmatrix} Z_1 & 0 & 0 \\ 0 & Z_2 & 0 \\ 0 & 0 & Z_0 \end{bmatrix} \begin{bmatrix} I_{ae1} \\ I_{ae2} \\ I_{ae0} \end{bmatrix} \qquad [7.48]$$

since negative- and zero-sequence e.m.f.s are not generated in synchronous machines.

Equations [7.48] are those of a single-phase system of sequence networks where the manner in which these single-phase networks are connected together depends upon the type of shunt fault as shown in Chapter 8 of Volume I. These interconnected networks can then be solved for I_{ae1}, I_{ae2} and I_{ae0} which may be synthesised to give the system phase currents, and for V_{ae1}, V_{ae2} and V_{ae0}, which can then be combined to give the potentials to earth of the phases at the point of fault. Similarly the symmetrical components of current and voltage at any other point between the fault and the generators can be calculated. This process (which is more simple and flexible than formulating the mesh impedance matrix and inverting it), has been illustrated in Volume 1 for some fault conditions which are symmetrical with respect to one phase. The difficulty in applying this method to a fault which is not symmetrical with respect to one phase is illustrated for a simple case in the next section.

7.6.3 FAULT WHICH IS ASYMMETRICAL WITH RESPECT TO THE REFERENCE PHASE

The normal method of symmetrical components is applied to a restricted class of faults, viz. those where the fault is symmetrical with respect to the reference phase a, because this restriction leads to simpler relationships between the currents and voltages of the three sequences, which correspond to the sequence networks being interconnected either directly or by means of $1:1$ ideal current transformers with no phase shift (see Chapter 8 of Volume 1).

If phase b in an unloaded system is connected to earth e by a fault impedance Z_f, then the potential difference between the faulted conductor and earth at the point of fault is

$$V_{be} = Z_f I_{be} \qquad [7.49]$$

where I_{be} is the current flowing to earth in phase conductor b. If I_{ae} and I_{ce} are similarly defined (as in Fig. 7.1) for the other two phases then

$$I_{ae} = 0 \qquad [7.50]$$

$$I_{ce} = 0 \qquad [7.51]$$

Writing these currents and voltages as the sum of their symmetrical components gives

$$h^2 V_{ae1} + h V_{ae2} + V_{ae0} = Z_f(h^2 I_{ae1} + h I_{ae2} + I_{ae0}) \qquad [7.52]$$

$$I_{ae1} + I_{ae2} + I_{ae0} = 0 \qquad [7.53]$$

$$h I_{ae1} + h^2 I_{ae2} + I_{ae0} = 0 \qquad [7.54]$$

From [7.53]

$$I_{ae0} = -I_{ae1} - I_{ae2}$$

and substituting this result in [7.54] gives

$$(h-1)I_{ae1} + (h^2-1)I_{ae2} = 0$$

so that

$$I_{ae2} = \frac{(h-1)}{(1-h^2)} I_{ae1} = h I_{ae1} \qquad [7.55]$$

Substitution of [7.55] into [7.53] gives

$$I_{ae0} = -(1+h)I_{ae1} = h^2 I_{ae1} \qquad [7.56]$$

[7.52], [7.55] and [7.56] taken together give

$$h^2 V_{ae1} + h V_{ae2} + V_{ae0} = 3 Z_f I_{ae0} \qquad [7.57]$$

[7.55], [7.56] and [7.57] are satisfied by the interconnection of the sequence networks shown in Fig. 7.2. Thus the sequence networks in which are flowing the sequence currents of one particular phase (the reference phase) when the fault is symmetrical with respect to a different phase, are connected by current transformers having equal numbers of primary and secondary turns but with 0°, 120° and 240°

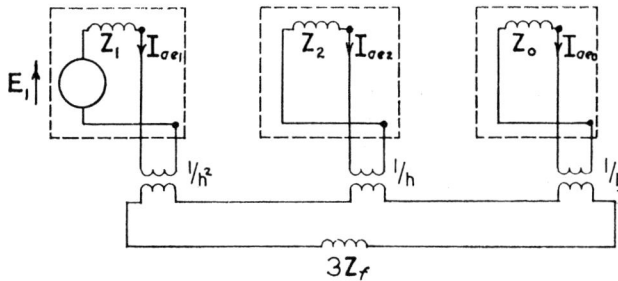

Fig. 7.2. Interconnection of sequence networks for an earth fault on a phase which is not taken as the reference.

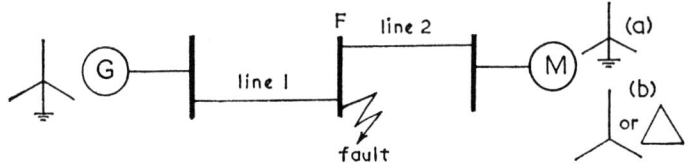

Fig. 7.3. Fault on a system with synchronous machines on both sides of the fault.

phase shifts. Obviously in this simple case, symmetry can be re-introduced merely by making phase *b* the reference phase but faults which have no symmetry at all would produce greater disturbance.

7.6.4 SYSTEM WITH SYNCHRONOUS MACHINES ON BOTH SIDES OF THE FAULT

A power system in which a shunt fault has occurred can be reduced to the type of circuit shown in Fig. 7.3, where the single equivalent synchronous machine G is generating and the single equivalent synchronous machine M may be either motoring or generating. The unconnected sequence networks for reference phase *a* are shown in Fig. 7.4, where z_{g1} is the sum of the positive-sequence impedances of G and line 1, z_{m1} is the sum of the positive-sequence impedances of M and line 2, and similarly for the other two sequence networks. In the zero-sequence network, switch S is closed for condition (*a*) in Fig. 7.3 or open for condition (*b*) in Fig. 7.3. Condition (*a*) is assumed here but if (*b*) applies than z_{m0} may be written as infinity.

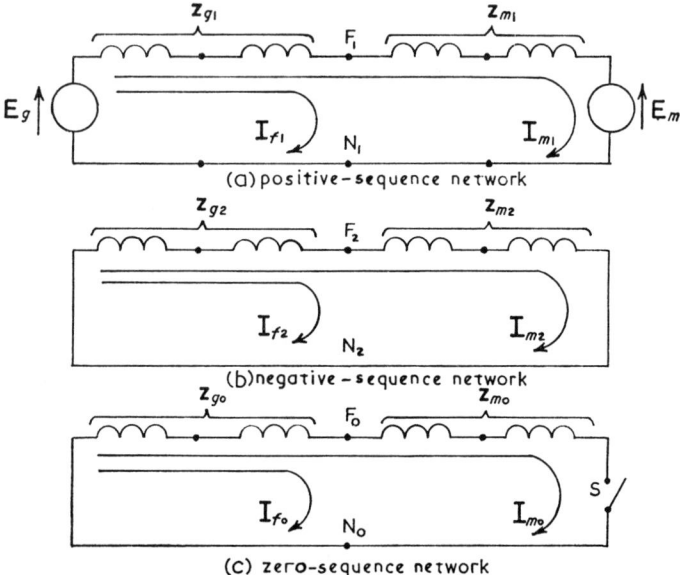

(a) positive-sequence network

(b) negative-sequence network

(c) zero-sequence network

Fig. 7.4. Unconnected sequence networks for system of Fig. 7.3.

The method is illustrated by considering the simplest case of an earth fault of zero impedance on phase *a* at point F in the system in Fig. 7.3. As shown in Chapter 8 of Volume 1, the three sequence networks for phase *a* are in series with each other for this fault so that F_1 is connected to N_2, F_2 to N_0 and F_0 to N_1. If the mesh currents are defined so that one of them, e.g. I_{f1}, is the fault current of that sequence, while the other, I_{m1}, is the current of that sequence in the single mesh of the circuit before the fault occurred, then a current $I_{f1} = I_{f2} = I_{f0}$ flows into these connections and in the impedances z_{g1}, z_{g2} and z_{g0}. The mesh equations are

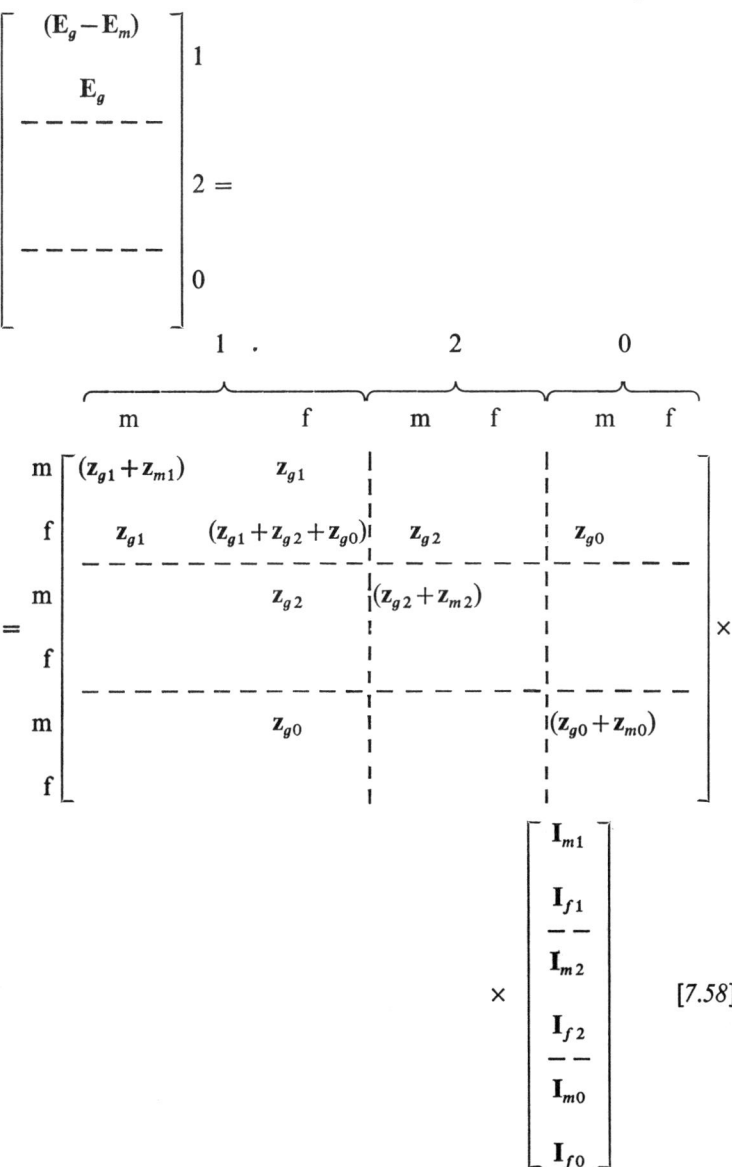

$$[7.58]$$

The rows in the impedance matrix of [7.58] corresponding to I_{f2} and I_{f0} are left unfilled because these currents are both equal to I_{f1}.

The solution to the actual fault current may be obtained by

(a) inverting the impedance matrix in [7.58] to give $[\mathbf{Z}_s]^{-1}$;

(b) computing the symmetrical components of the mesh currents from $[\mathbf{I}_s] = [\mathbf{Z}_s]^{-1}[\mathbf{E}_s]$;

(c) transforming these currents into the actual mesh currents by [7.33];

(d) obtaining the actual branch currents by inspection of Fig. 7.4 or by means of a connection matrix (see [8.3] of next chapter).

7.7 Transformation to rotor fixed axes (direct and quadrature axes)

In a synchronous machine the rotor usually revolves within a fixed stator winding but for convenience the rotor may be regarded as being stationary, while the three-phase stator winding rotates at an angular speed $\omega = d\theta/dt$. A transformation between rotating and fixed axes may then be made so that the three-phase stator winding can be replaced by two stationary fictitious coils, one on the direct axis and the other on the quadrature axis as shown in Fig. 7.5 (see Section 7.4 of Volume 1). Although physically the flux in the machine

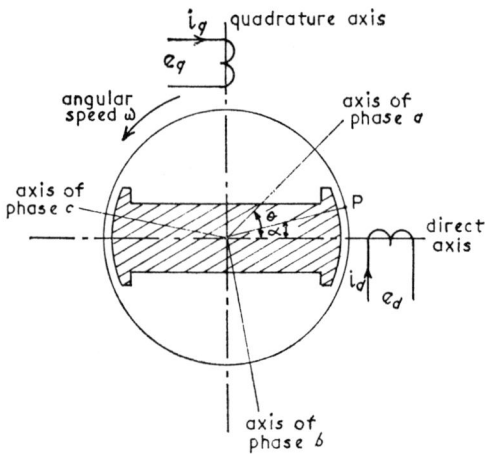

Fig. 7.5. Ideal two-pole synchronous machine.

is more significant than m.m.f. (see Harris, Lawrenson and Stephenson, 1970), and they are only proportional in an unsaturated cylindrical-rotor machine, it is simpler in the present context to use unit m.m.f. than unit flux. Unit m.m.f. may be taken to be the maximum value of the sinusoidal space wave of m.m.f. set up by unit balanced three-phase currents in the three stator phases (i.e. 1·5 times the peak of the sinusoidal m.m.f. due to full-load r.m.s. current in one phase). The two fictitious direct- and quadrature-axis coils are therefore assumed to have 1·5 times as many turns as each of the phase coils. For the instantaneous currents in the two coils i_d and i_q to be equivalent to those in the three stator phases i_a, i_b and i_c, the two sets of currents must at all times produce the same m.m.f. at any point. Equating the two m.m.f.s at any point P (Fig. 7.5) gives

$$i_d \cos \alpha + i_q \sin \alpha = \tfrac{2}{3}(i_a \cos (\theta - \alpha) + i_b \cos (\theta - 2\pi/3 - \alpha)$$
$$+ i_c \cos (\theta - 4\pi/3 - \alpha))$$

Equating coefficients of $\cos \alpha$ and $\sin \alpha$ gives

$$i_d = \tfrac{2}{3}(i_a \cos \theta + i_b \cos (\theta - 2\pi/3) + i_c \cos (\theta - 4\pi/3)) \qquad [7.59]$$

$$i_q = \tfrac{2}{3}(i_a \sin \theta + i_b \sin (\theta - 2\pi/3) + i_c \sin (\theta - 4\pi/3)) \qquad [7.60]$$

[7.59] and [7.60] cannot be inverted since only a square matrix can be inverted (Appendix A.1.8), and yet the phase currents must be expressed in terms of the fictitious primitive system currents. An additional equation is required to make the inversion possible and this is provided by the current $i_0 = \tfrac{1}{3}(i_a + i_b + i_c)$, which is an instantaneous current similar to the r.m.s. zero-sequence component of current (see Chapter 8, Volume 1). These three equal fictitious currents i_0 flowing in the three phase windings a, b and c do not produce any air-gap flux but only leakage flux. Therefore they do not appear in the relationships between the direct- and quadrature-axis currents and the actual three-phase currents. Combining this with [7.59] and [7.60] gives

$$\begin{bmatrix} i_d \\ i_q \\ i_0 \end{bmatrix} = \tfrac{2}{3} \begin{bmatrix} \cos \theta & \cos (\theta - 2\pi/3) & \cos (\theta - 4\pi/3) \\ \sin \theta & \sin (\theta - 2\pi/3) & \sin (\theta - 4\pi/3) \\ \tfrac{1}{2} & \tfrac{1}{2} & \tfrac{1}{2} \end{bmatrix} \begin{bmatrix} i_a \\ i_b \\ i_c \end{bmatrix} \qquad [7.61]$$

Inverting [7.61] (which may be done by students as an exercise) gives

$$
\begin{bmatrix} i_a \\ i_b \\ i_c \end{bmatrix} = \begin{bmatrix} \cos \theta & \sin \theta & 1 \\ \cos (\theta - 2\pi/3) & \sin (\theta - 2\pi/3) & 1 \\ \cos (\theta - 4\pi/3) & \sin (\theta - 4\pi/3) & 1 \end{bmatrix} \begin{bmatrix} i_d \\ i_q \\ i_0 \end{bmatrix} \qquad [7.62]
$$

[7.61] corresponds to the general equation [7.1] for the currents of the primitive system in terms of the actual currents. Thus [7.5] gives the voltage relation for invariance of power between the two systems as

$$
\begin{bmatrix} e_a \\ e_b \\ e_c \end{bmatrix} = \tfrac{2}{3} \begin{bmatrix} \cos \theta & \sin \theta & \tfrac{1}{2} \\ \cos (\theta - 2\pi/3) & \sin (\theta - 2\pi/3) & \tfrac{1}{2} \\ \cos (\theta - 4\pi/3) & \sin (\theta - 4\pi/3) & \tfrac{1}{2} \end{bmatrix} \begin{bmatrix} e_d \\ e_q \\ e_0' \end{bmatrix} \qquad [7.63]
$$

Comparison of [7.62] and [7.63] shows that the complete transformation matrices for currents and voltages are not the same. However, the relationships between the two-axis currents and voltages and the actual three-phase currents and voltages are exactly the same because i_0 and e_0' do not appear in them. Inverting [7.63] gives $e_0' = e_a + e_b + e_c$ instead of the value $e_0 = \tfrac{1}{3}(e_a + e_b + e_c)$ which corresponds with the zero-sequence component of voltage. The reason for this difference is that the condition of power invariance

$$
e_a i_a + e_b i_b + e_c i_c = e_d i_d + e_q i_q + e_0' i_0 \qquad [7.64]
$$

has been applied here whereas in the normal transformation of symmetrical components it is not applied (Section 7.6).

7.8 α, β and zero components (Clarke's components)

It is possible to analyse unbalanced conditions in three-phase systems by Clarke's α, β and zero components rather than by symmetrical components. This method appears perhaps relatively more often in textbooks than in practical applications, though it is occasionally used (e.g. Ching and Adkins, 1954), and is included here for completeness and to show its relationship with the transformation in the previous section. In place of the d and q coils of Fig. 7.5 which carry constant currents for given stator currents, two new fictitious coils, which are stationary with respect to the stator a, b and c coils, carrying instantaneous 2-phase system currents i_α and i_β, may be defined by putting $\theta = 0$, so that the first of these coils is on the axis of the reference phase a. Since the phase rotation is α, β, the latter coil must lag 90° behind the former, and is 30° ahead of the axis of phase b. Then equating m.m.f.s gives

$$\begin{bmatrix} i_\alpha \\ i_\beta \\ i_0 \end{bmatrix} = \begin{bmatrix} 2/3 & -1/3 & -1/3 \\ 0 & 1/\sqrt{3} & -1/\sqrt{3} \\ 1/3 & 1/3 & 1/3 \end{bmatrix} \begin{bmatrix} i_a \\ i_b \\ i_c \end{bmatrix} \qquad [7.65]$$

Thus the α-components in phases b and c are equal and they are opposite in sign, and of half the magnitude of the α-component in phase a. The β-components in phases b and c are equal in magnitude but opposite in sign while in phase a the β-component is zero.

REFERENCES

ADKINS, B. 1964. *The General Theory of Electrical Machines*. Chapman and Hall, London.

BANDYOPADHYAY, A. K. 1964. An introduction to the engineering philosophy of the generalized machine theory: Part II. A resumé of the basic assumptions of electrical machine analysis and the development of the fundamentals of generalized theory. *Matrix and Tensor Quarterly*, pp. 6–23.

BANDYOPADHYAY, A. K. 1969. Private communication.

CHING, Y. K. & ADKINS, B. 1954. Transient theory of synchronous generators under unbalanced conditions. *Journal I.E.E.*, **101**, Pt. IV, 166–170.

CLARKE, E. 1943. *Circuit Analysis of A.C. Power Systems*: Vol. 1, *Symmetrical and Related Components*. Wiley, New York.

HARRIS, M. R., LAWRENSON, P. J. & STEPHENSON, J. M. 1970. Per-unit systems: with special reference to electrical machines. *I.E.E. Monograph Series* No. 4. Cambridge University Press.

JONES, C. V. 1958. An analysis of commutation for the unified-machine theory. *Proc. I.E.E.*, Monograph No. 302U, 476–488.

LEWIS, W. E. & PRYCE, D. G. 1966. *The Application of Matrix Theory to Electrical Engineering*. Science Paperbacks. E. and F. Spon Ltd., London.

TROPPER, A. M. 1962. *Matrix Theory for Electrical Engineering Students*. George Harrap, London.

Chapter 8

POWER SYSTEM ANALYSIS

8.1 Introduction

From the late 1920s, analogue models of power systems called a.c. network analysers began to be developed, to study the increasingly large and complex systems which were requiring more and more detailed analysis. Digital computers began to be applied to this work in the late 1940s and their use increased when larger and faster machines become available in the next decade. They are now used for all the main analysis such as planning which of a number of alternatives to choose, designing and operating the new or extended system.

The problems of power system analysis which are solved by means of digital computers fall under three main headings:

(a) Fault or short-circuit calculations required to design protective systems and determine circuit breaker ratings. The calculations most commonly carried out are for 3-phase short-circuit and for an earth fault on one phase. The principles of these calculations as well as for the other unbalanced faults which are sometimes analysed are dealt with in Volume 1, Chapter 8.

(b) Transient stability studies which are required to ensure the capability of a power system to remain in synchronism during major disturbances such as faults, loss of generation or transmission plant, or sudden and sustained large changes in load. The principles involved in these calculations are dealt with in Chapter 3.

(c) Load-flow problems in which the steady-state busbar voltages and complex power flows are computed in an existing system so as to evaluate and regulate its performance, and are also used to analyse alternative plans for future expansion to meet new load demand.

In the present Chapter some of the basic principles of the analysis of power systems are developed from the foundations laid in Chapter 7, and their application to a systematic routine solution of load-flow problems is outlined.

8.2 Mesh current method and connection matrices

8.2.1 BASIC EQUATIONS

Mesh current analysis is one of the methods of solving network equations by transforming the equations of state of the network into another co-ordinate system (see Chapter 7). The method may be illustrated by means of the simple balanced 3-phase power system of Fig. 8.1(a), one phase of which is represented by Fig. 8.1(b) where z_{dd} and z_{ee} are the impedances of static loads (together with the shunt impedances of the nominal Π circuit representing the transmission line unless these are so much greater than the load impedances that they are negligible). All mutual impedances between branches such as z_{ad}, z_{dc}, etc. are assumed to be zero. The branch currents, driving voltages (e.m.f.s) and impedances are shown lower case, i.e. i_a, e_a, z_a, etc. in order to distinguish them from the mesh currents I_1, I_2 and I_3, the mesh driving voltages E_1, E_2 and E_3 and the mesh self-impedances Z_{11}, Z_{22} and Z_{33}. All currents and voltages are however r.m.s. values of sinusoidally-varying quantities of a single frequency.

For the three meshes 1, 2 and 3 of Fig. 8.1(b), the following three equations may be written by inspection,

$$\left.\begin{aligned}
e_a &= (z_{aa}+z_{dd})I_1 - z_{dd}I_2 \\
0 &= -z_{dd}I_1 + (z_{cc}+z_{dd}+z_{ee})I_2 - z_{ee}I_3 \\
-e_b &= -z_{ee}I_2 + (z_{bb}+z_{ee})I_3
\end{aligned}\right\} \qquad [8.1]$$

These equations may be re-written as

$$\left.\begin{aligned}
E_1 &= Z_{11}I_1 - Z_{12}I_2 - Z_{13}I_3 \\
E_2 &= -Z_{21}I_1 + Z_{22}I_2 - Z_{23}I_3 \\
E_3 &= -Z_{31}I_1 - Z_{32}I_2 + Z_{33}I_3
\end{aligned}\right\} \qquad [8.2]$$

In general

$$\mathbf{E}_n = -\mathbf{Z}_{n1}\mathbf{I}_1 - \mathbf{Z}_{n2}\mathbf{I}_2 \dots + \mathbf{Z}_{nn}\mathbf{I}_n - \mathbf{Z}_{n(n+1)}\mathbf{I}_{(n+1)} \dots$$

where for the nth mesh, $\mathbf{E}_n$ is the total driving voltage in a clockwise direction; $\mathbf{Z}_{nn}$ is the mesh self-impedance which is the total impedance

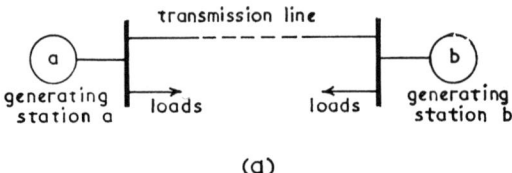

(a)

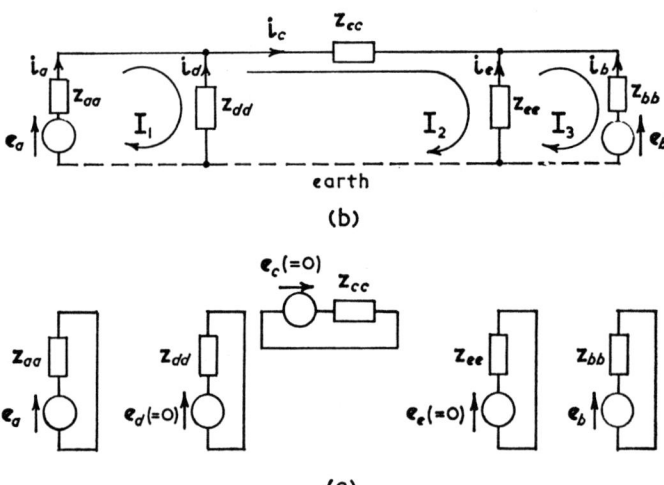

(b)

(c)

Fig. 8.1. Mesh currents:

(a) simple power system
(b) meshes
(c) primitive system of disconnected short-circuited branches.

around the mesh; $\mathbf{Z}_{nm}$ is the mutual impedance between meshes n and m (which for the case under consideration here of zero mutual coupling between branches reduces to the impedance common to

both meshes). For a bilateral network (and the analysis of power
systems containing nonbilateral elements such as rectifiers and
invertors is not dealt with in this book though convertors are considered in Chapter 6), $Z_{mn} = Z_{nm}$ and the impedance coefficients in
the mesh impedance matrix of [8.2] must be symmetrical about the
diagonal formed by the self-impedance terms. The non-diagonal
terms (mesh mutual impedances) are all negative when this cyclic
current convention is used.

The relations between the branch currents and the mesh currents
may be written by inspection of Fig. 8.1 as

$$i_a = I_1; \ i_b = -I_3; \ i_c = I_2; \ i_d = (I_2 - I_1); \ i_e = (I_3 - I_2)$$

These constraints which the connection of the branches shown in
Fig. 8.1 place upon the branch currents may be expressed by means
of a connection matrix [C] so that

$$[i] = [C][I] \qquad [8.3]$$

and since

$$
\begin{bmatrix} i_a \\ i_b \\ i_c \\ i_d \\ i_e \end{bmatrix}
=
\begin{bmatrix}
1 & 0 & 0 \\
0 & 0 & -1 \\
0 & 1 & 0 \\
-1 & 1 & 0 \\
0 & -1 & 1
\end{bmatrix}
\begin{bmatrix} I_1 \\ I_2 \\ I_3 \end{bmatrix}
$$

$$
[C] =
\begin{bmatrix}
1 & 0 & 0 \\
0 & 0 & -1 \\
0 & 1 & 0 \\
-1 & 1 & 0 \\
0 & -1 & 1
\end{bmatrix}
\qquad [8.4]
$$

It is shown in Chapter 7 (see [7.1] and [7.5]) that for power invariance
the mesh driving voltage matrix [E] is related to the branch driving
voltage matrix [e] by

$$[E] = [C^*]_t[e] \qquad [8.5a]$$

and since [C] contains only real terms

$$[E] = [C]_t[e] \qquad\qquad [8.5b]$$

The relationship between the branch driving voltage matrix [e] and the branch current matrix [i] of the primitive system formed when the branches are all in their respective locations in space, but are electrically (though not magnetically) isolated, and when all branches are short-circuited (see Fig. 8.1(c)), is given by

$$[e] = [z][i] \qquad\qquad [8.6]$$

where the branch impedance matrix [z] is

$$[z] = \begin{bmatrix} z_{aa} & z_{ab} & z_{ac} & z_{ad} & z_{ae} \\ z_{ba} & z_{bb} & z_{bc} & z_{bd} & z_{be} \\ z_{ca} & z_{cb} & z_{cc} & z_{cd} & z_{ce} \\ z_{da} & z_{db} & z_{dc} & z_{dd} & z_{de} \\ z_{ea} & z_{eb} & z_{ec} & z_{ed} & z_{ee} \end{bmatrix}$$

and since in the case being considered here there is no mutual coupling between the branches of the circuit of Fig. 8.1, then [z] reduces for this particular case to

$$[z] = \begin{bmatrix} z_{aa} & 0 & 0 & 0 & 0 \\ 0 & z_{bb} & 0 & 0 & 0 \\ 0 & 0 & z_{cc} & 0 & 0 \\ 0 & 0 & 0 & z_{dd} & 0 \\ 0 & 0 & 0 & 0 & z_{ee} \end{bmatrix} \qquad\qquad [8.7]$$

The currents in the separated short-circuited branches of the primitive network of Fig. 8.1(c) are related to the mesh currents in the circuit of Fig. 8.1(b) where the branches have been interconnected, by [8.3] using the connection matrix given by [8.4]. From [8.5] and [8.6]

$$[E] = [C]_t[z][i]$$

and using [8.3]

$$[E] = [C]_t[z][C][I] \qquad\qquad [8.8]$$

so that if the mesh driving voltage matrix [E] and the mesh current matrix [I] are related by a mesh impedance matrix [Z] then

$$[E] = [Z][I] \qquad\qquad [8.9]$$

This is the matrix form of [8.2] where

$$[Z] = [C]_t[z][C] \qquad\qquad [8.10]$$

[8.10] is an example of the application of [7.6].
    For the circuit of Fig. 8.1

$$[Z] = \begin{bmatrix} 1 & 0 & 0 & -1 & 0 \\ 0 & 0 & 1 & 1 & -1 \\ 0 & -1 & 0 & 0 & 1 \end{bmatrix} \begin{bmatrix} z_{aa} & 0 & 0 & 0 & 0 \\ 0 & z_{bb} & 0 & 0 & 0 \\ 0 & 0 & z_{cc} & 0 & 0 \\ 0 & 0 & 0 & z_{dd} & 0 \\ 0 & 0 & 0 & 0 & z_{ee} \end{bmatrix}$$

$$\times \begin{bmatrix} 1 & 0 & 0 \\ 0 & 0 & -1 \\ 0 & 1 & 0 \\ -1 & 1 & 0 \\ 0 & -1 & 1 \end{bmatrix}$$

It is left to the student to check that this gives the result

$$[Z] = \begin{bmatrix} (z_{aa}+z_{dd}) & -z_{dd} & 0 \\ -z_{dd} & (z_{cc}+z_{dd}+z_{ee}) & -z_{ee} \\ 0 & -z_{ee} & (z_{bb}+z_{ee}) \end{bmatrix} \qquad [8.11]$$

which is confirmed by [8.1] and [8.2].

8.2.2 INTERCONNECTION OF NETWORKS INTO WHICH A SYSTEM HAS BEEN SUBDIVIDED

In the previous section it was shown how the mesh impedance matrix [Z] of the whole network could be obtained from the impedance matrix [z] of a much simpler network (sometimes called the 'primitive network') consisting of each branch disconnected from all

others and short-circuited as shown in Fig. 8.1(c). It is not in fact
necessary to subdivide a network into its separate individual branches,
but instead it may be subdivided into a number of networks disconnected from each other. After analysing these, they may then be
interconnected to form the complete network.

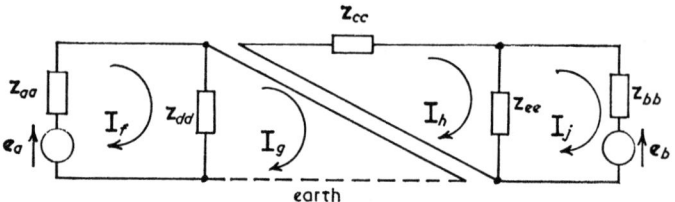

Fig. 8.2. System of Fig. 8.1 subdivided into two networks.

As a simple example of this process, the circuit of Fig. 8.1 may be
subdivided into two networks as in Fig. 8.2. For the left-hand network of Fig. 8.2, the driving voltage matrix $[\mathbf{E}']$, the mesh current
matrix $[\mathbf{I}']$, and the mesh impedance matrix $[\mathbf{Z}']$ where $[\mathbf{E}'] =
[\mathbf{Z}'][\mathbf{I}']$ are

$$[\mathbf{E}'] = \begin{bmatrix} \mathbf{e}_f \\ \mathbf{e}_g \end{bmatrix} = \begin{bmatrix} \mathbf{e}_a \\ 0 \end{bmatrix}$$

$$[\mathbf{I}'] = \begin{bmatrix} \mathbf{I}_f \\ \mathbf{I}_g \end{bmatrix}$$

$$[\mathbf{Z}'] = \begin{bmatrix} \mathbf{Z}_{ff} & \mathbf{Z}_{fg} \\ \mathbf{Z}_{gf} & \mathbf{Z}_{gg} \end{bmatrix} = \begin{bmatrix} (z_{aa}+z_{dd}) & -z_{dd} \\ -z_{dd} & z_{dd} \end{bmatrix}$$

Similarly for the right-hand network of Fig. 8.2

$$[\mathbf{E}''] = \begin{bmatrix} \mathbf{e}_h \\ \mathbf{e}_j \end{bmatrix} = \begin{bmatrix} 0 \\ -\mathbf{e}_b \end{bmatrix}$$

$$[I''] = \begin{bmatrix} \mathbf{I}_h \\ \mathbf{I}_j \end{bmatrix}$$

$$[\mathbf{Z}''] = \begin{bmatrix} \mathbf{Z}_{hh} & \mathbf{Z}_{hj} \\ \mathbf{Z}_{jh} & \mathbf{Z}_{jj} \end{bmatrix} = \begin{bmatrix} (z_{cc}+z_{ee}) & -z_{ee} \\ -z_{ee} & (z_{bb}+z_{ee}) \end{bmatrix}$$

The two separate networks of Fig. 8.2 form a primitive network for which the mesh driving voltage, current and impedance matrices are

$$[\mathbf{E}'''] = \begin{bmatrix} \mathbf{E}' \\ \mathbf{E}'' \end{bmatrix}; \quad [\mathbf{I}'''] = \begin{bmatrix} \mathbf{I}' \\ \mathbf{I}'' \end{bmatrix}; \quad [\mathbf{Z}'''] = \begin{bmatrix} \mathbf{Z}' & 0 \\ 0 & \mathbf{Z}'' \end{bmatrix}$$

A connection matrix [C] may again be used to relate the currents of the primitive network to those of the complete network (though it must be noted that it differs from the connection matrix of the previous section). Since $\mathbf{I}_f = \mathbf{I}_1$, $\mathbf{I}_g = \mathbf{I}_2$, $\mathbf{I}_h = \mathbf{I}_2$ and $\mathbf{I}_j = \mathbf{I}_3$, and $[\mathbf{I}''] = [\mathbf{C}][\mathbf{I}]$ (see [8.3]) then

$$[\mathbf{C}] = \begin{bmatrix} 1 & 0 & 0 \\ 0 & 1 & 0 \\ \hline 0 & 1 & 0 \\ 0 & 0 & 1 \end{bmatrix} = \begin{bmatrix} \mathbf{C}' \\ \mathbf{C}'' \end{bmatrix}$$

The mesh impedance matrix [Z] for the complete circuit of Fig. 8.1 may now be written using [8.10] as

$$[\mathbf{Z}] = [\mathbf{C}]_t[\mathbf{Z}'''][\mathbf{C}]$$

and using the statement of the transpose of a partitioned matrix given in Appendix A.1.5

$$[\mathbf{Z}] = [\mathbf{C}'_t \quad \mathbf{C}''_t] \begin{bmatrix} \mathbf{Z}' & 0 \\ 0 & \mathbf{Z}'' \end{bmatrix} \begin{bmatrix} \mathbf{C}' \\ \mathbf{C}'' \end{bmatrix}$$

$$[\mathbf{Z}] = [\mathbf{C}']_t[\mathbf{Z}'][\mathbf{C}'] + [\mathbf{C}'']_t[\mathbf{Z}''][\mathbf{C}''] \qquad [8.12]$$

Multiplying the components of the first term on the right-hand side of [8.12] for the circuit of Fig. 8.2 gives

$$[\mathbf{C}']_t[\mathbf{Z}'][\mathbf{C}'] = \begin{matrix} & 1 & 2 \\ 1 \\ 2 \end{matrix} \begin{bmatrix} (z_{aa}+z_{dd}) & -z_{dd} \\ -z_{dd} & z_{dd} \end{bmatrix}$$

The numbers opposite rows and columns of the last matrix indicate the meshes to which the terms belong. Similarly

$$[C'']_t[Z''][C''] = \begin{array}{c} \\ 2 \\ 3 \end{array}\begin{array}{cc} 2 & 3 \\ \left[\begin{array}{cc} (z_{cc}+z_{ee}) & -z_{ee} \\ -z_{ee} & (z_{bb}+z_{ee}) \end{array}\right] \end{array}$$

Addition of the elements of the two terms of [8.12] gives the result of [8.11], but it must be noted that the application of the addition rule of matrices (see Appendix A.1.2) must be done with care: it is as though both matrices were three by three with zeros for all elements in the first corresponding to mesh 3, and zeros in the second for all elements corresponding to mesh 1.

8.3 Nodal voltage method and connection matrices

8.3.1 BASIC EQUATIONS

Nodal voltage analysis is an alternative to mesh current analysis for solving network equations. The circuit of Fig. 8.1 has been redrawn in Fig. 8.3(a) with impedances replaced by the corresponding admittances, and with the nodes numbered, starting with the reference node 0 which for a 3-phase system will normally be earth. The total current flowing away from node 1 = zero = $-i_a-i_d+i_c$. Writing this in terms of the admittances, driving voltages and the potentials of one node with respect to another, and repeating this for node 2, gives for node 1

$$0 = y_{aa}(V_{10}-e_a)+y_{dd}V_{10}+y_{cc}V_{12}$$

and for node 2

$$0 = y_{cc}V_{21}+y_{ee}V_{20}+y_{bb}(V_{20}-e_b)$$

Rearranging terms and writing $V_{12} = V_{10}-V_{20}$ so that only potentials with respect to the reference node 0 are involved gives

$$\left.\begin{array}{l} y_{aa}e_a = (y_{aa}+y_{cc}+y_{dd})V_{10}-y_{cc}V_{20} \\ y_{bb}e_b = -y_{cc}V_{10}+(y_{bb}+y_{cc}+y_{ee})V_{20} \end{array}\right\} \qquad [8.13]$$

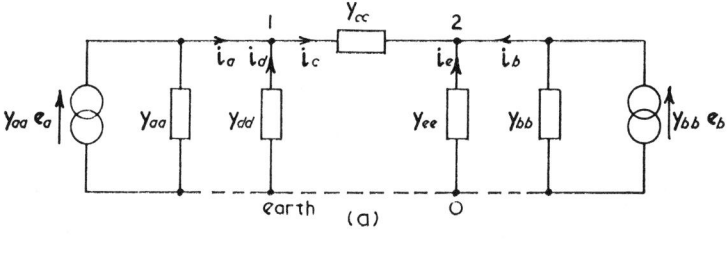

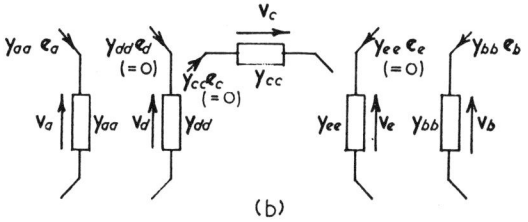

(b)

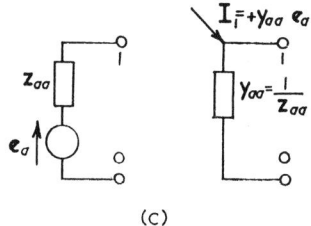

(c)

Fig. 8.3. Nodal voltages:

(a) simple power system
(b) primitive network of disconnected open-circuited branches.
(c) nodal current.

These equations may be rewritten as

$$(ye)_1 = I_1 = Y_{11}V_1 - Y_{12}V_2$$
$$(ye)_2 = I_2 = -Y_{21}V_1 + Y_{22}V_2$$

[8.14]

In general

$$(ye)_n = I_n = -Y_{n1}V_1 - Y_{n2}V_2 \cdots + Y_{nn}V_n - Y_{n(n+1)}V_{(n+1)} \cdots$$

For the node numbered n, $(ye)_n$ is the sum of the products of the driving voltages and the admittances of the branches containing them which are connected directly to node n, and it is effectively a current injected into the circuit at that node. This nodal current is positive, i.e. it is directed into the circuit at a node when the driving voltage is directed towards that node, as occurs in both equations of [8.13]. This is illustrated in Fig. 8.3(c). If the circuit contained a third node with a source of impedance z_{nn} having an e.m.f. e_n directed away from a node 3, then the current injected into node 3 would be $I_3 = -y_{nn}e_n$ which indicates that a current $+y_{nn}e_n$ leaves node 3 and returns to the circuit at the reference node. Y_{nn} is the nodal self-admittance at node n and is equal to the sum of all admittances connected *directly* to that node. Y_{mn} is the mutual admittance between nodes m and n and is equal to the sum of all admittances connected *directly* between these nodes. In a linear bilateral network $Y_{mn} = Y_{nm}$. V_n is written in place of V_{n0} as the potential of node n with respect to the reference node 0.

The relations between the branch voltages (i.e. potential differences across the branches) and the nodal voltages may be written by inspection of Fig. 8.3(a) and (b) as $v_a = V_{10}$; $v_b = V_{20}$; $v_c = V_{20} - V_{10}$; $v_d = V_{10}$; $v_e = V_{20}$. These constraints may be expressed by using a connection matrix [C] where

$$[v] = [C][V] \qquad [8.15]$$

$$[C] = \begin{bmatrix} 1 & 0 \\ 0 & 1 \\ -1 & 1 \\ 1 & 0 \\ 0 & 1 \end{bmatrix} \qquad [8.16]$$

The condition of power invariance (see [7.9] and [7.10]) gives the relationship between the nodal currents and the branch currents as

$$[I] = [C^*]_t[i]$$

and since [C] contains only real terms here

$$[I] = [C]_t[i] \qquad [8.17]$$

The relationship between the branch current matrix [i] and the branch voltage matrix [v] must now be considered. Using Norton's theorem, all active branches (i.e. those containing driving voltages e) may be replaced by a current source of (ye) in parallel with the branch admittance y, with the current source injecting current into the node towards which the driving voltage was directed (see [8.13]). Thus all branches in the network may be shown separately as their admittances, each active one being shunted by a current source (ye), and the branch currents must each be given by the product of the branch admittance and the branch voltage (i.e. potential difference across the branch) so that

$$[i] = [y][v] \qquad\qquad [8.18]$$

The branch admittance [y] is a diagonal matrix like the branch impedance matrix [z] shown in [8.7], since in the network under consideration there is no mutual coupling, but in general it takes the form of the matrix preceding [8.7]. [8.18] is the equation of a primitive network consisting of all the branches of Fig. 8.3 separated electrically from each other and on open-circuit, each with its injected nodal current (ye) as shown in Fig. 8.3(b). The branch voltages of this primitive network are related to the nodal voltages of the complete network of Fig. 8.3(a) by [8.15] and [8.16]. From [8.17] and [8.18]

$$[I] = [C]_t[y][v]$$

and using [8.15]

$$[I] = [C]_t[y][C][V] \qquad\qquad [8.19]$$

[8.19] shows that the nodal driving currents (currents injected into the network at nodes to which driving voltages are connected) are related to the nodal voltages by a nodal admittance matrix [Y], and this relation may be written as

$$[I] = [Y][V] \qquad\qquad [8.20]$$

which is the matrix form of [8.13], where

$$[Y] = [C]_t[y][C] \qquad\qquad [8.21]$$

[8.21] is an example of the application of [7.11].

For the circuit of Fig. 8.3(a)

$$[Y] = \begin{bmatrix} 1 & 0 & -1 & 1 & 0 \\ 0 & 1 & 1 & 0 & 1 \end{bmatrix} \begin{bmatrix} y_{aa} & 0 & 0 & 0 & 0 \\ 0 & y_{bb} & 0 & 0 & 0 \\ 0 & 0 & y_{cc} & 0 & 0 \\ 0 & 0 & 0 & y_{dd} & 0 \\ 0 & 0 & 0 & 0 & y_{ee} \end{bmatrix}$$

$$\times \begin{bmatrix} 1 & 0 \\ 0 & 1 \\ -1 & 1 \\ 1 & 0 \\ 0 & 1 \end{bmatrix}$$

It is left to the student to check that multiplying this out gives the result of [8.13], viz.:

$$[Y] = \begin{bmatrix} (y_{aa}+y_{cc}+y_{dd}) & -y_{cc} \\ -y_{cc} & (y_{bb}+y_{cc}+y_{ee}) \end{bmatrix} \qquad [8.22]$$

Worked example 8.1

In the circuit of Figs. 8.1 and 8.3, $e_a = 1.1\underline{/0°}$ p.u.; $e_b = 1 \cdot 2\underline{/30°}$ p.u.; $z_{aa} = j0 \cdot 11$ p.u., $z_{bb} = j0 \cdot 08$ p.u.; $z_{cc} = j0 \cdot 05$ p.u.; $z_{dd} = j1 \cdot 2$ p.u.; $z_{ee} = j1 \cdot 0$ p.u. All resistances may be neglected. Determine the equations relating the nodal currents and the nodal voltages.

SOLUTION

$$I_1 = y_{aa}e_a = \frac{-j1 \cdot 1}{0 \cdot 11}\underline{/0°} = -j10\underline{/0°} \text{ p.u.} = -j10 \text{ p.u.}$$

$$I_2 = y_{bb}e_b = \frac{-j1 \cdot 2}{0 \cdot 08}\underline{/30°} = -j15\underline{/30°} \text{ p.u.} = 7 \cdot 5 - j13 \text{ p.u.}$$

$$Y_{11} = \frac{1}{j0 \cdot 11} + \frac{1}{j0 \cdot 05} + \frac{1}{j1 \cdot 2} = -j29 \cdot 92 \text{ p.u.}$$

$$Y_{22} = \frac{1}{j0\cdot08} + \frac{1}{j0\cdot05} + \frac{1}{j1\cdot0} = -j33\cdot5 \text{ p.u.}$$

$$Y_{12} = Y_{21} = \frac{1}{j0\cdot05} = -j20 \text{ p.u.}$$

$$\begin{bmatrix} (0-j10) \\ (7\cdot5-j13) \end{bmatrix} = \begin{bmatrix} -j29\cdot92 & j20 \\ j20 & -j33\cdot5 \end{bmatrix} \begin{bmatrix} V_1 \\ V_2 \end{bmatrix}$$

8.3.2 NODE ELIMINATION BY MEANS OF SUB-MATRICES

In [8.20] [I] and [V] are column matrices and [Y] is a symmetrical square matrix. The latter has as many rows (and columns) as the original circuit has nodes (apart from the reference node). In order to reduce time and to ease the problem of handling a large power system on a computer, and to avoid unnecessary data being produced, it may be desirable to eliminate certain nodes. These can only be nodes at which current does not enter the system so that generator nodes must be retained. The process of node elimination described in this section is a systematic one which can be carried out as a part of the computer programme, and it amounts to circuit reduction, which when carried out by hand calculations involves reducing nodes one at a time by adding the impedances of two elements in series or by star-delta transformation (see Volume 1, section 8.2 and Appendix 2).

From [8.14] and [8.20] the nodal equation for a network with three nodes (apart from the reference node) may be written as

$$\begin{bmatrix} I_1 \\ I_2 \\ \hline I_3 \end{bmatrix} = \begin{bmatrix} Y_{11} & Y_{12} & \vdots & Y_{13} \\ Y_{21} & Y_{22} & \vdots & Y_{23} \\ \hline Y_{31} & Y_{32} & \vdots & Y_{33} \end{bmatrix} \begin{bmatrix} V_1 \\ V_2 \\ \hline V_3 \end{bmatrix} \qquad [8.23]$$

If current enters at only two of the three nodes then the third node can be eliminated, and this will be numbered as 3, so that the nodes are numbered starting with all of these (n in number) which must be retained and, when this group has been completely numbered, those which are to be eliminated are then given the last numbers. The nodal admittance matrix [Y] may then be partitioned by the dotted lines shown in [8.23], where the vertical line is between columns n and $(n+1)$, and the horizontal line is between rows n and $(n+1)$. Similarly the nodal current and voltage matrices are divided by a dotted line between elements n and $(n+1)$. Since all nodes below the dotted line are to be eliminated and no current can enter the network at them, all nodal currents below the dotted line must be zero. [8.23] with its matrices thus partitioned into sub-matrices (for which the square brackets have been omitted for convenience) may be written as

$$
\begin{bmatrix} I_r \\ -- \\ I_e \end{bmatrix} = \begin{bmatrix} A & \vdots & B \\ -- & \vdots & -- \\ C & \vdots & D \end{bmatrix} \begin{bmatrix} V_r \\ -- \\ V_e \end{bmatrix}
\qquad [8.24]
$$

where $[I_r]$ is the column matrix of currents at all nodes to be retained·
In the particular case of [8.23]

$$
[I_r] = \begin{bmatrix} I_1 \\ -- \\ I_2 \end{bmatrix}
$$

$[I_e]$ relates to all nodes to be eliminated and every element in it must be zero. For a bilateral network $[C] = [B]_t$ (as inspection of [8.23] shows for the simple case represented by it). The self and mutual nodal admittances in the sub-matrix A (omitting now the square brackets for convenience in writing matrix relations, as has been done in [8.24]) are those for the nodes to be retained, whereas those in D are for the nodes to be eliminated. B and B_t contain only mutual admittances between one node which is to be retained and another which is to be eliminated.

Expanding [8.24] gives

$$
I_r = AV_r + BV_e \qquad [8.25]
$$

$$
I_e = B_tV_r + DV_e = 0 \qquad [8.26]
$$

From [8.26]

$$
V_e = -D^{-1}B_tV_r
$$

and substituting this value in [8.25]

$$
I_r = AV_r - BD^{-1}B_tV_r \qquad [8.27]
$$

The circuit with its reduced number of nodes now has a nodal admittance matrix $[Y]$ to be used in [8.20], which is given by

$$
[Y] = A - BD^{-1}B_t \qquad [8.28]
$$

In order to remind the student that each of the terms in [*8.28*] is itself a matrix, it is repeated with the square brackets re-introduced

$$[\mathbf{Y}] = [\mathbf{A}] - [\mathbf{B}][\mathbf{D}]^{-1}[\mathbf{B}]_t \qquad [8.28]$$

In the nodal admittance [**Y**] which relates the nodal currents and voltages according to [*8.20*], there are elements which represent series branches which are admittances between two busbars, and other elements representing shunt branches which are admittances between a busbar and earth which is the reference node. Any diagonal term in [**Y**] such as Y_{22} in [*8.23*] is equal to the sum of all series and shunt admittances connected directly to busbar 2, so that it includes the series admittances of any equipment such as transformers and lines, the shunt susceptance of lines, shunt reactors or capacitors and the shunt elements of the equivalent circuit of a transformer (where this may need to be included, e.g. for tapped transformers). Any non-diagonal term such as Y_{32} in [*8.23*] is negative as [*8.14*] shows, and is equal to the sum of all series branch admittances connected directly between two busbars (3 and 2 in this case). It may include series reactors or capacitors as well as the series elements of a transformer equivalent circuit or transmission line.

Worked example 8.2

In the circuit of Fig. 8.4 the values of the per-unit e.m.f.s. and of the per-unit impedances (negligible resistance) are as marked. Eliminate nodes 3 and 4 from the network. Draw the reduced circuit.

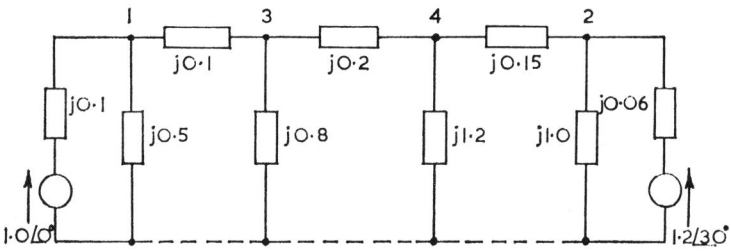

Fig. 8.4. System for worked example 8.2 showing per-unit e.m.f.'s and reactances.

SOLUTION

The nodal currents are $I_1 = -j10$ p.u. and $I_2 = -j20\underline{/30°} = 10 - j17·32$ p.u. The nodal admittance matrix of the circuit of Fig. 8.4 may be written by inspection as

$$
-j \left[
\begin{array}{cc|cc}
22·0 & 0 & -10·0 & 0 \\
0 & 24·33 & 0 & -6·67 \\
\hline
-10·0 & 0 & 16·25 & -5·0 \\
0 & -6·67 & -5·0 & 12·5
\end{array}
\right] \text{ p.u.}
$$

$$
= \left[
\begin{array}{c|c}
A & B \\
\hline
C & D
\end{array}
\right]
$$

$$
D^{-1} = \frac{j}{178} \begin{bmatrix} 12·5 & 5·0 \\ 5·0 & 16·25 \end{bmatrix}
$$

The nodal admittance matrix [Y] of the circuit with nodes 3 and 4 removed is given by [8·28] as

$$
[Y] = -j \begin{bmatrix} 22·0 & 0 \\ 0 & 24·33 \end{bmatrix} + j \begin{bmatrix} -10·0 & 0 \\ 0 & -6·67 \end{bmatrix} \frac{j}{178} \begin{bmatrix} 12·5 & 5·0 \\ 5·0 & 16·25 \end{bmatrix}
$$

$$
\times j \begin{bmatrix} 10·0 & 0 \\ 0 & 6·67 \end{bmatrix}
$$

$$
= -j \begin{bmatrix} 22·0 & 0 \\ 0 & 24·33 \end{bmatrix} - \frac{j}{178} \begin{bmatrix} -1250 & -333·3 \\ -333·3 & -722 \end{bmatrix}
$$

$$
= \begin{bmatrix} -j14·98 & j1·87 \\ j1·87 & -j20·27 \end{bmatrix}
$$

Inspection of [Y] enables the circuit shown in Fig. 8.5 to be drawn, where p.u. nodal currents and nodal admittances are marked. This network has now been reduced to the form of that of Worked example 8.1 and the nodal voltages may be evaluated by inversion

of [Y] as shown in Worked example 8.3. The student can check
if he wishes that by using the star-delta transformation twice the
network of Fig. 8.4 can be reduced to that of Fig. 8.7, but the method
of reduction by matrix partitioning using a digital computer is far
more convenient.

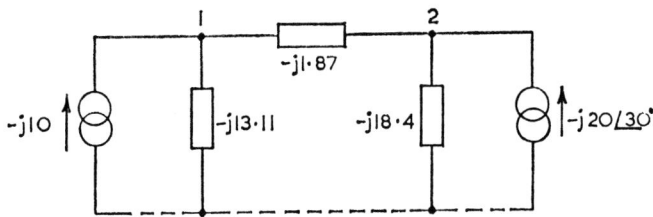

Fig. 8.5. Reduced circuit for system of Fig. 8.4 showing
per-unit currents and admittances.

8.4 Application of nodal voltage method to the solution of power-system load flow problems

In order to discover the best ways of operating an existing power
system, and to plan possible future extensions of the system, it is
necessary to investigate the steady-state solution of the network for
various generation and load requirements. It is necessary to check,
for example, that the system is stable, that no transmission lines or
cables are overloaded even when some others are out of circuit, that
all busbar voltages are within specified limits and that the flow of
reactive power is reasonable throughout the system. These load flow
studies are needed to give the magnitude and phase angle of each
busbar voltage, the power and reactive power in each transmission
line and the power and reactive power to be supplied at (at least)
one generator busbar which is called the floating bus (or 'swing'
or 'slack-take-up' bus in the U.S.A.). Although it is possible to
specify at the beginning the total load and the power input to the
network at nearly all busbars or all but one of them, the input at at
least one is unknown until the load flow study has been completed,
because the total power loss in the system can only then be determined.

Apart, therefore, from the information on the system impedances
and their interconnection, which can quickly be formed into a nodal

admittance matrix in a digital computer, the following data regarding node conditions must be specified at the outset, together with upper and lower limits of the magnitudes of busbar voltages and of reactive power, in order to carry out a particular load study.

(a) At some busbars the complex power $S = P + jQ$ taken from or injected into the network is specified. These are generally nodes at which loads are connected.

(b) At some busbars the power P and the magnitude of the voltage are specified (and limits may be placed to the reactive power flow at the busbar). These are usually generator nodes, though if the solution is obtained by breaking the network down into two or more subnetworks, a node connecting two subnetworks may also be specified in this way.

(c) At a generator busbar which is selected as the 'floating bus' (perhaps because it is the one most closely approaching an infinite busbar to ease the calculations), the busbar voltage is specified in magnitude and in phase. The complex power S fed into the network at this busbar is unknown until the load flow and thus the system losses have been determined, and this is therefore part of the solution.

These studies are now almost always made using digital computers rather than electrical models (network analysers), and only the former will be referred to here. The branch currents and voltages can be transformed into mesh form (see section 8.2) or nodal form (see section 8.3) (a third possibility of cut-set form is not discussed here). Mesh impedance matrices may sometimes be used for short-circuit calculations, particularly where mutual coupling has to be considered, but they are not often used for load-flow calculations, since nodal voltage equations have the following advantages:

(a) the number of equations is generally less.

(b) There is no need to combine parallel branches before forming equations, in order to reduce their number.

(c) There is no difficulty in dealing with non-planar networks, i.e. where in a diagram of the network there are some cross-over branches.

(d) Off-nominal turns ratios of transformers can more easily be represented (see section 8.6.3).

The disadvantages of the nodal method which arise in hand calculations, of the conversion of impedances into admittances, and the necessity of calculating nodal voltages to a relatively large number of significant figures because branch currents depend upon small differences between them, are no longer disadvantages when a digital computer is used.

At any node r, if the injected nodal current is I_r, the nodal voltage is V_r (potential difference between that node and the reference node which can generally be earthed since 3-phase systems are involved) and the complex power/phase fed into the system at that node is S_r, then (see Volume 1, Appendix 4)

$$S_r = V_r I_r^*$$ [8.29]

so that

$$I_r = S_r^*/V_r^*$$ [8.30]

If there are n nodes in the network (apart from the reference node) then [8.23] and [8.30] show that the load flow study involves the solution of the set of quadratic equations given by

$$\sum_{k=1}^{n} Y_{rk} V_k = S_r^*/V_r^*$$ [8.31]

where $r = 1, 2, 3, \ldots n$ in turn.

The difficulty in solving the set of equations given by [8.31] is the non-linearity caused by the complex powers and the voltages being interdependent. Whatever method of solution is adopted must therefore involve making successive approximations.

The methods available are usually classified as

(a) Direct.

(b) Iterative.

(c) Hybrid (combination of direct and iterative methods).

The principles involved in the direct methods where the nodal admittance matrix is inverted are given in the next section, (8.5), and two of these methods are then outlined in sections 8.5.1 and 8.5.2. Section 8.6 shows that for a given system, the nodal admittance matrix only needs to be inverted once, because modifications can then be made to the inverse matrix to take account of various changes in the system. Two examples of iterative methods are outlined in sections 8.7.1 and 8.7.2, and hybrid methods are referred to briefly in section 8.7.4. It is not possible here to compare the merits of the three types of method, and to discuss the relative problems of inverting the nodal matrix with least time and computer storage space in direct methods which may be ameliorated by such methods as partitioning (section 8.5.2) or tearing (section 8.8), and of obtaining the quickest possible convergence in iterative methods (see section 8.7.3). Such comparisons as have been made, show that it is difficult to generalise, but that the size of the network and the size of the computer are important factors (see Laughton and Humphrey Davies, 1964, and Stagg and El-Abiad, 1968), and that usually the larger the network the greater the relative advantages of iterative methods over direct or hybrid ones.

8.5 Direct methods involving inversion of the nodal admittance matrix

Direct methods involve computing the nodal impedance matrix [Z] by inverting the nodal admittance matrix [Y], so that, in place of [8.20], the following equation can be written (provided that the determinant of $[Y] = |Y| \neq 0$, i.e that [Y] is square and non-singular as is true of a power system nodal admittance matrix, see Appendix A.1)

$$[V] = [Y]^{-1}[I] = [Z][I] \qquad [8.32]$$

(The nodal impedance matrix in [8.32] must not be confused with the mesh impedance matrix which appears in [8.9] though the same symbol is used here for both, as this is common practice.)

Appendix A.1 shows how the inverse of a matrix may be obtained from the original matrix by hand calculation, but in power system analysis where the matrices can be very large, digital computers are used where the time taken to invert a matrix is approximately proportional to the cube of the number of equations. Because of the large number of equations involved in a power system study, a variety of methods have been developed to conserve computer time and storage. A discussion of these methods, together with further references, is given for example by Laughton and Humphrey Davies (1964), and two of them are outlined in sections 8.5.1 and 8.5.2. The following section (8.6) then shows how the inverse matrix [Z] may be modified in order to determine the effect upon the load flow of making certain circuit changes, such as adding a new load, interconnecting two places by a new transmission line or taking a transmission line out of circuit.

After the admittance matrix has been inverted to give the [Z] matrix and the latter has been modified if necessary to incorporate circuit changes, it is still necessary to arrive at the solution via some form of iterative process. These processes (see Laughton and Humphrey Davies, 1964) are not discussed in detail here, but the basic procedure is as follows. For any busbar the voltage and complex power (where not already specified by the nature of the problem as discussed in section 8.4) are estimated. Using these voltages and complex powers a first estimate of the current at each node can be made using [8.30]. Using the nodal impedance matrix, obtained by a method such as one of those outlined in section 8.5.1 and 8.5.2, a corrected set of nodal voltages may be obtained from [8.32]. This new set of voltages when used with the power requirements gives a second set of currents from [8.30], which leads from [8.32] to a further set of voltages. The computer programme cycle terminates when the difference between two successive sets of voltages lies within pre-determined limits.

Worked example 8.3

Determine the nodal voltages for the system in Worked example 8.1.

SOLUTION

The result of example 8.1 may be written

$$\begin{bmatrix} V_1 \\ V_2 \end{bmatrix} = \begin{bmatrix} -j29 \cdot 92 & j20 \\ j20 & -j33 \cdot 5 \end{bmatrix}^{-1} \begin{bmatrix} (0 & -j10) \\ (7 \cdot 5 - j13) \end{bmatrix}$$

$$= -\frac{1}{600} \begin{bmatrix} -j33 \cdot 5 & -j20 \\ -j20 & -j29 \cdot 92 \end{bmatrix} \begin{bmatrix} (0 & -j10) \\ (7 \cdot 5 - j13) \end{bmatrix}$$

$$\begin{bmatrix} V_1 \\ V_2 \end{bmatrix} = \begin{bmatrix} (0 \cdot 992 + j0 \cdot 25) \\ (0 \cdot 978 + j0 \cdot 373) \end{bmatrix} \text{p.u.}$$

8.5.1 GAUSSIAN ELIMINATION

The nodal admittance matrix [Y] and its inverse the nodal impedance matrix $[Z] = [Y]^{-1}$ are related by

$$[Y][Z] = [U] \qquad [8.33]$$

where [U] is the unit matrix (see Appendix A.1.7). [8.33] may be written for the simplest case of a network with only two nodes (apart from the reference node) as

$$\begin{bmatrix} Y_{11} & Y_{12} \\ Y_{21} & Y_{22} \end{bmatrix} \begin{bmatrix} Z_{11} & Z_{12} \\ Z_{21} & Z_{22} \end{bmatrix} = \begin{bmatrix} 1 & 0 \\ 0 & 1 \end{bmatrix}$$

which gives the equations

$$Y_{11}Z_{11} + Y_{12}Z_{21} = 1 \qquad [8.34]$$

$$Y_{21}Z_{11} + Y_{22}Z_{21} = 0 \qquad [8.35]$$

$$Y_{11}Z_{12} + Y_{12}Z_{22} = 0 \qquad [8.36]$$

$$Y_{21}Z_{12} + Y_{22}Z_{22} = 1 \qquad [8.37]$$

The nodal self and mutual admittances are therefore known coefficients for a given network in a set of linear simultaneous equations in which the elements of the [Z] matrix are the unknowns which are being determined. The method of Gaussian elimination for a set of n equations with unknowns $x_1, x_2 \ldots x_n$ (here the number of nodes $n = 2$ and there are two equations in Z_{11} and Z_{21} and two equations in Z_{12} and Z_{22}) is to add to another of the equations in the set, that multiple of the first equation which makes the coefficient of x_1 become zero in the new equation resulting from the addition. There are then $(n-1)$ equations in the $(n-1)$ unknowns $x_2, x_3 \ldots x_n$. This is repeated for each of the unknowns in turn until finally $a_{nn}x_n = b_n$ is reached, where a_{nn} and b_n are known so that x_n is found. Substitution of this value of x_n back into the $(n-1)$th equation yields x_{n-1} and so on.

Applying this method to [8.34] and [8.35] involves adding the former multiplied by $(-Y_{21}/Y_{11})$ to the latter, which gives

$$(Y_{22} - Y_{21}Y_{12}/Y_{11})Z_{21} = -Y_{21}/Y_{11}$$

so that

$$Z_{21} = -Y_{21}/(Y_{22}Y_{11} - Y_{21}Y_{12}) \qquad [8.38]$$

Z_{11} can then be found by substitution, and in a similar way Z_{12} and Z_{22} may be found from [8.36] and [8.37]. There are a number of special arrangements of this method which reduce the time taken in a computer to carry out the process.

8.5.2 PARTITIONING OR FACTORISED INVERSE MATRIX

The nodal admittance matrix may be partitioned into submatrices with the square brackets omitted here for convenience

$$[Y] = \begin{bmatrix} Y_1 & \vdots & Y_2 \\ -- & -\vdots- & -- \\ Y_3 & \vdots & Y_4 \end{bmatrix} \qquad [8.39]$$

Comparison of [8.23] and [8.39] shows that in this particular case of a 3-node network

$$[Y_1] = \begin{bmatrix} Y_{11} & Y_{12} \\ Y_{21} & Y_{22} \end{bmatrix}, [Y_2] = \begin{bmatrix} Y_{13} \\ Y_{23} \end{bmatrix} \text{ etc.}$$

In general

$$\begin{bmatrix} I_a \\ -- \\ I_b \end{bmatrix} = \begin{bmatrix} Y_1 & \vdots & Y_2 \\ -- & -\vdots- & -- \\ Y_3 & \vdots & Y_4 \end{bmatrix} \begin{bmatrix} V_a \\ -- \\ V_b \end{bmatrix} \qquad [8.40]$$

The notation I_a, I_b, V_a and V_b has been used here instead of I_1, I_2, etc. for the submatrices of currents and voltages to avoid confusion with the nodal currents and voltages such as I_1 and V_1 at node 1 and I_2 and V_2 at node 2. Again the square brackets around I_a, I_b, V_a and V_b are omitted.

$$I_a = Y_1 V_a + Y_2 V_b \qquad [8.41]$$

$$I_b = Y_3 V_a + Y_4 V_b \qquad [8.42]$$

From [8.41]

$$V_a = Y_1^{-1}(I_a - Y_2 V_b)$$

and substituting this expression for V_a in [8.42] gives

$$I_b = Y_3 Y_1^{-1}(I_a - Y_2 V_b) + Y_4 V_b$$

$$I_b = Y_3 Y_1^{-1} I_a + (Y_4 - Y_3 Y_1^{-1} Y_2) V_b$$

$$V_b = (Y_4 - Y_3 Y_1^{-1} Y_2)^{-1}(I_b - Y_3 Y_1^{-1} I_a) \qquad [8.43]$$

If the inverse of [Y] is [Z] and the latter is partitioned into four submatrices corresponding to those in [8.39], then

$$[\mathbf{Z}] = \begin{bmatrix} \mathbf{Z}_1 & | & \mathbf{Z}_2 \\ --&|&-- \\ \mathbf{Z}_3 & | & \mathbf{Z}_4 \end{bmatrix} \qquad [8.44]$$

and

$$\begin{bmatrix} \mathbf{V}_a \\ -- \\ \mathbf{V}_b \end{bmatrix} = \begin{bmatrix} \mathbf{Z}_1 & | & \mathbf{Z}_2 \\ --&|&-- \\ \mathbf{Z}_3 & | & \mathbf{Z}_4 \end{bmatrix} \begin{bmatrix} \mathbf{I}_a \\ -- \\ \mathbf{I}_b \end{bmatrix} \qquad [8.45]$$

so that [8.43] shows that

$$\mathbf{Z}_3 = (\mathbf{Y}_3\mathbf{Y}_1^{-1}\mathbf{Y}_2 - \mathbf{Y}_4)^{-1}\mathbf{Y}_3\mathbf{Y}_1^{-1} \qquad [8.46]$$

$$\mathbf{Z}_4 = (\mathbf{Y}_4 - \mathbf{Y}_3\mathbf{Y}_1^{-1}\mathbf{Y}_2)^{-1} \qquad [8.47]$$

Similarly, from [8.42]

$$\mathbf{V}_b = \mathbf{Y}_4^{-1}(\mathbf{I}_b - \mathbf{Y}_3\mathbf{V}_a)$$

and substituting this value in [8.41] gives

$$\mathbf{I}_a = \mathbf{Y}_2\mathbf{Y}_4^{-1}\mathbf{I}_b + (\mathbf{Y}_1 - \mathbf{Y}_2\mathbf{Y}_4^{-1}\mathbf{Y}_3)\mathbf{V}_a$$

so that

$$\mathbf{V}_a = (\mathbf{Y}_1 - \mathbf{Y}_2\mathbf{Y}_4^{-1}\mathbf{Y}_3)^{-1}(\mathbf{I}_a - \mathbf{Y}_2\mathbf{Y}_4^{-1}\mathbf{I}_b) \qquad [8.48]$$

From [8.45] and [8.48]

$$\mathbf{Z}_1 = (\mathbf{Y}_1 - \mathbf{Y}_2\mathbf{Y}_4^{-1}\mathbf{Y}_3)^{-1} \qquad [8.49]$$

$$\mathbf{Z}_2 = -(\mathbf{Y}_1 - \mathbf{Y}_2\mathbf{Y}_4^{-1}\mathbf{Y}_3)^{-1}\mathbf{Y}_2\mathbf{Y}_4^{-1} \qquad [8.50]$$

Thus

$$[\mathbf{Z}] = \begin{bmatrix} (\mathbf{Y}_1 - \mathbf{Y}_2\mathbf{Y}_4^{-1}\mathbf{Y}_3)^{-1} & | & -(\mathbf{Y}_1 - \mathbf{Y}_2\mathbf{Y}_4^{-1}\mathbf{Y}_3)^{-1}\mathbf{Y}_2\mathbf{Y}_4^{-1} \\ -------------&|&----------- \\ (\mathbf{Y}_3\mathbf{Y}_1^{-1}\mathbf{Y}_2 - \mathbf{Y}_4)^{-1}\mathbf{Y}_3\mathbf{Y}_1^{-1} & | & (\mathbf{Y}_4 - \mathbf{Y}_3\mathbf{Y}_1^{-1}\mathbf{Y}_2)^{-1} \end{bmatrix}$$

$$[8.51]$$

Since the matrices which have to be inverted in [8.51] are of lower order than [Y], and the computer time in inversion is approximately proportional to the cube of the number of equations (i.e. to the cube of the matrix order), partitioning can be advantageous. This is particularly so when the equations of a large power system are being solved on a relatively small computer, and submatrices can be contained within the fast store, so that relatively slow transfers to and from the back-up store are avoided. The student should note (and check for himself) that the submatrices of [Z] can take several forms, some of which involve less computation time than others, e.g. Z_2 may be written as $-Z_1 Y_2 Y_4^{-1}$, and if V_a is found from [8.42] and this result is substituted in [8.41], then Z_3 takes the form of $(Y_2 - Y_1 Y_3^{-1} Y_4)^{-1}$.

8.6 Modification of the inverse of the nodal admittance matrix

When information for a particular power system has been stored in a computer in the form of a nodal admittance matrix, and this has been inverted to give its inverse which is a nodal impedance matrix, the solution can be found for a particular set of load and generation conditions. Solutions will, however, also be required for a number of other network conditions, e.g. if one transmission line is out of circuit, if a new transmission line links two nodes or if a new load admittance is added at a particular node. For each of these and other circuit changes there is a new nodal admittance matrix and a new inverse matrix. In order to obtain the latter without the relatively lengthy procedure of forming the new admittance matrix and performing a new complete inversion of this large matrix, it is possible to modify the inverse of the original admittance matrix.

In the discussion of this modification, the original n by n admittance matrix for a network with n nodes is denoted by [Y]. The square brackets are omitted here for convenience, so that it appears as Y, and its original inverse (or nodal impedance matrix) is Y^{-1}, and this is done for other matrices in this section. The new admittance matrix Y_m for the modified network is also a square n by n matrix, if no new node has been set up, and this is the sum of the original matrix Y and a new modification matrix. The modification matrix, which must also be n by n, can readily be obtained for the usual circuit changes for which it is required, e.g. if a new equivalent load admittance y is added to the network between node r and earth (reference node 0), then this constitutes only an increase of y in the self-admittance at node r, so that the modification matrix consists entirely of zeros except for element rr which is $\mathbf{y}$. This modification matrix may be written as a scalar α multiplied by the product of a column vector of order n (matrix with only one column and n rows) and a row vector of order n (matrix with one row and n columns), so that the modification matrix is $n \times n$ like the original matrix $\mathbf{Y}$ (see Appendix A.1). If the column vector is denoted by $\mathbf{G}$ and the row vector is the transpose of another column vector $\mathbf{H}$, then the new modified nodal admittance matrix $\mathbf{Y}_m$ is given by the sum of the original matrix and the modification matrix

$$\mathbf{Y}_m = \mathbf{Y} + \alpha \mathbf{G} \mathbf{H}_t \qquad [8.52]$$

and the inverse of the modified admittance matrix which it is required to obtain from $\mathbf{Y}^{-1}$ without inverting $\mathbf{Y}_m$ directly is given by

$$\mathbf{Y}_m^{-1} = (\mathbf{Y} + \alpha \mathbf{G} \mathbf{H}_t)^{-1} \qquad [8.53]$$

The inverse of a modified matrix may be obtained by using a formula due to Woodbury, but for the restricted conditions which apply here of α being a scalar and $\mathbf{G}$ and $\mathbf{H}$ being column vectors, the following formula by Kron may be used

$$\mathbf{Y}_m^{-1} = (\mathbf{Y} + \alpha \mathbf{G} \mathbf{H}_t)^{-1} = \mathbf{Y}^{-1} - \frac{(\mathbf{Y}^{-1}\mathbf{G})(\mathbf{H}_t\mathbf{Y}^{-1})}{1/\alpha + \mathbf{H}_t\mathbf{Y}^{-1}\mathbf{G}} \qquad [8.54]$$

It may be verified as follows that [8.54] is indeed an identity, though this treatment is not mathematically rigorous (e.g. there are conditions as regards determinants to be satisfied, but for power system problems these are in fact satisfied and are not discussed here). Pre-multiplying both sides of [8.54] by $(\mathbf{Y} + \alpha \mathbf{G} \mathbf{H}_t)$, rewriting the right-hand side on a common scalar denominator and multiplying both sides by it gives

$$(\mathbf{Y} + \alpha \mathbf{G} \mathbf{H}_t)(\mathbf{Y}^{-1}/\alpha + \mathbf{Y}^{-1}\mathbf{H}_t\mathbf{Y}^{-1}\mathbf{G} - \mathbf{Y}^{-1}\mathbf{G}\mathbf{H}_t\mathbf{Y}^{-1}) = (1/\alpha + \mathbf{H}_t\mathbf{Y}^{-1}\mathbf{G})\mathbf{U}$$

where $\mathbf{U}$ is the unit matrix of order n (Appendix A.1.7).

The product $\mathbf{H}_t\mathbf{G}$ is a scalar (see Appendix A.1) and as $\mathbf{H}_t\mathbf{Y}^{-1}$ is another row vector of order n then $\mathbf{H}_t\mathbf{Y}^{-1}\mathbf{G}$ is also a scalar β (as shown by dimensional inspection of [8.54]). The terms on the left-hand side may now be considered: $YY^{-1}/\alpha = (1/\alpha)U$ and cancels the first term on the right-hand side; $YY^{-1}H_tY^{-1}G = \beta U$ and cancels the second term on the right-hand side; the remaining terms on the left-hand side are

$$-YY^{-1}GH_tY^{-1}+GH_tY^{-1}+\alpha GH_tY^{-1}\beta-\alpha GH_tY^{-1}GH_tY^{-1}$$

The first two terms cancel and writing β in place of the middle three components of the last term shows that it cancels with the previous term, which verifies the identity of [8.54].

Sections 8.6.1 and 8.6.2 show respectively the form to which [8.54] is reduced, for changes in shunt admittance (i.e. admittance between a node and the reference node) and changes in branch impedance between two nodes. The former modification may be required for example (i) to ascertain the load flow if changes are made in load admittance (this also requires a change in the power data specified as discussed in section 8.4) or (ii) to simulate the shunt admittance change when a load change or fault causes the synchronous machine rotors to swing. In this latter case where the swing curve is determined by iteration process, it is also necessary to modify the inverse matrix to change the synchronous machine reactances from the synchronous values to the transient reactances, and also to change the machine e.m.f.s from a 'voltage behind synchronous reactance' to one 'behind transient reactance' (see Chapter 3 of this volume and Chapter 7 of Volume 1). The second modification, that of an impedance between two nodes, may be required, for example, to find the effect of a new transmission line linking two points, or to check that under steady state flow conditions, no other part of the transmission system is overloaded if each in turn of a number of transmission lines is out of circuit due to maintenance or a fault. In the latter case if the line linking node f to node k has a series impedance Z_{fk}, then the modification required for line outage is to connect a second impedance of $-Z_{fk}$ between nodes f and k. Transformer tap changes can also be dealt with by modifying the inverse matrix as shown in section 8.6.3. In this case and also where transmission line capacitance can not be neglected, the modification involves changes in a π circuit. It is possible to make simultaneous modification of the series and shunt parameters, but it is often found more convenient to make these changes separately by the results given in sections 8.6.1 and 8.6.2.

8.6.1 APPLICATION TO THE ADDITION OF A NEW LOAD ADMITTANCE AT ONE NODE

If a new load admittance **y** is to be added to the network between node k and the reference node, the modification matrix which must be added to **Y** to give the new modified matrix $\mathbf{Y}_m$ is

$$
\alpha\mathbf{GH}_t =
\begin{array}{c}
1 \\ 2 \\ \vdots \\ k \\ \vdots \\ n
\end{array}
\begin{array}{cccc}
1 & 2 & k & n \\
\end{array}
\begin{bmatrix}
0 & 0\ldots\ldots0\ldots\ldots0 \\
0 & 0\ldots\ldots0\ldots\ldots0 \\
\vdots & \vdots \quad \vdots \quad \vdots \\
0 & 0 \quad y \quad 0 \\
\vdots & \vdots \quad \vdots \quad \vdots \\
0 & 0\ldots\ldots0\ldots\ldots0 \\
\end{bmatrix}
\qquad [8.55]
$$

In this case it is convenient to choose $\alpha = 1$ and $\mathbf{G} = \mathbf{H}$ so that

$$
\mathbf{G} = \mathbf{H} =
\begin{array}{c}
1 \\ 2 \\ \vdots \\ k \\ \vdots \\ n
\end{array}
\begin{bmatrix}
0 \\
0 \\
\vdots \\
\sqrt{y} \\
\vdots \\
0
\end{bmatrix}
\qquad [8.56]
$$

The terms on the right-hand side of [8.54] can now be examined. $\mathbf{Y}^{-1}$, the inverse of the original admittance matrix, has already been obtained (possibly by direct methods such as those outlined in section 8.5). For the rest

$\mathbf{Y}^{-1}\mathbf{G}$ gives column k of $\mathbf{Y}^{-1}$ multiplied by $\sqrt{y}$

$\mathbf{H}_t\mathbf{Y}^{-1}$ gives row k of $\mathbf{Y}^{-1}$ multiplied by $\sqrt{y}$

$\mathbf{H}_t\mathbf{Y}^{-1}\mathbf{G}$ gives element kk of $\mathbf{Y}^{-1}$ multiplied by **y**

Thus

$$
\mathbf{Y}_m^{-1} = \mathbf{Y}^{-1} - \frac{(\text{column } k \text{ of } \mathbf{Y}^{-1})(\text{row } k \text{ of } \mathbf{Y}^{-1})}{1/y + \text{element } kk \text{ of } \mathbf{Y}^{-1}}
\qquad [8.57]
$$

Clearly there is the alternative of choosing $\alpha = y$ with the only non-zero element in $\mathbf{G}$, and $\mathbf{H}$ as 1. It may appear confusing to write $\alpha = y$ but α is termed a scalar as opposed to a matrix, though in this case it represents a complex admittance $y = 1/z = 1/(R + jX)$.

8.6.2 APPLICATION TO THE ADDITION OF A NEW IMPEDANCE BETWEEN TWO NODES

If a new impedance z is added to the network to interconnect nodes f and k, then the nodal self admittances at nodes f and k are increased by $1/z$ and the nodal mutual admittances between nodes f and k are reduced by $1/z$, so that the modification matrix is given by

$$\alpha \mathbf{GH}_t = \quad \begin{matrix} & f & k \\ f & \\ k & \end{matrix} \begin{bmatrix} \vdots & \vdots \\ \cdots\cdots 1/z \cdots\cdots -1/z \cdots\cdots \\ \vdots & \vdots \\ \cdots\cdots -1/z \cdots\cdots 1/z \cdots\cdots \\ \vdots & \vdots \end{bmatrix} \qquad [8.58]$$

and all other elements are zero. If α is chosen to be $1/z$ and $\mathbf{G} = \mathbf{H}$ then the latter are column vectors with element $f = 1$ and element $k = -1$ and all other elements are zero. (The student should satisfy himself that these conditions do fulfil [8.58].)

$\mathbf{Y}^{-1}\mathbf{G}$ gives a column vector with each element consisting of an element of column f minus the corresponding element of column k of $\mathbf{Y}^{-1}$

$\mathbf{H}_t\mathbf{Y}^{-1}$ gives a row vector with each element consisting of an element of row f minus the corresponding element of row k of $\mathbf{Y}^{-1}$.

$\mathbf{H}_t\mathbf{Y}^{-1}\mathbf{G}$ gives the sum of elements $ff + kk - fk - kf$ of $\mathbf{Y}^{-1}$ (denoted by $\mathbf{Y}_{ff}^{-1}$ etc.).

Thus

$$\mathbf{Y}_m^{-1} = \mathbf{Y}^{-1} - \frac{(\text{column } f - \text{column } k \text{ of } \mathbf{Y}^{-1})\,(\text{row } f - \text{row } k \text{ of } \mathbf{Y}^{-1})}{z + \mathbf{Y}_{ff}^{-1} + \mathbf{Y}_{kk}^{-1} - \mathbf{Y}_{fk}^{-1} - \mathbf{Y}_{kf}^{-1}}$$

$$[8.59]$$

8.6.3 APPLICATION TO TRANSFORMER TAP CHANGING

In Appendix 1 of Volume 1, it is shown that when the voltages on both sides of a two-winding transformer are the base (rated or nominal) voltages, then the per-unit series impedance $\mathbf{Z}$ and thus the per-unit series admittance $\mathbf{Y} = 1/\mathbf{Z}$ for the combination of primary and secondary windings has the same value on both sides of the transformer. The shunt admittance associated with magnetisation and iron losses may be neglected in most power system studies.

Some power transformers however, have a ratio which may be varied from the nominal value by perhaps $\pm 10\%$ (per-unit tap $t = \pm 0.1$ per unit) by off-load or on-load tap changing (section 6.7, Volume 1). The variation of series impedance with tap position is kept to a minimum in the design and may be neglected here. The tap-change transformer may be represented by its series per-unit admittance $\mathbf{Y}$ for the MVA base chosen and for nominal ratio, in series with a fictitious ideal transformer which for a per-unit tap t has a per-unit turns ratio $N:1$ where $N = 1+t$. This is shown in Fig. 8.6(a) where the admittance of the power transformer which is connecting nodes f and k is shown on the non-tapped side, and an extra node x is introduced temporarily. The base value (1·0 per unit) of the voltage with respect to the reference node 0 (earth) of node x is related to the base voltage (1·0 per unit) of node k, when both are expressed in kV by the nominal turns ratio, and the per-unit impedance is the same at either of these voltages. If, however, the impedance is transferred to the tapped-side of the transformer its ohmic value is N^2 times the value corresponding to the voltage level at x, and its per-unit value is $N^2\mathbf{Z}$. This is because the per-unit impedance is proportional to base MVA and ohmic impedance and inversely proportional to (base line kV)^2 (see [A.1.5], page 395 of Volume 1) and the base voltages at x and f are the same, whereas the ohmic impedances differ by the factor N^2. The per-unit admittance represented on the tapped side is therefore $\mathbf{Y}/N^2$ and a new auxiliary node a is then introduced temporarily as shown in Fig. 8.6(b).

From Fig. 8.6(a) it can be seen that applying [8,14] to node k on the non-tapped side gives

$$\mathbf{I}_k = (\mathbf{Y}+\mathbf{Y}_{ks}+\mathbf{Y}_{k0})\mathbf{V}_k - \mathbf{Y}\mathbf{V}_x - \mathbf{Y}_{ks}\mathbf{V}_s$$

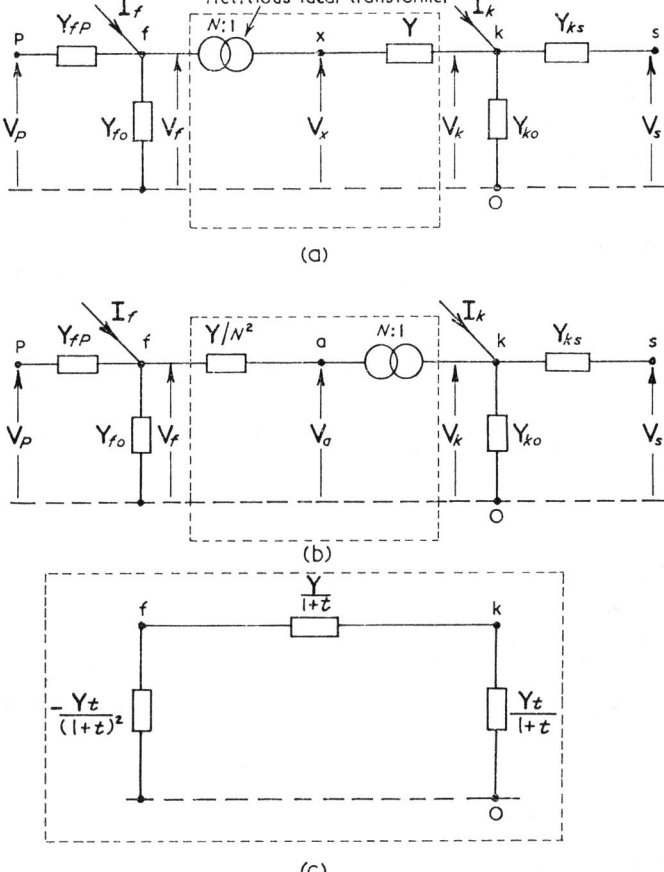

Fig. 8.6. Transformer tap changing:

   (a) series admittance represented on the non-tapped
       side
   (b) series admittance represented on the tapped side
   (c) π-equivalent circuit.

and since

$$\mathbf{V}_x = \mathbf{V}_f/N$$

$$\mathbf{I}_k = (\mathbf{Y}+\mathbf{Y}_{ks}+\mathbf{Y}_{k0})\mathbf{V}_k-\mathbf{Y}\mathbf{V}_f/N-\mathbf{Y}_{ks}\mathbf{V}_s$$

Thus the nodal self admittance at the node on the non-tapped side of the transformer contains its normal series admittance $\mathbf{Y}$, while the nodal mutual admittance between the nodes on either side of it is $\mathbf{Y}/N = \mathbf{Y}/(1+t)$.

For node f on the tapped side, [8.14] may be written by inspection of Fig. 8.6(b) as

$$\mathbf{I}_f = (\mathbf{Y}/N^2+\mathbf{Y}_{fp}+\mathbf{Y}_{f0})\mathbf{V}_f-\mathbf{Y}\mathbf{V}_a/N^2-\mathbf{Y}_{fp}\mathbf{V}_p$$

and since

$$\mathbf{V}_a = N\mathbf{V}_k$$

$$\mathbf{I}_f = (\mathbf{Y}/N^2 + \mathbf{Y}_{fp} + \mathbf{Y}_{f0})\mathbf{V}_f - \mathbf{Y}\mathbf{V}_k/N - \mathbf{Y}_{fp}\mathbf{V}_p$$

The nodal self admittance on the tapped side is therefore $\mathbf{Y}/N^2 = \mathbf{Y}/(1+t)^2$. These nodal admittances show that in general the tap-changing transformer may be represented by the π-equivalent circuit shown in Fig. 8.6(c), which reduces to the series admittance only on nominal ratio. These nodal admittances appear in the nodal admittance matrix as

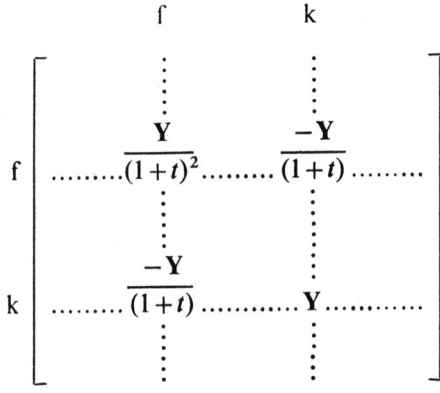

so that the modification matrix required to change from a per-unit tap of t_1 to a tap of t_2 is

$$\alpha\,\mathbf{GH}_t = \begin{array}{cc} & \begin{array}{cc} \text{f} & \qquad\qquad\qquad \text{k} \end{array} \\ \begin{array}{c} \text{f} \\[4em] \text{k} \end{array} & \left[\begin{array}{c:c} \ldots\mathbf{Y}\!\left(\dfrac{1}{(1+t_2)^2}-\dfrac{1}{(1+t_1)^2}\right)\ldots & \mathbf{Y}\!\left(\dfrac{1}{(1+t_1)}-\dfrac{1}{(1+t_2)}\right)\ldots \\[2em] \ldots\mathbf{Y}\!\left(\dfrac{1}{(1+t_1)}-\dfrac{1}{(1+t_2)}\right)\ldots\ldots\ldots & \qquad 0 \ldots\ldots\ldots \end{array}\right] \end{array}$$

Worked example 8.4

In the network of Worked example 8.1, a further load of impedance j0·6 p.u. is to be connected to node 2. Obtain the new nodal impedance matrix by using Kron's formula and verify the result by writing down the new admittance matrix and inverting it.

SOLUTION

From the solution to Worked example 8.3, the inverse of the original admittance matrix is

$$[\mathbf{Y}]^{-1} = \frac{\text{j}}{600}\begin{bmatrix} 33\cdot5 & 20 \\ 20 & 29\cdot92 \end{bmatrix} = \text{j}\begin{bmatrix} 0\cdot0558 & 0\cdot0333 \\ 0\cdot0333 & 0\cdot0499 \end{bmatrix}$$

From [8.57] the matrix to be subtracted from $[\mathbf{Y}]^{-1}$ to give $\mathbf{Y}_m^{-1}$ is

$$\frac{\left(\dfrac{\text{j}}{600}\right)^2\begin{bmatrix} 20 \\ 29\cdot92 \end{bmatrix}\begin{bmatrix} 20 & 29\cdot92 \end{bmatrix}}{\text{j}0\cdot6+\dfrac{\text{j}29\cdot92}{600}}$$

$$= \text{j}\begin{bmatrix} 0\cdot00176 & 0\cdot00264 \\ 0\cdot00264 & 0\cdot00394 \end{bmatrix}$$

so that subtracting this from $[\mathbf{Y}]^{-1}$ gives

$$[\mathbf{Y}_m]^{-1} = j \begin{bmatrix} 0 \cdot 05404 & 0 \cdot 03066 \\ 0 \cdot 03066 & 0 \cdot 04596 \end{bmatrix}$$

(It should be noted that these values were obtained with a slide rule and the student should observe that this is not adequate to deal with such small differences which are typical of nodal voltage analysis, but which are no problem at all when a digital computer is used).

8.7 Iterative methods

Iterative methods are classed as those which solve [8.20] and [8.29] together, so as to obtain the voltages by successive approximation. This is in contrast to direct methods (discussed in section 8.5), where the equations which are solved together are [8.32] and [8.29].

It would be beyond the scope of this book to deal with more than two of the iterative methods which are available, and the basic principles of the Gauss–Seidel and Newton–Raphson methods have been selected to illustrate iterative methods. There are improvements and variations of these methods as there are of other iterative methods, but again these are beyond the present scope. For a Bibliography on these the reader is referred to Laughton and Humphrey Davies, 1964 and to Stagg and El-Abiad, 1969.

8.7.1 GAUSS–SEIDEL METHOD

At any node r where the complex power $\mathbf{S}_r$ entering or leaving a network is specified from the start (see section 8.4), then using [8.31] (which was obtained from $[\mathbf{I}] = [\mathbf{Y}][\mathbf{V}]$ and $\mathbf{S}_r = \mathbf{V}_r \mathbf{I}_r^*$) it follows that the final solution to the busbar voltages must conform with

$$\sum_{k=1}^{n} \mathbf{Y}_{rk}^* \mathbf{V}_k^* = \frac{\mathbf{S}_r}{\mathbf{V}_r} \qquad [8.60]$$

where n is the total number of nodes apart from the reference node. The set of equations similar to [8.60] for all nodes at which the complex power has been specified for the given load flow problem may be written in matrix form as

$$[\mathbf{Y}^*][\mathbf{V}^*] = [\mathbf{B}] \qquad [8.61]$$

which, provided that $[Y^*]$ is non-singular may be written as

$$[V^*] = [Y^*]^{-1}[B]$$

$[Y^*]$ may be written as the algebraic sum of three submatrices

$$[Y^*] = [D] - [L] - [T]$$

where $[D]$ is the diagonal matrix of $[Y^*]$ which may be assumed to be composed of finite complex elements, and $[L]$ and $[T]$ are its lower and upper triangular matrices (with the signs of the elements changed).

[8.61] now becomes

$$[D][V^*] = ([L]+[T])[V^*]+[B] \qquad [8.62]$$

so that

$$[V^*] = [D]^{-1}[L][V^*]+[D]^{-1}[T][V^*]+[D]^{-1}[B] \qquad [8.63]$$

When the Gauss–Seidel method (or method of successive displacements) is applied to [8.63] it becomes

$$[V^*]^{p+1} = [D]^{-1}[L][V^*]^{p+1}+[D]^{-1}[T][V^*]^{p}+[D]^{-1}[B]$$
$$[8.64]$$

where the superscript p denotes the iteration cycle number and not a power, with elements of $[V^*]^{p+1}$ replacing elements of $[V^*]^{p}$ obtained in the previous cycle as soon as they have been calculated, and with one iteration cycle corresponding to one displacement for every element. (This may be seen more clearly after the relevant equations [8.68], [8.69] and [8.70] have been obtained below).

If there are three nodes 1, 2 and 3 for each of which the complex power is specified from the beginning, then inspection of [8.23] shows that

$$[D]^{-1} = \frac{1}{Y_{11}^* Y_{22}^* Y_{33}^*} \begin{bmatrix} Y_{22}^* Y_{33}^* & 0 & 0 \\ 0 & Y_{11}^* Y_{33}^* & 0 \\ 0 & 0 & Y_{11}^* Y_{22}^* \end{bmatrix}$$

$$= \begin{bmatrix} 1/Y_{11}^* & 0 & 0 \\ 0 & 1/Y_{22}^* & 0 \\ 0 & 0 & 1/Y_{33}^* \end{bmatrix} \qquad [8.65]$$

$$[D]^{-1}[L] = \begin{bmatrix} 1/Y_{11}^* & 0 & 0 \\ 0 & 1/Y_{22}^* & 0 \\ 0 & 0 & 1/Y_{33}^* \end{bmatrix} \begin{bmatrix} 0 & 0 & 0 \\ -Y_{21}^* & 0 & 0 \\ -Y_{31}^* & -Y_{32}^* & 0 \end{bmatrix}$$

$$[D]^{-1}[L] = \begin{bmatrix} 0 & 0 & 0 \\ -Y_{21}^*/Y_{22}^* & 0 & 0 \\ -Y_{31}^*/Y_{33}^* & -Y_{32}^*/Y_{33}^* & 0 \end{bmatrix} \qquad [8.66]$$

similarly

$$[D]^{-1}[T] = \begin{bmatrix} 0 & -Y_{12}^*/Y_{11}^* & -Y_{13}^*/Y_{11}^* \\ 0 & 0 & -Y_{23}^*/Y_{22}^* \\ 0 & 0 & 0 \end{bmatrix} \qquad [8.67]$$

and the general form of each term of $[D]^{-1}[B] = S_r/Y_{rr}^*V_r$. Substituting this result, together with those of [8.65], [8.66] and [8.67] into [8.64] gives

$$V_1^{*p+1} = -\frac{Y_{12}^*}{Y_{11}^*} V_2^{*p} - \frac{Y_{13}^*}{Y_{11}^*} V_3^{*p} + \frac{S_1}{Y_{11}^*} \frac{1}{V_1^p} \qquad [8.68]$$

$$V_2^{*p+1} = -\frac{Y_{21}^*}{Y_{22}^*} V_1^{*p+1} - \frac{Y_{23}^*}{Y_{22}^*} V_3^{*p} + \frac{S_2}{Y_{22}^*} \frac{1}{V_2^p} \qquad [8.69]$$

$$V_3^{*p+1} = -\frac{Y_{31}^*}{Y_{33}^*} V_1^{*p+1} - \frac{Y_{32}^*}{Y_{33}^*} V_2^{*p+1} + \frac{S_3}{Y_{33}^*} \frac{1}{V_3^p} \qquad [8.70]$$

Inspection of [8.68], [8.69] and [8.70] shows how new values obtained in one equation are then used at once in the next; e.g. [8.69] uses the new value of V_1^{*p+1} obtained from [8.68]; [8.70] uses V_2^{*p+1} from [8.69] and so on through a complete iteration cycle which in this simple case is a correction to each of only t hree voltages).

[8.68], [8.69] and [8.70] apply to those nodes at which the complex power is specified at the outset, and these may be nodes at which loads are connected.

For a node g (probably a generator node) at which $|V_g|$ and P_g are specified (possibly with limits placed to Q_g) then $P_g + jQ_g$ may be written in place of S_g in the appropriate equation; e.g. if in the above three-node example, node 2 is a generator bus with $|V_2|$ and P_2 specified and the unknown Q_2 is to be determined then

$$V_2^{*p+1} = \frac{-Y_{21}^*}{Y_{22}^*} V_1^{*p+1} - \frac{Y_{23}^*}{Y_{22}^*} V_3^{*p} + \frac{P_2}{Y_{22}^*} \frac{1}{V_2^p} + \frac{jQ_2}{Y_{22}^*} \frac{1}{V_2^p}$$

[8.71]

Writing $(V_2^{*p+1})'$ for the sum of the first three terms on the right hand side of [8.71] gives

$$V_2^{*p+1} = (V_2^{*p+1})' + j \frac{Q_2}{Y_{22}^*} \frac{1}{V_2^p}$$

In general, therefore, for a node g with $|V_g|$ and P_g specified

$$V_g^{*p+1} = (V_g^{*p+1})' + j \frac{Q_g}{Y_{gg}^*} \frac{1}{V_g^p}$$

[8.72]

One possible approximation which may be employed because Q_g is not specified, is to add the second term on the right hand side of [8.72] containing an estimated value of Q_g to the first term as though they were in phase, and then to adjust the magnitudes of these two components of V_g^{*p+1} by multiplying them by a ratio which makes $|V_g^{*p+1}| = |V_g^*|$. Thus the magnitudes of the voltages at these generator busbars are readjusted to the correct specified values in each iteration cycle.

Since both the magnitude and phase angle of the voltage at the floating bus is specified at the start, no equation is required for this, and the iteration cycle only deals in turn with each of the other voltages, while the floating bus voltage is maintained constant in magnitude and in phase in each equation in which it appears.

8.7.2 NEWTON–RAPHSON METHOD

If $\bar{S}_r = \bar{P}_r + j\bar{Q}_r$ is the complex power at any node r specified from the outset of a load flow study, the discrepancy between this and the complex power estimated at any stage of the iteration process may be written from [8.60] as

$$\Delta S_r = \bar{S}_r - V_r \sum_{k=1}^{n} Y_{rk}^* V_k^*$$

[8.73]

If $\mathbf{V}_r = e_r + jf_r$ and at an iteration cycle numbered p

$$\Delta S_r^p = \Delta P_r^p + j\Delta Q_r^p$$

then by using Taylor's theorem

$$\Delta S_r^p = \left(\frac{\partial P_r}{\partial e_r} + j\frac{\partial Q_r}{\partial e_r}\right)^p \Delta e_r^p + \left(\frac{\partial P_r}{\partial f_r} + j\frac{\partial Q_r}{\partial f_r}\right)^p \Delta f_r^p \qquad [8.74]$$

There are n busbars, at $(n-1)$ of which the real power $\bar{P}_r$ is specified and the reactive power $\bar{Q}_r$ is specified either definitely or as a maximum value, but where e_r and f_r are unknown. At the nth busbar which is the floating busbar, both of the voltage components are specified but the complex power is unknown. Thus in the Newton–Raphson method $2(n-1)$ equations of the type of [8.74] are to be solved and these may be written in the form

$$
\begin{bmatrix}
\Delta P_1 \\
\vdots \\
\Delta P_{n-1} \\
\hline
\Delta Q_1 \\
\vdots \\
\Delta Q_{n-1}
\end{bmatrix}
$$

$$
=
\left[
\begin{array}{ccc|ccc}
\partial P_1/\partial e_1 & \cdots\cdots & \partial P_1/\partial e_{n-1} & \partial P_1/\partial f_1 & \cdots\cdots & \partial P_1/\partial f_{n-1} \\
\cdots\cdots\cdots\cdots & & \cdots\cdots\cdots\cdots & \cdots\cdots\cdots\cdots & & \cdots\cdots\cdots\cdots \\
\partial P_{n-1}/\partial e_1 & \cdots & \partial P_{n-1}/\partial e_{n-1} & \partial P_{n-1}/\partial f_1 & \cdots & \partial P_{n-1}/\partial f_{n-1} \\
\hline
\partial Q_1/\partial e_1 & \cdots\cdots & \partial Q_1/\partial e_{n-1} & \partial Q_1/\partial f_1 & \cdots\cdots & \partial Q_1/\partial f_{n-1} \\
\cdots\cdots\cdots\cdots & & \cdots\cdots\cdots\cdots & \cdots\cdots\cdots\cdots & & \cdots\cdots\cdots\cdots \\
\partial Q_{n-1}/\partial e_1 & \cdots & \partial Q_{n-1}/\partial e_{n-1} & \partial Q_{n-1}/\partial f_1 & \cdots & \partial Q_{n-1}/\partial f_{n-1}
\end{array}
\right]
$$

$$
\times
\begin{bmatrix}
\Delta e_1 \\
\vdots \\
\Delta e_{n-1} \\
\hline
\Delta f_1 \\
\vdots \\
\Delta f_{n-1}
\end{bmatrix}
\qquad [8.75]
$$

or more briefly as

$$\left[\frac{\Delta \mathbf{P}}{\Delta \mathbf{Q}}\right] = \left[\frac{\mathbf{J}_1 | \mathbf{J}_2}{\mathbf{J}_3 | \mathbf{J}_4}\right]\left[\frac{\Delta \mathbf{e}}{\Delta \mathbf{f}}\right]$$

The elements in the matrix of coefficients (known as the Jacobian) in [8.75] may be determined from the busbar power equations. It is sufficient for this brief outline to illustrate this for the elements in $[\mathbf{J}_1]$ only. For more detailed treatment and references to load flow studies using this method, the reader is referred to Laughton and Humphrey Davies, 1964, and to Stagg and El-Abiad, 1968.

Writing $Y_{rk} = G_{rk} + jB_{rk}$, then the power flow at a busbar r at any stage of the iteration process which when ΔS_r has been reduced to zero will finally equal $\bar{P}_r + j\bar{Q}_r$, may be written from [8.60] as

$$P_r + jQ_r = (e_r + jf_r) \sum_{k=1}^{n} (G_{rk} - jB_{rk})(e_k - jf_k)$$

so that

$$P_r = \sum_{k=1}^{n} \{e_r(e_k G_{rk} - f_k B_{rk}) + f_r(e_k B_{rk} + f_k G_{rk})\} \qquad [8.76]$$

$$Q_r = \sum_{k=1}^{n} \{f_r(e_k G_{rk} - f_k B_{rk}) - e_r(e_k B_{rk} + f_k G_{rk})\} \qquad [8.77]$$

From [8.76]

$$P_r = e_r(e_r G_{rr} - f_r B_{rr}) + f_r(e_r B_{rr} + f_r G_{rr})$$

$$+ \sum_{\substack{k=1 \\ k \neq r}}^{n} \{e_r(e_k G_{rk} - f_k B_{rk}) + f_r(e_k B_{rk} + f_k G_{rk})\}$$

The non-diagonal elements of $[\mathbf{J}_1]$ for $k \neq r$ are

$$\frac{\partial P_r}{\partial e_k} = e_r G_{rk} + f_r B_{rk} \qquad [8.78]$$

The diagonal elements of $[\mathbf{J}_1]$ are

$$\frac{\partial P_r}{\partial e_r} = 2e_r G_{rr} - f_r B_{rr} + f_r B_{rr} + \sum_{\substack{k=1 \\ k \neq r}}^{n} (e_k G_{rk} - f_k B_{rk})$$

This expression may be simplified by substitution of the real part c_r of the current at busbar r

$$\mathbf{I}_r = c_r + jd_r = (G_{rr} + jB_{rr})(e_r + jf_r) + \sum_{\substack{k=1 \\ k \neq r}}^{n} (G_{rk} + jB_{rk})(e_k + jf_k)$$

$$[8.79]$$

so that

$$\frac{\partial P_r}{\partial e_r} = e_r G_{rr} + f_r B_{rr} + c_r \qquad [8.80]$$

Similarly the elements in $[\mathbf{J}_2]$, $[\mathbf{J}_3]$ and $[\mathbf{J}_4]$ may be obtained.

There are several variants of the Newton–Raphson method but it may broadly be outlined by the following steps:

(a) Assign initial estimated values to the busbar voltages

(b) Compute P_r and Q_r for each busbar from [8.76] and [8.77]

(c) Compute ΔQ_r and ΔP_r from [8.73]

(d) Compute the elements of the matrix of coefficients of [8.75] by means of such equations as [8.78] and [8.80] using the components of the busbar currents from [8,79] (using the estimated voltages and calculated powers) or otherwise

(e) Obtain Δe_r and Δf_r from [8.75]

(f) Using these values, modify the voltages at all busbars and begin the next iteration cycle at step (b)

(g) Continue until ΔP_r^p and ΔQ_r^p for all busbars are within a specified very small tolerance.

8.7.3 CONVERGENCE AND ACCELERATION FACTORS

Computer time can be saved if means are employed to reduce the number of iteration cycles taken to obtain the desired convergence, and many methods have been developed for accelerating the iteration procedure so as to obtain faster convergence.

One common method of acceleration is to multiply both the real and the imaginary component of the difference in the voltage between one iteration cycle and the next, by the same acceleration factor. Using an estimated voltage $\mathbf{V}^p$ during the $(p+1)$th iteration cycle of any method such as the Gauss–Seidel method gives a new estimate $(\mathbf{V}^{p+1})'$. If now an acceleration factor ω is used, the voltage used in the next iteration cycle $(p+2)$ will be $(\mathbf{V}^{p+1})$ instead of $(\mathbf{V}^{p+1})'$ where

$$(\mathbf{V}^{p+1}) = \mathbf{V}^p + \omega((\mathbf{V}^{p+1})' - \mathbf{V}^p) \qquad [8.81]$$

ω has so far been limited to real values and may be about 1.6.

If the Gauss–Seidel method (section 8.7.1) is accelerated then convergence of equations such as [8.68], [8.69] and [8.70], can be accelerated by applying [8.64] and [8.81] which gives

$$[\mathbf{V}^*]^{p+1} = [\mathbf{V}^*]^p + \omega([\mathbf{D}]^{-1}[\mathbf{L}][\mathbf{V}^*]^{p+1} + [\mathbf{D}]^{-1}[\mathbf{T}][\mathbf{V}^*]^p$$
$$+ [\mathbf{D}]^{-1}[\mathbf{B}] - [\mathbf{V}^*]^p) \qquad [8.82]$$

Writing [8.82] in the form

$$[\mathbf{V}^*]^{p+1} = [\mathbf{H}][\mathbf{V}^*]^p + [\mathbf{M}][\mathbf{D}]^{-1}[\mathbf{B}]$$

it can be shown (Laughton and Humphrey Davies, 1964) that the optimum acceleration factor may be determined in terms of the eigenvalues (see Appendix A.1.10) of the iteration matrix [H].

8.7.4 HYBRID METHODS

If the nodal admittance equation [8.20] for the network is partitioned to give

$$\begin{bmatrix} \mathbf{I}_a \\ -- \\ \mathbf{I}_b \end{bmatrix} = \begin{bmatrix} \mathbf{Y}_{aa} & \mathbf{Y}_{ab} \\ -- & -- \\ \mathbf{Y}_{ba} & \mathbf{Y}_{bb} \end{bmatrix} \begin{bmatrix} \mathbf{V}_a \\ -- \\ \mathbf{V}_b \end{bmatrix} \qquad [8.83]$$

then

$$\mathbf{I}_a = \mathbf{Y}_{aa}\mathbf{V}_a + \mathbf{Y}_{ab}\mathbf{V}_b \qquad [8.84]$$

$$\mathbf{I}_b = \mathbf{Y}_{ba}\mathbf{V}_a + \mathbf{Y}_{bb}\mathbf{V}_b \qquad [8.85]$$

and from [8.85]

$$\mathbf{V}_b = \mathbf{Y}_{bb}^{-1}\mathbf{I}_b - \mathbf{Y}_{bb}^{-1}\mathbf{Y}_{ba}\mathbf{V}_a \qquad [8.86]$$

Substituting this in [8.84] gives together with [8.86]

$$\begin{bmatrix} \mathbf{I}_a \\ -- \\ \mathbf{V}_b \end{bmatrix} = \begin{bmatrix} (\mathbf{Y}_{aa} - \mathbf{Y}_{ab}\mathbf{Y}_{bb}^{-1}\mathbf{Y}_{ba}) & \mathbf{Y}_{ab}\mathbf{Y}_{bb}^{-1} \\ ------- & ---- \\ -\mathbf{Y}_{bb}^{-1}\mathbf{Y}_{ba} & \mathbf{Y}_{bb}^{-1} \end{bmatrix} \begin{bmatrix} \mathbf{V}_a \\ -- \\ \mathbf{I}_b \end{bmatrix} \qquad [8.87]$$

The matrix in [8.87] relating the system currents and voltages is a mixed one composed of impedance elements, admittance elements and dimensionless elements. In this way it is possible to use a hybrid method, i.e. one combining direct methods to form the inverse matrix Y_{bb}^{-1}, followed by an iterative method to obtain the solution.

8.8 Tearing

As was mentioned at the end of the introductory section 8.4, direct methods may be at a disadvantage compared with iterative methods for solving equations of large power networks. This is because of the time taken to invert the large admittance matrix and the large storage space required, so that the size and speed of the computer are vital to the choice of method. In order to reduce computer time and storage so that apart from the advantage of these reductions, it becomes possible to solve problems with more equations without faster and larger computers, the methods of diakoptics or tearing developed by Kron may be employed. The use of tearing in solving power system problems has been examined in recent years (e.g. Brammeller, 1964 and Day and Parton, 1965).

Tearing consists essentially of dividing the power system into a number of component networks and inverting the admittance matrices of these smaller or sub-networks separately, and then obtaining the solution to the whole system by using connection matrices. Tearing is not an alternative to other methods of solution but is an extension which may be used when these methods become unsatisfactory due to the limitations of the computers which are available. Because many power systems consist of load and generation centres interconnected by relatively few transmission lines, they are suited to the technique of tearing where the number of cut branches should be much less than the total number of system nodes, as is indicated by the following comparison of computation times. Since the time taken to invert a matrix is approximately proportional to n^3 where n is the number of nodes, then tearing the network into N subnetworks involves N inversions, each taking a time which is about $1/N^3$ of the time T to invert the complete network matrix. Then if there are m lines which are cut to form the N subnetworks the time t taken for inversion using tearing is given by Brameller, 1964 as

$$t \simeq (1/N^2 + m/n)T \qquad [8.88]$$

so that the smaller the number of cut lines relative to the total number of nodes the more economical in inversion time is the technique of tearing. The effect of the term m/n is to make t exceed T/N^2 and it may in some cases be as high as between $2T/N^2$ and $3T/N^2$. The term involving the number of cut lines m appears in [8.88] because new unknowns, viz. the cut line currents, appear at each tear, and the column vector of the currents $[\mathbf{I}]$ must now include these, together with the nodal currents in the equations $[\mathbf{I}] = [\mathbf{Y}][\mathbf{V}]$ and $[\mathbf{V}] = [\mathbf{Y}]^{-1}[\mathbf{I}] = [\mathbf{Z}][\mathbf{I}]$. The column vector of the voltages similarly includes not only the nodal voltages but also the voltages (usually zero) in series with the cut branches.

Since the method of partitioning or factorizing the inverse matrix is, as shown in section 8.5.2, a way of reducing the order of the matrices needing to be inverted in direct methods of solution, there is some similarity between this and tearing. These two methods are therefore used to solve a simple example after the process of tearing has been outlined in section 8.8.1.

8.8.1 SIMPLE APPLICATION OF TEARING

The treatment in this section follows that adopted by Brameller (1964). Fig. 8.7 shows a simple power system with two generating stations interconnected by transmission lines and with loads at various places. Fig. 8.8 shows the same system with the corresponding branch admittances and the currents injected into the system at nodes 1 and 4 marked in. Fig. 8.8 also shows a jagged line across it indicating where it is being torn into two subnetworks, and in Fig. 8.9 these subnetworks are shown separately, with the branches of the original network through which the cuts were made shown isolated from the subnetworks. Thus there are four primitive networks in Fig. 8.9 for which the nodal admittance matrices can be formed and inverted. The networks of Fig. 8.9 must be related (and are sometimes called the related network) to that of Fig. 8.8 by the condition of power invariance.

A prime will be used to denote quantities in the related or primitive network of Fig. 8.9, as distinct from the corresponding unprimed quantities in the actual power system of Fig. 8.7 or 8.8. The actual nodal currents are $\mathbf{I}_1(= \mathbf{y}_{aa}\mathbf{e}_a)$, $\mathbf{I}_2(= 0)$, $\mathbf{I}_3(= 0)$ and $\mathbf{I}_4(=\mathbf{y}_{jj}\mathbf{e}_j)$. The relationships between the related circuit currents and those of the

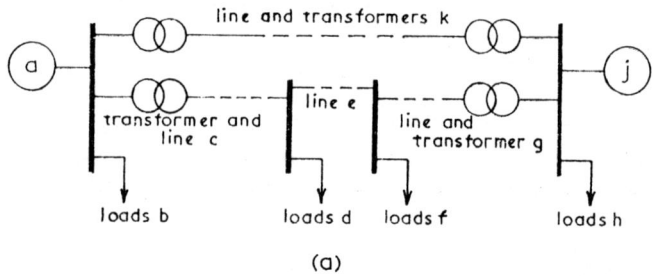

(a)

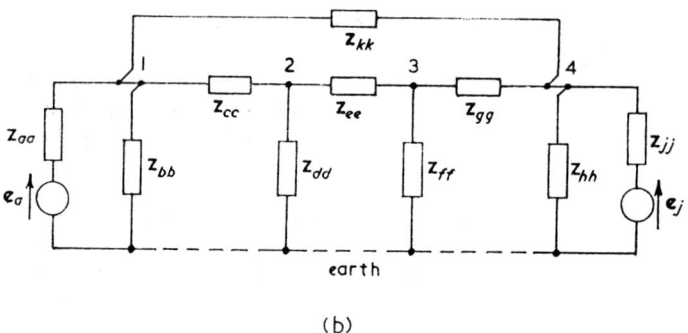

(b)

Fig. 8.7. (a) system
(b) network of voltage sources and per-unit impedances.

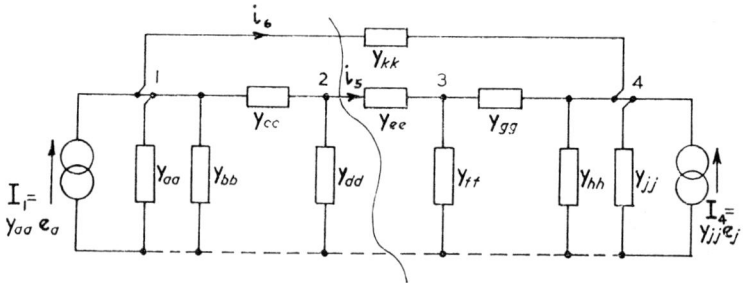

Fig. 8.8. Network of current sources and per-unit admittances.

actual system may be expressed by means of a connection matrix
[C'] as

$$[I'] = [C'][I] \qquad\qquad [8.89]$$

The symbol [C'] has been used here in place of [C] because the
latter has been used in section 7.2.2 and in section 8.3 in nodal
voltage analysis to pre-multiply the nodal voltage matrix to give the
branch voltages of the primitive network. In [8.89] the connection
matrix [C'] has for convenience been used to relate the corresponding

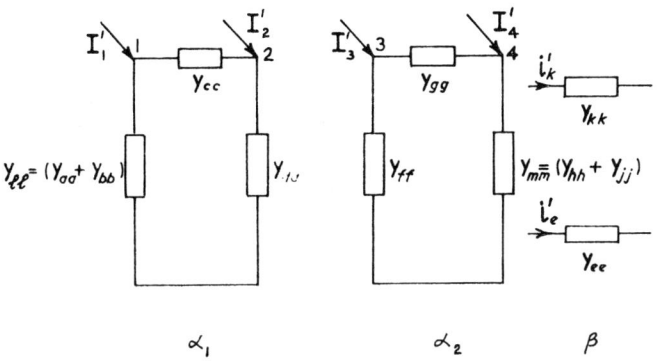

Fig. 8.9. Primitive system.

currents instead of the voltages. Comparison of [8.89] and [7.18]
shows that $[C'] = [C^*]_t^{-1}$ so that

$$[C']^{-1} = [C^*]_t \qquad\qquad [8.90]$$

and

$$[C'^*]_t^{-1} = [C] \qquad\qquad [8.91]$$

The column vector [I] of currents in the actual system of Fig. 8.8
consists of two parts; $[I_\alpha]$ which contains the nodal currents I_1,
$I_2 (= 0)$, $I_3 (= 0)$ and I_4, and $[I_\beta]$ which contains the currents i_5 and
i_6 flowing in the branches of Fig. 8.8 which are then cut to form the
primitive system of Fig. 8.9. Similarly the column vector [I'] of
currents in the primitive system consists of two parts $[I'_\alpha]$ and $[I'_\beta]$.
By inspection of Figs. 8.8 and 8.9 the relationships between these
sets of currents may be written; e.g. equating the current I_1 flowing
towards node 1 in Fig. 8.8 to the current $(I'_1 - i'_k)$ flowing towards

node 1 when the cut branch $\mathbf{y}_{kk}$ is connected to the α_1 network of Fig. 8.9, and using the fact that $\mathbf{i}'_k$ must equal $\mathbf{i}_6$, gives the first of the equations of [8.92], which are the full statement of [8.89].

$$\left.\begin{aligned}
\mathbf{I}'_1 &= \mathbf{I}_1 + \mathbf{i}_6 \\
\mathbf{I}'_2 &= \mathbf{I}_2 + \mathbf{i}_5 \\
\mathbf{I}'_3 &= \mathbf{I}_3 - \mathbf{i}_5 \\
\mathbf{I}'_4 &= \mathbf{I}_4 - \mathbf{i}_6 \\
\mathbf{i}'_e &= \mathbf{i}_5 \\
\mathbf{i}'_k &= \mathbf{i}_6
\end{aligned}\right\} \qquad [8.92]$$

[8.89] and [8.92] may be partitioned so that the currents on the left hand sides of the first four equations of [8.92], which relate to the α_1 and α_2 networks of Fig. 8.9 are the column vector $[\mathbf{I}'_\alpha]$, and similarly the last two equations have currents $[\mathbf{I}'_\beta]$. If

$$\begin{bmatrix} \mathbf{I}_1 \\ \mathbf{I}_2 \\ \mathbf{I}_3 \\ \mathbf{I}_4 \end{bmatrix} = [\mathbf{I}_\alpha] \quad \text{and} \quad \begin{bmatrix} \mathbf{i}_5 \\ \mathbf{i}_6 \end{bmatrix} = [\mathbf{I}_\beta]$$

then [8.89] and [8.92] may be written

$$\begin{bmatrix} \mathbf{I}'_\alpha \\ \hline \mathbf{I}'_\beta \end{bmatrix} = \begin{bmatrix} \mathbf{U} & \mathbf{C}'' \\ \hline 0 & \mathbf{U} \end{bmatrix} \begin{bmatrix} \mathbf{I}_\alpha \\ \hline \mathbf{I}_\beta \end{bmatrix} \qquad [8.93]$$

where $\mathbf{U}$ denotes a unit matrix (Appendix A.1.7). In this case the first unit matrix is of order 4 and the second of order 2 and

$$[\mathbf{C}''] = \begin{bmatrix} 0 & 1 \\ 1 & 0 \\ -1 & 0 \\ 0 & -1 \end{bmatrix}$$

If the column vector of the nodal voltages and the cut-branch voltages in the primitive network of Fig. 8.9 is [V'] then

$$[\mathbf{I}'] = [\mathbf{Y}'][\mathbf{V}']$$ [8.94]

Pre-multiplying both sides of [8.94] by $[\mathbf{Y}']^{-1} = [\mathbf{Z}']$ gives

$$[\mathbf{V}'] = [\mathbf{Z}'][\mathbf{I}']$$ [8.95]

Since the admittance matrix [Y'] consists of the admittance matrices for the separate parts of the primitive network with no mutual coupling between them, these matrices are the diagonal elements of the compound matrix [Y'] with all non-diagonal terms zero.

$$[\mathbf{Y}'] = \begin{bmatrix} \mathbf{Y}'_\alpha & \vdots & 0 \\ -- & -- \\ 0 & \vdots & \mathbf{Y}'_\beta \end{bmatrix} = \begin{bmatrix} \mathbf{Y}'_{\alpha 1} & 0 & \vdots & 0 \\ 0 & \mathbf{Y}'_{\alpha 2} & \vdots & 0 \\ -- & -- & -- \\ 0 & 0 & \vdots & \mathbf{Y}'_\beta \end{bmatrix}$$ [8.96]

Similarly the inverse or impedance matrix [Z'] consists of diagonal elements each of which is the inverse of the corresponding element in [Y'], since each relates to one of the separate parts of the network of Fig. 8.9, i.e. $[\mathbf{Z}'_{\alpha 1}] = [\mathbf{Y}'_{\alpha 1}]^{-1}$, $[\mathbf{Z}'_{\alpha 2}] = [\mathbf{Y}'_{\alpha 2}]^{-1}$ and $[\mathbf{Z}'_\beta] = [\mathbf{Y}'_\beta]^{-1}$ where

$$[\mathbf{Z}'] = \begin{bmatrix} \mathbf{Z}'_\alpha & \vdots & 0 \\ -- & -- \\ 0 & \vdots & \mathbf{Z}'_\beta \end{bmatrix} = \begin{bmatrix} \mathbf{Z}'_{\alpha 1} & 0 & \vdots & 0 \\ 0 & \mathbf{Z}'_{\alpha 2} & \vdots & 0 \\ -- & -- & -- \\ 0 & 0 & \vdots & \mathbf{Z}'_\beta \end{bmatrix}$$ [8.97]

The inverse admittance matrix $[\mathbf{Y}]^{-1} = [\mathbf{Z}]$ for the complete system of Fig. 8.7, together with the cut-branch terms is related to [Z'] by the condition of power invariance. This condition gives rise to a relationship between the nodal admittance matrix [Y] for the whole system plus the cut branches and that of the primitive network [Y'], which is stated in [7.11] in terms of the connection matrix [C]. Substituting the results of [8.90] and [8.91] into [7.11] defines the following relationship in terms of the connection matrix [C'] which is being used here

$$[\mathbf{Y}] = [\mathbf{C}']^{-1}[\mathbf{Y}'][\mathbf{C}'^*]_t^{-1}$$ [8.98]

so that using the rule given in Appendix A.1.8.1

$$[\mathbf{Z}] = [\mathbf{Y}]^{-1} = [\mathbf{C}'^*]_t [\mathbf{Y}']^{-1} [\mathbf{C}']$$

$$[\mathbf{Z}] = [\mathbf{C}'^*]_t [\mathbf{Z}'] [\mathbf{C}'] \qquad [8.99]$$

Substituting for $[\mathbf{C}']$ from [8.89] and [8.93], and for $[\mathbf{Z}']$ from [8.97], into [8.99] and omitting the conjugate signs for convenience since $[\mathbf{C}']$ will normally contain only real terms.

$$[\mathbf{Z}] = \begin{bmatrix} 1 & \vdots & 0 \\ ---- \\ \mathbf{C}_t'' & \vdots & 1 \end{bmatrix} \begin{bmatrix} \mathbf{Z}_\alpha' & \vdots & 0 \\ ---- \\ 0 & \vdots & \mathbf{Z}_\beta' \end{bmatrix} \begin{bmatrix} 1 & \vdots & \mathbf{C}'' \\ ---- \\ 0 & \vdots & 1 \end{bmatrix}$$

$$[\mathbf{Z}] = \begin{bmatrix} \mathbf{Z}_\alpha' & \vdots & \mathbf{Z}_\alpha' \mathbf{C}'' \\ ----------- \\ \mathbf{C}_t'' \mathbf{Z}_\alpha' & \vdots & (\mathbf{C}_t'' \mathbf{Z}_\alpha' \mathbf{C}'' + \mathbf{Z}_\beta') \end{bmatrix} \qquad [8.100]$$

The nodal impedance matrix $[\mathbf{Z}]$ for the whole network of Fig. 8.7, together with the cut-branch terms, partitioned as shown in [8.100] can have its submatrices designated as $[\mathbf{Z}_1]$, $[\mathbf{Z}_2]$, $[\mathbf{Z}_3]$ and $[\mathbf{Z}_4]$ so that

$$\begin{bmatrix} \mathbf{V}_\alpha \\ -- \\ \mathbf{V}_\beta \end{bmatrix} = \begin{bmatrix} \mathbf{Z}_1 & \vdots & \mathbf{Z}_2 \\ ---- \\ \mathbf{Z}_3 & \vdots & \mathbf{Z}_4 \end{bmatrix} \begin{bmatrix} \mathbf{I}_\alpha \\ -- \\ \mathbf{I}_\beta \end{bmatrix} \qquad [8.101]$$

where $[\mathbf{V}_\alpha]$ is a column vector of the nodal voltages $[\mathbf{V}_1]$, $[\mathbf{V}_2]$, $[\mathbf{V}_3]$ and $[\mathbf{V}_4]$ of the circuit of Fig. 8.7, and $[\mathbf{V}_\beta] = 0$ since it is a column vector of the voltages in series with the cut branches and these are both zero. Expanding [8.101] and temporarily omitting the square brackets around matrices for convenience gives

$$\mathbf{V}_\alpha = \mathbf{Z}_1 \mathbf{I}_\alpha + \mathbf{Z}_2 \mathbf{I}_\beta \qquad [8.102]$$

$$0 = \mathbf{Z}_3 \mathbf{I}_\alpha + \mathbf{Z}_4 \mathbf{I}_\beta \qquad [8.103]$$

From [8.103]

$$\mathbf{I}_\beta = -\mathbf{Z}_4^{-1} \mathbf{Z}_3 \mathbf{I}_\alpha$$

and substituting this value in [8.102] gives

$$\mathbf{V}_\alpha = (\mathbf{Z}_1 - \mathbf{Z}_2 \mathbf{Z}_4^{-1} \mathbf{Z}_3) \mathbf{I}_\alpha \qquad [8.104]$$

Comparing [8.100] and [8.101] and substituting the values of $\mathbf{Z}_1$, $\mathbf{Z}_2$, etc. in [8.104] gives

$$\mathbf{V}_\alpha = \{\mathbf{Z}'_\alpha - \mathbf{Z}'_\alpha \mathbf{C}''(\mathbf{C}''_t \mathbf{Z}'_\alpha \mathbf{C}'' + \mathbf{Z}'_\beta)^{-1} \mathbf{C}''_t \mathbf{Z}'_\alpha\}\mathbf{I}_\alpha \qquad [8.105]$$

Since for the circuit of Fig. 8.7 the nodal currents and voltages are $[\mathbf{I}_\alpha]$ and $[\mathbf{V}_\alpha]$, the required nodal impedance matrix is that given by [8.105].

An example will now be worked through directly, i.e. by matrix inversion, using two methods which reduce the order of the matrices to be inverted, (a) tearing and (b) partitioning. Reference to the solutions of this example will illustrate some differences between these two methods, even though the usual difficulty arises here that methods of analysis cannot be properly illustrated by examples which are relatively trivial and which could more easily be solved by simpler means. Partitioning, though reducing the time of computation, requires the full nodal admittance matrix, which has relatively few zeros, and does not reduce the demand on computer storage, whereas the compound nodal admittance and impedance matrices used in tearing contain many zeros, so that storage demand as well as time is reduced. In partitioning, matrix multiplication occurs many times while in tearing the multiplications by $[\mathbf{C}'']$ or $[\mathbf{C}'']_t$ are only additions since their elements are ± 1. The inversion involved in the two methods is similar, e.g. in the following example four inversions of 2×2 matrices occur in each method.

Worked Example 8.5

In the system of Fig. 8.7 the per-unit nodal currents are $\mathbf{I}_1 = 2{\cdot}0$, $\mathbf{I}_2 = 0$, $\mathbf{I}_3 = 0$ and $\mathbf{I}_4 = 3{\cdot}0$. The per-unit admittances (to a common base) are $y_{aa} = 0{\cdot}5$, $y_{bb} = 1{\cdot}5$, $y_{cc} = 2{\cdot}5$, $y_{dd} = 0{\cdot}5$, $y_{ee} = 5{\cdot}0$, $y_{ff} = 0{\cdot}25$, $y_{gg} = 4{\cdot}0$, $y_{hh} = 0{\cdot}75$, $y_{jj} = 0{\cdot}25$ and $y_{kk} = 4{\cdot}0$. (The operator $-j$ has been omitted for convenience). Calculate the voltages at nodes 1, 2, 3 and 4, (a) by tearing and (b) by partitioning.

Solution

(a) *By tearing*

The nodal current matrix for the system of Fig. 8.7 is designated $[\mathbf{I}_\alpha]$ (see [8.93]), and this is partitioned to correspond with tearing the network into the two component networks α_1 and α_2 as in Fig. 8.9. Thus

$$[I_\alpha] = \begin{bmatrix} I_{\alpha 1} \\ -- \\ I_{\alpha 2} \end{bmatrix} = \begin{bmatrix} 2 \\ 0 \\ - \\ 0 \\ 3 \end{bmatrix}$$

The connection matrix $[C'']$ (see $[8.93]$) is

$$[C''] = \begin{bmatrix} 0 & 1 \\ 1 & 0 \\ ---- \\ -1 & 0 \\ 0 & -1 \end{bmatrix}$$

so that

$$[C'']_t = \begin{bmatrix} 0 & 1 & : -1 & 0 \\ 1 & 0 & : 0 & -1 \end{bmatrix}$$

The compound matrix $[Y'_\alpha]$ for the separate networks of Fig. 8.9 into which the original network has been torn is

$$[Y'_\alpha] = \begin{bmatrix} Y'_{\alpha 1} & 0 \\ 0 & Y'_{\alpha 2} \end{bmatrix} = \begin{bmatrix} 4 \cdot 5 & -2 \cdot 5 & 0 & 0 \\ -2 \cdot 5 & 3 & 0 & 0 \\ ----- & & ----- \\ 0 & 0 & 4 \cdot 25 & -4 \\ 0 & 0 & -4 & 5 \end{bmatrix}$$

The admittance matrix for the cut branches is

$$[Y'_\beta] = \begin{bmatrix} 5 & 0 \\ 0 & 4 \end{bmatrix}$$

The inverse admittance matrices for the separate component net-works and the cut branches $[\mathbf{Z}'_{\alpha 1}]$, $[\mathbf{Z}'_{\alpha 2}]$ and $[\mathbf{Z}'_{\beta}]$ are given by

$$[\mathbf{Z}'_{\alpha 1}] = [\mathbf{Y}'_{\alpha 1}]^{-1} = \frac{1}{7 \cdot 25} \begin{bmatrix} 3 & 2 \cdot 5 \\ 2 \cdot 5 & 4 \cdot 5 \end{bmatrix} = \begin{bmatrix} 0 \cdot 415 & 0 \cdot 345 \\ 0 \cdot 345 & 0 \cdot 621 \end{bmatrix}$$

$$[\mathbf{Z}'_{\alpha 2}] = [\mathbf{Y}'_{\alpha 2}]^{-1} = \frac{1}{5 \cdot 25} \begin{bmatrix} 5 & 4 \\ 4 & 4 \cdot 25 \end{bmatrix} = \begin{bmatrix} 0 \cdot 952 & 0 \cdot 762 \\ 0 \cdot 762 & 0 \cdot 809 \end{bmatrix}$$

$$[\mathbf{Z}'_{\beta}] = [\mathbf{Y}'_{\beta}]^{-1} = \frac{1}{20} \begin{bmatrix} 4 & 0 \\ 0 & 5 \end{bmatrix} = \begin{bmatrix} 0 \cdot 2 & 0 \\ 0 & 0 \cdot 25 \end{bmatrix}$$

$$[\mathbf{Z}'_{\alpha}] = \begin{bmatrix} \mathbf{Z}'_{\alpha 1} & 0 \\ 0 & \mathbf{Z}'_{\alpha 2} \end{bmatrix} = \begin{bmatrix} 0 \cdot 415 & 0 \cdot 345 & 0 & 0 \\ 0 \cdot 345 & 0 \cdot 621 & 0 & 0 \\ 0 & 0 & 0 \cdot 952 & 0 \cdot 762 \\ 0 & 0 & 0 \cdot 762 & 0 \cdot 809 \end{bmatrix}$$

$$[\mathbf{Z}'_{\alpha}][\mathbf{C}''] = \begin{bmatrix} 0 \cdot 345 & 0 \cdot 415 \\ 0 \cdot 621 & 0 \cdot 345 \\ -0 \cdot 952 & -0 \cdot 762 \\ -0 \cdot 762 & -0 \cdot 809 \end{bmatrix}$$

$$[\mathbf{C}'']_t[\mathbf{Z}'_{\alpha}][\mathbf{C}''] = \begin{bmatrix} 1 \cdot 573 & 1 \cdot 107 \\ 1 \cdot 107 & 1 \cdot 224 \end{bmatrix}$$

$$[\mathbf{C}'']_t[\mathbf{Z}'_{\alpha}][\mathbf{C}''] + [\mathbf{Z}'_{\beta}] = \begin{bmatrix} 1 \cdot 773 & 1 \cdot 107 \\ 1 \cdot 107 & 1 \cdot 474 \end{bmatrix}$$

$$([\mathbf{C}'']_t[\mathbf{Z}'_{\alpha}][\mathbf{C}''] + [\mathbf{Z}'_{\beta}])^{-1} = \frac{1}{1 \cdot 38} \begin{bmatrix} 1 \cdot 474 & -1 \cdot 107 \\ -1 \cdot 107 & 1 \cdot 773 \end{bmatrix}$$

$$= \begin{bmatrix} 1 \cdot 07 & -0 \cdot 803 \\ -0 \cdot 803 & 1 \cdot 29 \end{bmatrix}$$

$$[Z'_\alpha][C'']([C'']_t[Z'_\alpha][C'']+[Z'_\beta])^{-1} = \begin{bmatrix} 0\cdot036 & 0\cdot258 \\ 0\cdot387 & -0\cdot055 \\ -0\cdot41 & -0\cdot219 \\ -0\cdot167 & -0\cdot427 \end{bmatrix}$$

$$[C'']_t[Z'_\alpha] = \begin{bmatrix} 0\cdot345 & 0\cdot621 & -0\cdot952 & -0\cdot762 \\ 0\cdot415 & 0\cdot345 & -0\cdot762 & -0\cdot809 \end{bmatrix}$$

$$[Z'_\alpha][C'']([C'']_t[Z'_\alpha][C'']+[Z'_\beta])^{-1}[C'']_t[Z'_\alpha]$$

$$= \begin{bmatrix} 0\cdot119 & 0\cdot11 & -0\cdot231 & -0\cdot235 \\ 0\cdot111 & 0\cdot221 & -0\cdot326 & -0\cdot250 \\ -0\cdot232 & -0\cdot329 & 0\cdot557 & 0\cdot489 \\ -0\cdot235 & -0\cdot251 & 0\cdot485 & 0\cdot472 \end{bmatrix}$$

and subtracting this impedance matrix from $[Z'_\alpha]$ (see [8.105]) gives the nodal impedance matrix for the network of Fig. 8.7, which relates the required nodal voltages to the given nodal currents as follows

$$\begin{bmatrix} V_1 \\ V_2 \\ V_3 \\ V_4 \end{bmatrix} = \begin{bmatrix} 0\cdot296 & 0\cdot235 & 0\cdot231 & 0\cdot235 \\ 0\cdot234 & 0\cdot400 & 0\cdot326 & 0\cdot250 \\ 0\cdot232 & 0\cdot329 & 0\cdot395 & 0\cdot273 \\ 0\cdot235 & 0\cdot251 & 0\cdot277 & 0\cdot337 \end{bmatrix} \begin{bmatrix} 2 \\ 0 \\ 0 \\ 3 \end{bmatrix}$$

from which the nodal voltages are

$V_1 = 1\cdot297$ p.u., $V_2 = 1\cdot218$ p.u., $V_3 = 1\cdot283$ p.u., and $V_4 = 1\cdot481$ p.u.

Solution

(b) By partitioning

For the network of Fig. 8.7 with four nodes, it is convenient to partition in such a way that the submatrices of the four by four admittance matrix are both two by two, so that by inspection [8.40] may be written as

$$[\mathbf{I}] = \begin{bmatrix} 2 \\ 0 \\ -- \\ 0 \\ 3 \end{bmatrix} = \begin{bmatrix} 8\cdot5 & -2\cdot5 & 0 & -4 \\ -2\cdot5 & 8 & -5 & 0 \\ ---- & ---- & ---- & ---- \\ 0 & -5 & 9\cdot25 & -4 \\ -4 & 0 & -4 & 9 \end{bmatrix} \begin{bmatrix} \mathbf{V}_1 \\ \mathbf{V}_2 \\ -- \\ \mathbf{V}_3 \\ \mathbf{V}_4 \end{bmatrix}$$

The solution may be written in partitioned form as

$$\begin{bmatrix} \mathbf{V}_1 \\ \mathbf{V}_2 \\ -- \\ \mathbf{V}_3 \\ \mathbf{V}_4 \end{bmatrix} = \begin{bmatrix} \mathbf{Z}_1 & \mathbf{Z}_2 \\ -- & -- \\ \mathbf{Z}_3 & \mathbf{Z}_4 \end{bmatrix} \begin{bmatrix} 2 \\ 0 \\ - \\ 0 \\ 3 \end{bmatrix}$$

where $\mathbf{Z}_1$, $\mathbf{Z}_2$, $\mathbf{Z}_3$ and $\mathbf{Z}_4$ are defined in terms of the four submatrices in the nodal admittance matrix by [8.51].

$[\mathbf{Z}_1] = ([\mathbf{Y}_1] - [\mathbf{Y}_2][\mathbf{Y}_4]^{-1}[\mathbf{Y}_3])^{-1}$ will be evaluated first.

$$[\mathbf{Y}_4]^{-1} = \frac{1}{67\cdot25} \begin{bmatrix} 9 & 4 \\ 4 & 9\cdot25 \end{bmatrix}$$

$$[\mathbf{Y}_2][\mathbf{Y}_4]^{-1} = \frac{1}{67\cdot25} \begin{bmatrix} 0 & -4 \\ -5 & 0 \end{bmatrix} \begin{bmatrix} 9 & 4 \\ 4 & 9\cdot25 \end{bmatrix} = \frac{1}{67\cdot25} \begin{bmatrix} -16 & -37 \\ -45 & -20 \end{bmatrix}$$

$$[\mathbf{Y}_2][\mathbf{Y}_4]^{-1}[\mathbf{Y}_3] = \frac{1}{67\cdot25} \begin{bmatrix} -16 & -37 \\ -45 & -20 \end{bmatrix} \begin{bmatrix} 0 & -5 \\ -4 & 0 \end{bmatrix} = \begin{bmatrix} 2\cdot2 & 1\cdot19 \\ 1\cdot19 & 3\cdot35 \end{bmatrix}$$

$$[\mathbf{Y}_1] - [\mathbf{Y}_2][\mathbf{Y}_4]^{-1}[\mathbf{Y}_3] = \begin{bmatrix} 6\cdot3 & -3\cdot69 \\ -3\cdot69 & 4\cdot65 \end{bmatrix}$$

$$[\mathbf{Z}_1] = \frac{1}{15\cdot7} \begin{bmatrix} 4\cdot65 & 3\cdot69 \\ 3\cdot69 & 6\cdot3 \end{bmatrix} = \begin{bmatrix} 0\cdot296 & 0\cdot235 \\ 0\cdot235 & 0\cdot401 \end{bmatrix}$$

$$[Z_2] = -[Z_1][Y_2][Y_4]^{-1} = \frac{1}{67 \cdot 25} \begin{bmatrix} 0 \cdot 296 & 0 \cdot 235 \\ 0 \cdot 235 & 0 \cdot 401 \end{bmatrix} \begin{bmatrix} 16 & 37 \\ 45 & 20 \end{bmatrix}$$

$$[Z_2] = \begin{bmatrix} 0 \cdot 228 & 0 \cdot 233 \\ 0 \cdot 324 & 0 \cdot 248 \end{bmatrix}$$

$[Z_4] = ([Y_4] - [Y_3][Y_1]^{-1}[Y_2])^{-1}$ will be evaluated before $[Z_3]$ for convenience beginning with the inversion of $[Y_1]$

$$[Y_1]^{-1} = \frac{1}{61 \cdot 75} \begin{bmatrix} 8 & 2 \cdot 5 \\ 2 \cdot 5 & 8 \cdot 5 \end{bmatrix}$$

$$[Y_1]^{-1}[Y_2] = \frac{1}{61 \cdot 75} \begin{bmatrix} -12 \cdot 5 & -32 \\ -42 \cdot 5 & -10 \end{bmatrix}$$

$$[Y_3][Y_1]^{-1}[Y_2] = \frac{1}{61 \cdot 75} \begin{bmatrix} 212 \cdot 5 & 50 \\ 50 & 128 \end{bmatrix} = \begin{bmatrix} 3 \cdot 44 & 0 \cdot 81 \\ 0 \cdot 81 & 2 \cdot 07 \end{bmatrix}$$

$$[Y_4] - [Y_3][Y_1]^{-1}[Y_2] = \begin{bmatrix} 5 \cdot 81 & -4 \cdot 81 \\ -4 \cdot 81 & 6 \cdot 93 \end{bmatrix}$$

$$([Y_4] - [Y_3][Y_1]^{-1}[Y_2])^{-1} = \frac{1}{17 \cdot 2} \begin{bmatrix} 6 \cdot 93 & 4 \cdot 81 \\ 4 \cdot 81 & 5 \cdot 81 \end{bmatrix} = \begin{bmatrix} 0 \cdot 403 & 0 \cdot 279 \\ 0 \cdot 279 & 0 \cdot 338 \end{bmatrix}$$

$$= Z_4$$

$$[Z_3] = ([Y_3][Y_1]^{-1}[Y_2] - [Y_4])^{-1}[Y_3][Y_1]^{-1}$$

$$[Y_3][Y_1]^{-1} = \frac{1}{61 \cdot 75} \begin{bmatrix} -12 \cdot 5 & -42 \cdot 5 \\ -32 & -10 \end{bmatrix} = \begin{bmatrix} -0 \cdot 203 & -0 \cdot 689 \\ -0 \cdot 518 & -0 \cdot 163 \end{bmatrix}$$

$$([Y_3][Y_1]^{-1}[Y_2] - [Y_4])^{-1} = \begin{bmatrix} -0 \cdot 403 & -0 \cdot 279 \\ -0 \cdot 279 & -0 \cdot 338 \end{bmatrix}$$

by inspection from the negative of it evaluated in finding $[Z_4]$.

Thus

$$[Z_3] = \begin{bmatrix} 0 \cdot 228 & 0 \cdot 325 \\ 0 \cdot 233 & 0 \cdot 248 \end{bmatrix}$$

Assembling the complete nodal impedance matrix gives

$$
\begin{bmatrix} V_1 \\ V_2 \\ V_3 \\ V_4 \end{bmatrix} = \begin{bmatrix} 0{\cdot}296 & 0{\cdot}235 & 0{\cdot}228 & 0{\cdot}233 \\ 0{\cdot}235 & 0{\cdot}401 & 0{\cdot}324 & 0{\cdot}248 \\ 0{\cdot}228 & 0{\cdot}325 & 0{\cdot}403 & 0{\cdot}279 \\ 0{\cdot}233 & 0{\cdot}251 & 0{\cdot}279 & 0{\cdot}338 \end{bmatrix} \begin{bmatrix} 2 \\ 0 \\ 0 \\ 3 \end{bmatrix}
$$

Comparison with the results of the tearing method shows discrepancies of about $\pm 1\%$ due to the use of a slide rule. These differences have been left here deliberately to call attention to them, although students cannot always use a hand calculating machine or programme a computer during their course of study.

REFERENCES

BRAMELLER, A. 1965. Tensors, diakoptics and power system load flow calculations. *Conference on the Application of Digital Methods to Electrical Power Systems*. University of Newcastle-upon-Tyne.

BRAMELLER, A., JOHN, M. N. & SCOTT, M. R. 1969. *Practical Diakoptics For Electrical Networks*. Chapman and Hall, London.

DAY, J. E. & PARTON, K. C. 1965. Generalised computer program for power-system analysis. *Proc. I.E.E.*, **112**, No. 12, 2261–2274.

LAUGHTON, M. A. & DAVIES, M. W. H. 1964. Numerical techniques in solution of power system load flow problems. *Proc. I.E.E.*, **111**, No. 9, 1575–1588.

MORTLOCK, J. R. & DAVIES, M. W. H. 1952. *Power System Analysis*. Chapman and Hall, London.

PARTON, K. C., ALLEN, J. W. & TEOH, C. K. 1963. Power-system analysis using the digital computer. *G.E.C. Journal*, **30**, No. 3, 3–13.

STAGG, G. W. & EL-ABIAD, A. H. 1968. *Computer Methods in Power System Analysis*. McGraw-Hill, New York.

STEVENSON, W. D. 1962. *Elements of Power System Analysis*. McGraw-Hill, New York.

Examples

1. Show how a nodal admittance matrix can be inverted by partitioning to give a factorized inverse matrix, and use this method to calculate the nodal voltage at node 1 in the following balanced 3-phase power system with 3 generating stations. All resistances may be neglected and the per-unit reactances are all expressed at the same base MVA. At node 1 a generator transformer unit $G_1 T_1$ has a phase e.m.f. of $1{\cdot}1\underline{/0°}$ p.u. and a reactance of $0{\cdot}55$ p.u. and there is

a local phase-to-neutral load of 2·0 p.u. reactance. At node 2, G_2T_2 has an e.m.f. of $1·16\underline{/0°}$ p.u. and reactance of 0·4 p.u. and there is a load of 2·5 p.u. reactance. At node 3, G_3T_3 has an e.m.f. of $1·14\underline{/0°}$ p.u. and reactance of 0·6 p.u. and there is a load of 3·0 p.u. reactance. The reactances of the transmission lines linking the generating stations are; 0·02 p.u. between 1 and 2, 0·04 p.u. between 2 and 3 and 0·025 p.u. between 1 and 3.

(University of Leeds) (0·934 p.u.)

2. In the system shown in Fig. 8.10 the e.m.f.s are $E_1 = 1·2\underline{/0°}$ p.u. $E_2 = 1·4\underline{/0°}$ p.u. and the per unit reactances are: line between 1 and 2 = 0·05, line between 1 and 3 = 0·02, line between 3 and 2 = 0·02, phase-to-neutral load at 1 = 3·0, phase-to-neutral load at 2 = 1·0, phase-to-neutral load at 3 = 1·0, generator 1 = 0·3, generator 2 = 0·2. Resistances may be neglected.

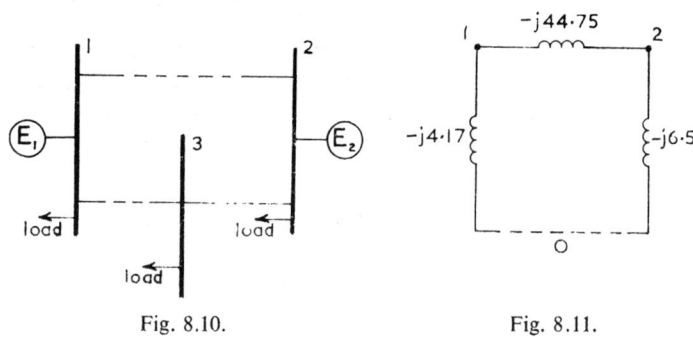

Fig. 8.10.   Fig. 8.11.

Draw the circuit with per-unit admittances and current generators and write down the nodal admittance matrix. Eliminate node 3 by matrix partitioning, justifying the method. Show the reduced circuit with the values of per unit admittance in it. Obtain the voltages at the busbars of the two generating stations.

(University of Leeds)

(Fig. 8.11 shows solution with p.u. admittances; 1·027 p.u., 1·036 p.u.)

3. In a 3-phase power system, generator 1 having an e.m.f. of $1 \cdot 2\underline{/0°}$ p.u., a reactance of $0 \cdot 1$ p.u. and a local phase-to-neutral load of $\overline{1 \cdot 0}$ p.u. reactance is linked by a transmission line of $0 \cdot 04$ p.u. reactance to a busbar 3 where there is a load of $2 \cdot 5$ p.u. reactance. A second generator 2 having an e.m.f. of $1 \cdot 1\underline{/0°}$ p.u., a reactance of $0 \cdot 2$ p.u. and a local load of $4 \cdot 0$ p.u. reactance is linked to busbar 3 by a line of $0 \cdot 05$ p.u. reactance. All resistances may be neglected and all reactances are given to a common MVA base.

Write down the nodal admittance matrix and hence determine the nodal voltages by matrix inversion.

If a new load of reactance $1 \cdot 0$ p.u. is connected in parallel with the existing load at the busbar of generator 2, obtain the new inverse matrix from [8.54].

(University of Leeds)

$$\left(1 \cdot 05 \text{ p.u., } 1 \cdot 04 \text{ p.u., } 1 \cdot 037 \text{ p.u.: } \frac{j}{11130} \begin{bmatrix} 726 & 458 & 601 \\ 458 & 926 & 660 \\ 601 & 660 & 866 \end{bmatrix} \right)$$

4. Use the method of matrix partitioning to eliminate as many nodes as possible in the network shown in Fig. 8.12. Give a diagram showing the admittances which remain after the circuit has been reduced in this way. Indicate very briefly how the potentials with

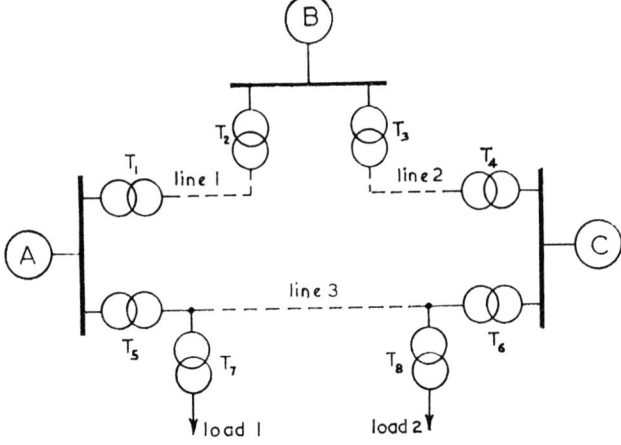

Fig. 8.12.

respect to earth of the remaining nodes could then be determined. In the circuit of Fig. 8.12 the resistance of all branches may be neglected. The per-unit reactances for each phase are: generators A, B and C each 0·20, transformers T_1 to T_8 each 0·10, line 1 = 0·04, line 2 = 0·05, line 3 = 0·10, load 1 (phase-to-neutral) = 1·90, and load 2 (phase-to-neutral) = 2·40.

The e.m.f.s of corresponding phases of generators A, B and C respectively are $1·4\underline{/0°}$, $1·2\underline{/0°}$ and $1·3\underline{/0°}$ per unit.

(University of Leeds)

(Fig. 8.13 shows solution with p.u. currents and admittances)

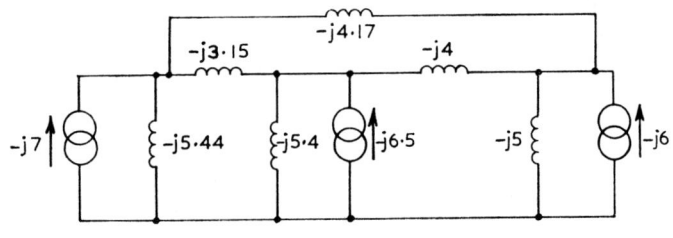

Fig. 8.13.

5. In the power system shown in Fig. 8.14 the per-unit reactances of the equipment (all to the same MVA base) are: Generator/ transformer $G_1T_1 = 1·0$, $G_2T_2 = 1·25$, $G_3T_3 = 2·0$, Load 1 = 2·0,

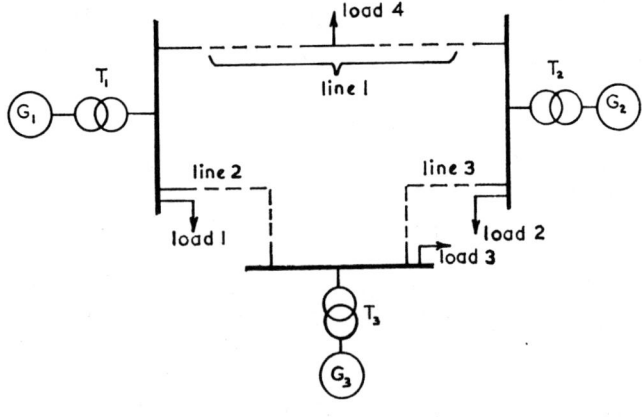

Fig. 8.14.

Load $2 = 1\cdot0$, Load $3 = 1\cdot25$, Load $4 = 2\cdot5$, Line $1 = 0\cdot05$, Line $2 = 0\cdot02$, and Line $3 = 0\cdot025$. The generator phase e.m.f.s are $E_1 = 2\cdot2\underline{/0°}$ p.u., $E_2 = 2\cdot5\underline{/0°}$ p.u., and $E_3 = 2\cdot0\underline{/0°}$ p.u. The load 4 is situated halfway along Line 1. Set up the nodal admittance matrix for this system, and using matrix algebra, eliminate one node to give a reduced nodal admittance matrix. Show how a nodal admittance matrix may be inverted by partitioning, and apply the derived results to the reduced admittance matrix obtained for the system of Fig. 8.14, and thus calculate the voltage at the busbars to which load 2 is connected.

(University of Leeds)

$$\left(-j\begin{bmatrix} 2\cdot2 \\ 2\cdot0 \\ 1\cdot0 \end{bmatrix} = -j\begin{bmatrix} 71\cdot6 & -19\cdot9 & -50 \\ -19\cdot9 & 61\cdot9 & -40 \\ -50 & -40 & 91\cdot3 \end{bmatrix}\begin{bmatrix} V_1 \\ V_2 \\ V_3 \end{bmatrix}; \right)$$

$$V_2 = 1\cdot034 \text{ p.u.}$$

6. Write down the general nodal voltage equations $I = YV$, for three nodes and from the data given below determine the value of each of the following complex quantities in per-unit values based on 100 MVA: I_3, Y_{13}, Y_{11}. Three identical power stations 1, 2 and 3 each contain four identical generator-transformer sets. Each generator has a synchronous reactance of $1\cdot9$ per unit, and each transformer has a leakage reactance of $0\cdot1$ per unit. All machines are running under steady-state conditions. The open-circuit source e.m.f.s at the stations are respectively $2\cdot1\underline{/20°}$, $2\cdot05\underline{/10°}$ and $2\cdot0\underline{/0°}$ per unit. The stations are inter-connected by a closed ring of 275-kV lines having lengths:—1 to 2, 160 km; 2 to 3, 160 km; and 1 to 3, 320 km. The line reactance is $0\cdot47$ ohms/km, and resistance is negligible. The only load is connected to station 3. All per-unit data are based on 100 MVA.

(L.C.T.) $(-j4\cdot0, j5\cdot0, -j17\cdot0$ per unit)

7. A transmission system is represented by a symmetrical-T having a total series reactance of $j1\cdot2$ p.u. and a total shunt reactance of $j2$ p.u. The supply alternator is represented by an equivalent series circuit having an e.m.f. of $(1\cdot1 + j0\cdot1)$ p.u. and a reactance of $j0\cdot4$ p.u. The corresponding data for the load (synchronous motor) is a back

e.m.f. of $(0.9 - j0.2)$ p.u. and a reactance of 1.4 p.u. The data refers to the equivalent line-to-neutral values and the per-unit values are all on the same base. Using the mesh current method and matrix inversion, calculate the input and output currents of the system and hence the volt-amperes from the alternator and to the load in the form $(P + jQ)$ where lagging vars are positive.

(L.C.T.) ($0.1 - j0.325$ p.u., $0.1 + j0.0625$ p.u., $0.0775 + j0.368$ p.u. /phase, $0.0775 - j0.0762$ p.u./phase)

Chapter 9

ASPECTS OF SYSTEM INTEGRATION AND DEVELOPMENT

9.1 Introduction

The design and operation of a particular item of equipment in a power system cannot be dealt with by considering the characteristics of the device in isolation. The equipment operates as a part of an integrated system where each component must interact satisfactorily with other equipment under steady-state and transient conditions, and improvements in one area are constantly affecting other devices. However, before an appreciation of the problems involved in the interaction of system components is possible, an introduction to their individual characteristics is necessary and this was given in Volume 1. Even there, attempts were made to indicate some aspects of system operation, e.g. when synchronous generators were dealt with the effects of improvements in voltage regulators on machine stability at leading power factor were discussed.

This volume has discussed principles in the design and operation of a power system as a whole or of particular devices in relation to the rest of the system. This Chapter is intended to emphasize this systems approach in the context of present developments in the supply system, and to give up-to-date references to some of the most significant areas of current development.

Generation and transmission systems, plant and equipment, and techniques of their operation and maintenance are still under continuous development. By the judicious application of these developments, supply authorities can improve the quality of supply given to consumers, either in terms of its cost or its continuity, or both. Today, consumers in England and Wales experience, on average, only some 90 minutes a year without supply. Although some improvement in continuity is likely through greater use of automatic circuit breaker reclosure and of computers at major control centres and large stations, developments in plant design are more likely to result in present standards of continuity being achieved with greater economy.

Not only is it necessary to pursue the development of improved designs of plant which offer performances comparable with current designs at lower capital cost, but it is also important to consider new developments which offer, perhaps at small additional cost, performances which allow considerable economies in capital costs to be made on other plant. Thus technical development can be exploited to realise maximum gain in overall economy. The best examples of interaction between the design of different items of plant and equipment lie in the field of system transient stability which is dependent on a large number of factors.

9.2 Transient stability

As discussed in Chapter 3, the factors involved in transient stability include the load angles of the machines prior to the disturbance, the duration of the disturbance, the synchronising power which can flow between machines after the disturbance and the control exerted on steam flow and excitation to the machines during and after the disturbance. As indicated in sections 7.8 and 7.9 of Volume 1, considerable saving in alternator rotor m.m.f. and hence machine size, weight and cost is made by reducing the air-gap and thus the short-circuit ratio (section 7.3 of Volume 1), but this increases the load angle and reduces the machine stability. At times when the active demand on the system is low, reactive power may have to be absorbed either by the provision of shunt reactors or by operating generators at leading power factor, as discussed in section 9.3. Stability considerations will decide whether expenditure on shunt reactors is necessary and this will depend, to some extent, on the machine short-circuit ratio adopted.

The duration of a disturbing effect of a system fault is the fault clearance time which is the sum of the circuit breaker and protection operating times. In Britain, these operating times have now been reduced to 0·08 and 0·06 s at high fault currents, but designs will soon become available which will bring both operating times down to 0·04 s. Although these reduced times will add to circuit breaker and protection costs, they will allow greater savings to be made in the costs of generators or shunt reactors. The development of vacuum interruptors with small contact separation and short arc duration times holds out the possibility of reducing circuit breaker operating times to 0·02 s or even less. Such short circuit breaker operating times with comparable protection times would not only allow major financial savings to be made on other plant but would also much reduce fault damage and limit consumer disturbance to negligible proportions.

The synchronising power which can flow between machines after a disturbance is dependent on the forcing factor of the excitation system, and on the sophistication of control which can be exerted on the excitation current and steam flow during and after the disturbance. Up to the present, excitation and steam flow have been controlled by reference only to stator voltage and machine speed, respectively, and voltage regulators and governors have functioned independently. The introduction of faster excitation control systems using thyristors, and of electronic governors and steam interceptor valves between reheat furnaces and intermediate pressure turbines, opens the way to faster control of excitation and steam input to turbines, to electronic coupling between excitation regulators and governors and to suitable application of first and second derivatives of rotor angle, voltage, speed and possibly MW output and low pressure turbine steam pressure to ensure that excitation and steam control are at all times exerting maximum stabilising effects on the machine. Thus it can be seen that, ultimately, there would be a single electronic machine controller supplied with full information about the instantaneous conditions influencing the machine and their rates of change and the necessary stabilising feedback circuits.

A method of excitation control, known as 'bang-bang' control, has been proposed to alleviate the effects of a generator pole-slipping during or after a major system disturbance. With normal excitation control, the swing of a machine rotor is either arrested and reversed by forced excitation or accelerated into a pole-slip. In the latter event, the speed rise is detected by the governor which reduces steam input and MW output and, by throwing more load on to other machines, causes them to slow down. Such undesirable oscillations and exchanges in active power between machines can be minimised by arranging for excitation to be removed from a machine when it starts to accelerate into a pole-slip and restoring excitation when a suitable rotor angle has been reached after a pole-slip has taken place. This bang-bang control of excitation has now become possible with the advent of fast-acting thyristor excitation control, but it is yet too early to forecast whether the overall effects on a supply system of one or more large machines pole-slipping without excitation can be tolerated.

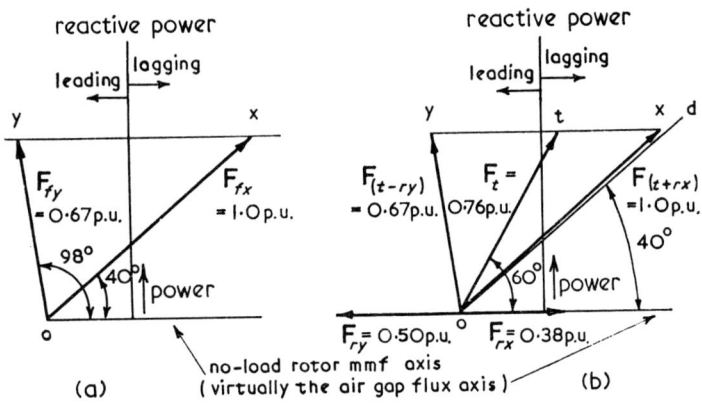

Fig. 9.1. Field m.m.f. diagrams:
(a) conventionally-wound rotor
(b) divided-winding rotor.

Another development of interest is the divided-winding rotor which allows the alternator rotor or load angle to be kept constant, irrespective of the power factor of the load. Such a development allows a machine to absorb reactive power up to the limit of its stator thermal rating without impairing its transient stability. Fig. 9.1(a) is a field m.m.f. diagram of a conventionally-wound rotor drawn for a machine with a short-circuit ratio of 0·4. (The student may find it helpful to refer to Figs. 3.1(a), 3.13 and possibly also Fig. 7.21 of Volume 1). Points x and y are at rated power factor lagging and leading at rated load. As the alternator reactive output changes from x to y, its load angle increases from 40° to 98° and the rotor direct axis moves from ox to oy. Fig. 9.1(b) is an m.m.f. diagram for a similar alternator with a divided-winding rotor which has the same ampere turns and is wound in the same slots as the conventional winding referred to in Fig. 9.1(a). As before, x and y are at rated lagging and leading outputs. The leading section of the rotor winding in the direction of rotation provides all the torque necessary for full MW load and is shown acting along ot at an angle of about 60° to the no-load rotor direct axis, which is virtually the same as the air-gap flux axis over the load range (because the stator leakage reactance voltage drop is only about 0·1 to 0·2 p.u. at full load). The rear or reactive section of the rotor winding exerts an m.m.f. F_r along the air gap flux and zero torque axis, and controls the alternator reactive output by operating with continuous positive or negative excitation. At full load, the torque winding m.m.f. F_t is 0·76 p.u. of F_{fx} (the rated m.m.f. of the conventional rotor winding shown in Fig. 9.1(a)), F_{rx} is 0·38 p.u. for rated power factor lagging and F_{ry} is $-0·5$ p.u. for rated power factor leading. Reactive winding m.m.f. F_{rx} or F_{ry} when added to F_t total 1·0 p.u. or 0·67 p.u. respectively, exactly as F_{fx} or F_{fy} in Fig. 9.1(a). Thus the conventional and divided-rotor-winding arrangements are equivalent under steady state conditions. The torque winding excitation is controlled by a regulator responsive to rotor angle which maintains the angle at about 40° (see od in Fig. 9.1(b)) throughout the load range of the machine. The reactive winding excitation is controlled by a separate regulator responsive to stator terminal voltage.

The important feature of the divided-winding arrangement is that a change in machine reactive output does not require a corresponding change in rotor position relative to the air gap flux, i.e. there is no change in load angle. This is illustrated in Fig. 9.1(b), which shows that the rotor position remains at od for a change in reactive output from rated lagging power factor at x to rated leading power factor at y for full MW load. The improved transient stability performance of the divided-winding rotor arises from the fact that, when the rotor angle increases during a system fault and F_t moves forward, the forward movement of the total rotor m.m.f. is arrested by the voltage regulator applying full positive excitation to the reactive winding. After fault clearance, the reverse action takes place.

A prototype divided-winding rotor is in manufacture for a standard 500-MW turbo-alternator. A conventional a.c. exciter/static rectifier will be used to excite the torque winding. A separately driven d.c. generator, with two separate fields supplied from a shaft-driven pilot exciter through two thyristor bridges, will provide forward and reverse excitation of the turbo-generator reactive winding.

A separate cost analysis has shown that reactive power absorption by high-merit turbo-alternators with divided-winding rotors is half the cost of alternative compensation equipment. Against this must be set the disadvantages of an additional excitation control system with its slip rings and brushgear or rotating diodes or thyristors, and more serious expansion and contraction problems with the rotor windings due to the torque and reactive sections not carrying the same current. It may prove desirable to preheat the windings before reaching full speed of the rotor to give a differential expansion of the two sections equal to the mean of the extremes possible due to the differential heating which will occur in service.

9.3 Comparison of system conditions at high and low loads

Many of the problems involved in the design and operation of large interconnected supply systems are highlighted by comparing the conditions encountered at high and low loads. The National Supply System in England, Wales and South Scotland will be described as a typical example of a system serving a comparatively compact and densely loaded area. This system is of special interest because the generation and transmission section in England and Wales, for which the Central Electricity Generating Board is responsible, is the largest system of its kind in the world which is designed and operated as a single entity to give maximum economy.

The main voltage levels of the system are shown in Fig. 9.2. The larger generators, which are the most modern and efficient, are connected to 400-kV and 275-kV networks which are closely interconnected to give a nationwide Supergrid. Smaller, older and less efficient generators are connected to individual 132-kV and 66-kV grid networks, the majority of which are coupled to the Supergrid at only one point. The 33-kV and 11-kV voltage levels shown in the figure represent local distribution networks to some of which are connected the smallest and least efficient generators.

The capacity of the transmission system is sufficient to allow economic pooling of generation. Maximum economy is achieved by running and loading generating plant according to its incremental fuel cost (which is least for nuclear fuel and, for coal and oil is roughly in inverse proportion to generator MW rating) and paying due regard to transmission losses. Some departure from this ideal may be necessary to give adequate security of supply, avoid circuit overloading and provide control of voltage. It follows that, as the system load falls, the less efficient generators on local networks are shut down and the load, when light, is left supplied by large base-load machines connected to the Supergrid.

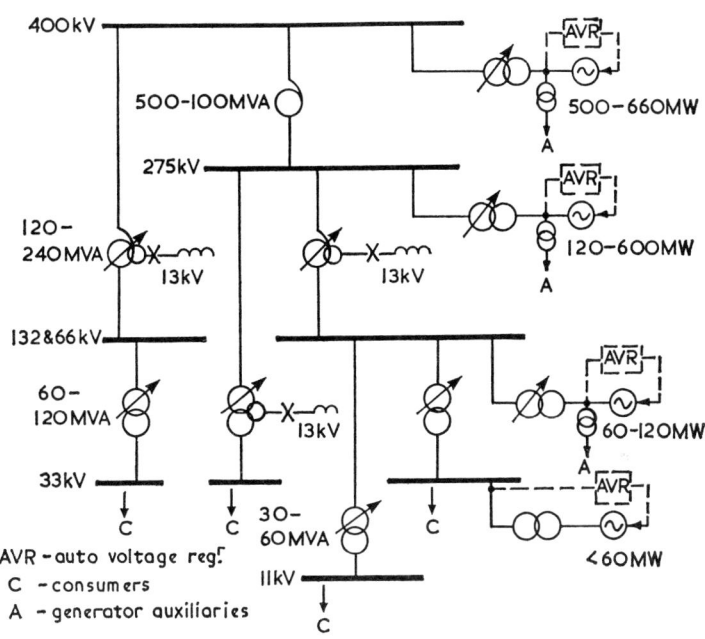

Fig. 9.2. Voltage levels, generator ratings and methods of voltage control of the British transmission system.

Control of voltage on the various parts of the system is obtained by an automatic regulator on each generator, set to hold the stator terminal voltage at rated value, and on-load tap-changers (section 6.7, Volume 1) on all transformers except those coupling the 275-kV and 400-kV sections of the Supergrid, those supplying generator unit auxiliaries and step-up transformers for generators connected to 66-kV and lower voltage networks.

Although the voltages at the supply end of 33-kV and 11-kV distribution networks are kept between 100 and 105% of nominal value, the 132-kV and 66-kV grid voltages can, with advantage, be varied over a wider range. Grid voltages are kept as high as practicable at times of high load to reduce I^2R losses and voltage regulation due to load reactive power, as was stated in section 1.5 of Volume 1 where reactive power compensation was considered. Conversely, grid voltages are kept as low as practicable at times of low load to reduce the reactive power produced by the shunt capacitance of the networks. This is particularly important for networks containing underground cables. Since the 400-kV and 275-kV Supergrid networks are coupled by fixed ratio transformers, independent control of their working voltages is not possible. At times of high system load, Supergrid voltages are kept as high as practicable, the top limits, as set by insulation, being 420-kV and 300-kV. At times of low system load, the Supergrid, by virtue of operating well below its natural load (see section 1.6 of Volume 1), is a large producer of reactive power, much of which can be absorbed by the directly-connected base load generators. Optimum Supergrid voltages are then only slightly below the values prevalent at high loads because, with lower voltages, the saving in reactive power produced by the Supergrid is less than the reduction in absorption capacity of the generators.

Figure 9.3 shows, for the early 1970s, the reactive powers required by the C.E.G.B. transmission system and the consumers (area board loads and distribution networks) plotted against the active power demands of the consumers. The maximum reactive power production and absorption capacities of the generation connected to meet the active load are also shown. These capacities are based on generators designed to deliver rated MW at 0·8 p.f. lagging and capable, when under-excited, of operating continuously at load angles of up to 75 degrees (see sections 7.8 and 7.9 of Volume 1). It is seen that the total reactive power variation of the consumers and system is about half the active power variation. At times of high system load, the reactive power requirements of the consumers are met partly by the system, but chiefly by over-exciting the generators. Thus, the reactive power flows are from the generators and transmission system to the consumers, and the transformer tap-changers are adjusted to allow these flows to occur within acceptable voltage limits. At times of low

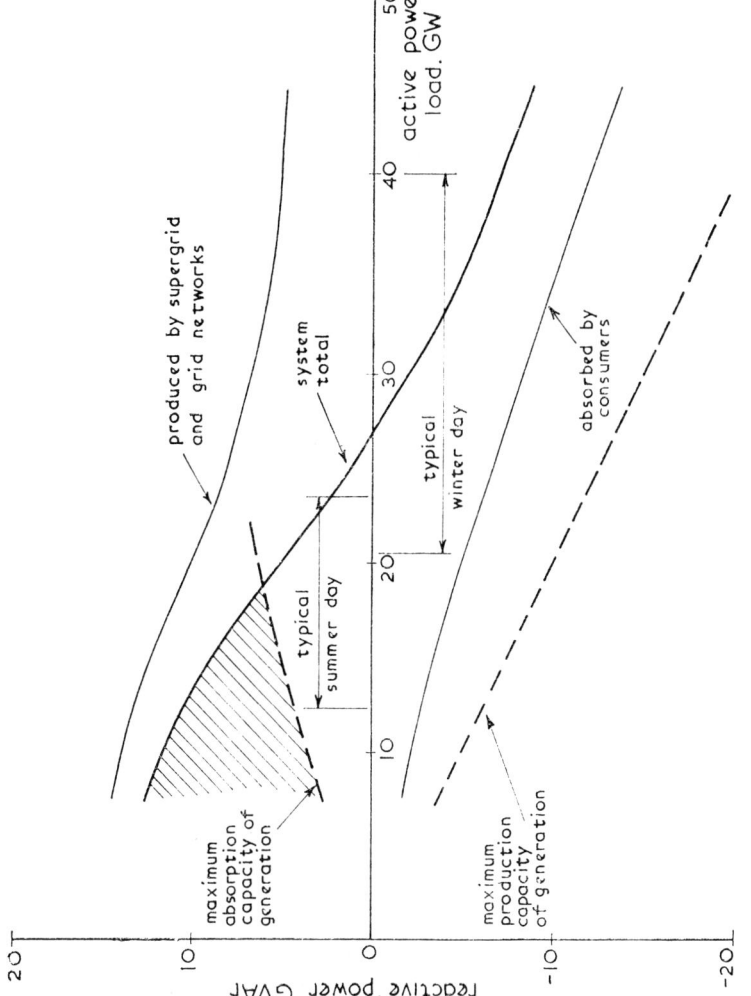

Fig. 9.3. **System reactive power requirements and capacity of generation in England and Wales.**

system load, the reactive power produced by the transmission system is many times that absorbed by the consumers, and the surplus exceeds the absorption capacity of the base load generators obtainable by under-excitation. The shaded area shows the minimum absorption capacity which must be provided by other means at low loads. In fact, considerably more than this amount of additional absorption capacity must be arranged for in the planning stage, to limit voltage gradients across the fixed ratio 400/275-kV transformers, cover unforeseen eventualities such as non-availability of particular generators for absorption and under-estimation of future cable mileage, and to provide a dynamic reserve.

The means available for providing the additional absorption capacity are, in rising order of overall cost:—switching out unnecessary circuits (particularly cables), staggering taps on paralleled transformers, installing shunt reactors and running out-of-merit generation at minimum active load. Although switching out unnecessary circuits involves no direct financial cost, it does impair the transient stability of the system by increasing the impedance between generators, as shown in Chapter 3. The staggering of taps on transformers in parallel absorbs reactive power by virtue of the circulation of current through the reactance of the transformers (see section 6.4.2 of Volume 1). This method is much more effective with groups of three or four transformers than with groups of two, because for a given step-change in consumer voltage following loss of a transformer, larger tap staggers are permissible on the larger groups. This method has the attraction of involving no capital cost and of making maximum use of transformer MVA capacity for reactive power absorption, at times when the active load requirement for this capacity is at a minimum. Shunt reactors, which were discussed in section 1.5.1 of Volume 1, although they involve additional capital cost, have the advantage of very low running cost and simplicity of control. Normally, 60 MVAr units are connected to Supergrid transformer 13-kV tertiary windings through circuit breakers, although units of 100 MVAr or more may be connected in parallel with long 275-kV cable circuits. These sizes of reactor are such that they can be switched without producing changes in voltage exceeding about 3% in normal circumstances and 6% under fault conditions.

There are situations where, for short periods, it is cheapest to provide the required absorption capacity by running out-of-merit generators under-excited at the minimum active load which can be supplied by the boilers. Substantial dynamic reserves of reactive power must, at all times, be carried on rotating plant to compensate for the additional reactive demand arising from the loss of one of two or more parallel circuits at times of high system load, or of the loss of an under-excited generator or shunt reactor at times of low system load.

Figure 9.3 shows the variation in active and reactive power over typical summer and winter days. In both cases, the active power load falls to a minimum during the night and is maintained near the maximum between 8.00 a.m. and 10.00 p.m. The change from minimum to near maximum load occurs in about two hours and, during this period, additional generators must be synchronised and loaded up and steps taken to keep the network voltages within acceptable limits by switching out shunt reactors, strengthening generator excitation and adjusting transformer tap-changers. The reverse process takes place between 10.00 p.m. and 2.00 a.m.

It is clear from what has been said, that system voltage control by the balance of reactive power production and absorption is more complex, at times of low system load, than when the load is high. It should also be noted that generators tend to have less dynamic stability at times of low system load, because of the higher impedance to the rest of the system after fault clearance (due partly to switching out unnecessary circuits), and because generators are then operating with weaker excitation. Critical fault clearance times of protection and circuit breakers are, therefore, generally least at times of low load.

The complex activities involved in designing and operating an integrated supply system to meet, with due economy and reliability, active power loads which vary over a ratio of about two to one every 24 hours, open up many interesting possibilities for computer application, particularly in dealing with the rapid changes in both active and reactive power requirements which occur twice daily.

9.4 Computer applications in power systems

Digital computers are employed, both off-line and on-line, in the design and operation of power transmission systems. As commented on in section 8.1, power system design problems became too complex for manual calculation in the 1920s, and network analysers were used until the advent of digital computers towards the end of the 1940s, caused design calculations to shift more and more from models to computers. The operating problems have similarly become so complex that it has become increasingly necessary to aid the operating engineers by means of digital computers and to aim at greater economy. This has taken the form initially of providing the operator with information based upon computation of data transmitted to a control centre from a number of points in the system. In the longer term it will lead to the provision of closed-loop automatic optimisation of more and more aspects of system control. The British supply system differs in a number of respects from that of the United States or the interconnected national systems in Europe. It is a fully integrated system with virtually no hydroelectric plant available for regulation purposes, in which the load must be generated in the most economical way with a safe minimum of spinning reserve (spare capacity of sets already connected to the system). An economic balance has to be struck, for example, between having too much generating capacity and not having enough margin against frequency drop and even power failure. One of the prime necessities is that the load shall not only be distributed economically, but in a way which allows load changes to be met sufficiently quickly, so that methods of load prediction (which are commented upon in the next section) must be improved, and so that as progress is made towards on-line closed-loop control the predictions must be made more quickly for short intervals ahead. Greater accuracy in prediction can enable a reduction to be made in the spinning reserve which caters for errors in load forecasting as well as actual load increases which materialise, and for the loss of the largest infeed of power in the area.

In the off-line or predictive mode, computers are used to calculate for large networks, the circuit load flows, switchgear fault levels, voltage profiles and generator stability. Load flows, the principles in the calculation of which are discussed in Chapter 8, are necessary to determine whether any circuits are overloaded with the network normal, and when one or more circuits are out of service for maintenance or due to faults. Section 8.6.2 shows how readily a computer programme enables this latter check to be made. The fault levels to which switchgear may be subjected, (see sections 8.4.8 and 8.4.9 of

Volume 1), are compared with the switchgear rating and, if the latter is exceeded, some rearrangement of the network may be necessary. Attention to voltage profiles is particularly important on the highest voltage networks which are generally designed to work within comparatively narrow limits of voltage to avoid over-stressing of insulation or excessive copper loss at low voltage. As generator stability, which is discussed in Chapter 3, is affected by network arrangement, fault clearance time, machine inertia, and machine reactances and loading conditions, a check may be necessary where normal margins are impaired by, for instance, a special network arrangement or the risk of an abnormally long fault clearance time.

Other off-line applications which have been, or are being, developed are economic scheduling of generating plant, circuit outage costs, demand estimation and information storage, retrieval and analysis. The most economic pattern of generation connected to a network at any time must take account of the actual and predicted loads, the cost of production of power from each generator, avoidance of circuit overloads and the effect of faults on network security. A circuit outage may have to be planned to avoid increasing the overall cost of generation, either because the outage would restrict output from low-cost generation or because it would necessitate high-cost generation being run to secure supply to parts of the network. Demand estimation must take account of such meteorological factors as temperature, wind, illumination and precipitation of moisture, in addition to the time of day and the day of the week. Network data is often stored on magnetic tape as it is then more readily available when required for computer surveys of load flows, fault levels, voltage profiles and scheduling of generating plant, and can be classified or analysed by the computer on demand. Computers are also being used to analyse network fault data and protection performance. In this case, full information about each fault incident is fed into the computer which supplies an analysis of the information stored at regular intervals or on demand.

Computers are also being increasingly employed in design and operating problems in distribution systems. Two examples of this are (i) in the optimum design of branched radial low-voltage distributors using standard cable size, taking account of voltage-drop limitation, thermal ratings, capitalisation of losses and factors for diversity and unbalanced loading (Snelson and Carson, 1970); and

(ii) in the security assessment of distribution systems in determining alternative supply routes under emergency conditions (Pragnell and Cory, 1970).

On-line or real-time applications of computers to transmission system operation are a more recent development. These applications include data logging, alarm analysis and computer-assisted system control. These are open-loop systems which supply control engineers with up-to-date key information in an easily digestible form, leaving the engineer to decide on the corrective action to take. For instance, the new C.E.G.B. National Control Centre in London is equipped with two high-speed digital computers in which on-line information from about 1000 power flow meters and 5000 devices (circuit breakers, isolators and current overload relays) is stored. This information is displayed on demand on 12 cathode ray tubes, the necessary symbols to make up the diagrams being generated by the computers. Although up-to-the-minute data for the whole Supergrid system is held as a complete mosaic picture in the computer store, only about one sixth of the whole picture is shown on any tube. A 'joystick' is provided with each tube so that it can be steered progressively to display that part of the complete mosaic required, complete with on-line power flows and any current-overloaded indications. The following information on system conditions can be displayed on demand:—network-voltage profile, national and area generation and demand, area transfers, group generation incremental and decremental costs, continuous summation of flows on groups of up to 20 feeders and automatic annunciation of switchgear operation, current overloads, etc. The on-line data is stored at minute intervals for up to 30 minutes, so that animated pictures of system conditions leading up to an incident can be studied and permanently recorded for subsequent reproduction. On-line security calculations of power flow and short-circuit levels under actual and credible outage conditions are now made cyclically for the entire Supergrid system every thirty minutes. If any circuit overload or circuit breaker over-stressing is revealed, an alarm is initiated and, on acknowledgement, details of the abnormal conditions are displayed on a cathode-ray tube. The need is recognised of improving security assessment in optimisation studies (see discussion on paper by Shen and Laughton).

Examples of closed-loop on-line computer applications are automatic switching at substations, automatic starting, loading and shutting down of boiler-turbo-generator units in power stations and automatic system control. Automatic switching without computers is already common practice on high-voltage networks, but the introduction of small computers at large and important substations allows a much greater variety of emergency conditions to be catered for and can include other facilities such as selective load shedding and alarm analysis. Automatic system control involves the provision of a machine controller connected to the turbine governor of each generating set controlled. The controllers are themselves automatically controlled by 'teletransmission from a central process control computer. The computer controls the output of each generator to merit order and checks the power flow on the transmission circuits to ensure secure operation. The computer can be used to calculate the expected demand, say 1 hour ahead, and schedule the loading on each generator to meet the demand with maximum economy subject to secure operation. An operational test of such an automatic system of load dispatching has been successfully carried out on a section of the British Grid System involving 31 generators in six power stations with a total capacity of 1600 MW, and this is described in references listed by Farmer *et al.*, 1969. In this initial experiment, automatic switching was excluded as inappropriate at this stage and so was voltage control, because the economic benefits of optimising reactive power flow were less than those of power loading. It seems likely that automatic system control will be extended to include the optimum provision of spinning reserve to cover errors in demand prediction and sudden loss of the largest generator on the system, and the control of reactive power flows and system voltage by suitable adjustment of generator excitation and transformer on-load tap changers. Another possibility is to use the computer to select optimum sectioning of the system in the event of a fault not being cleared by normal main or back-up protection. On large highly-meshed systems, the pattern of generation is constantly changing and only by taking account of the network flows immediately prior to the fault, is it possible to choose sectioning points so that the sustained fault is removed from the system with the minimum of disturbance to consumers. To sum up, the field of computer application in the design and operation of power systems is widening and several exciting possibilities are under consideration.

9.5 Load prediction

An essential part of the developing computer control schemes is the forecasting of demand both for the calculation of an economic schedule of loading and for the assessment of the security of the system. 'Short-term' prediction can provide a forecast every few minutes upon which load changes in the power stations can be made. 'Long-term' prediction up to 2 hours ahead, and revised every half-hour, can give advanced warning to the stations of load changes expected, which would also enable action to be taken for a while in the event of a failure of the control system, and this allows time to plan loading to meet peak demands without exceeding maximum permissible rates of load increase. Load prediction 8 to 12 hours ahead is important in a system in which almost all generating sets are thermal, so as to provide information for the starting up and shutting down of sets to meet the daily demand cycle. The further ahead the demand is forecast, the more vital it is to include data on the predicted weather.

The various methods applied at present to short-term load prediction (Farmer and Potton, 1968; Matthewman and Nicholson, 1968; Toyoda *et al.*, 1970), do not include the use of meteorological data because of the difficulty of obtaining in the right form and feeding into the computer the short-term variations in such data, (which may be unreliable), sufficiently quickly to give the frequent and detailed loading estimates required. Methods based upon input data on past load variations which are readily available in the form required for the computer, have been found to give relatively accurate short-term estimates, though they cannot predict special demands such as those which arise occasionally, for example from the reaction of large numbers of consumers to a television broadcast during the hours of darkness. Improvements in on-line telemetering equipment should give more accurate data on previous loading, and the conversion of graphical weather-correlation techniques to computer input is being examined. Progress is therefore likely to be made in short-term load prediction by a combination of estimating from past loading, together with weather-based data, perhaps even with psychological allowance added if necessary by an experienced forecaster. There is great need for the smallest possible tolerance in the estimated demand, because an error of even a small percentage in the estimated load for a given area may be a large fraction of the output of one alternator. It is more difficult to obtain satisfactory accuracy in the case when the prediction is part of an on-line programme for load dispatching, than it is when relatively smooth off-line data are provided in the form of half-hour averages over a larger area (Farmer and Potton, 1968). It will therefore be necessary for considerable work to be done in on-line prediction techniques as a part of the development of automatic control of power systems. Nevertheless, it will never be possible to predict with absolute certainty what the short-term variations will be in the system demand, so that anticipatory regulation will never be able to replace closed-loop regulation with correction of observed effects.

9.6 Problems associated with E.H.V. A.C. transmission

No decision has yet been reached as to whether a new primary transmission system operating at a voltage above 400-kV will be needed to meet the load growth in Great Britain in the forseeable future. Every effort will be made to find alternative more economic solutions, but if a new system has to be constructed it is likely to be at a voltage of the order of 1100-kV, and it is not likely to be in service before about the mid 1980s. In Great Britain, the use of overhead lines at this voltage would be restricted because of their severe impact on amenity, and special steps would be needed to obtain satisfactory distribution of voltage across the long insulator strings required under conditions of atmospheric pollution (see section 4.1.1 of Volume 1).

As discussed in section 2.8 and in Chapter 5, lightning is less of a problem in E.H.V. systems than it is at lower voltages, because lightning surges rarely reach the impulse withstand strength of the system insulation, whereas switching surges which are related to system voltage become very important. Also, as discussed in section 5.4, the flashover voltages of long air gaps and of system insulators may be appreciably lower for a surge wavefront of the order of a few hundred μs (which is typical of many switching surges), than for the relatively short wavefront time of lightning, or for the longer wavefront times of system-frequency over-voltages. This reduction in the breakdown strength of long gaps and insulators under switching-surge conditions may produce an ultimate limit to the transmission voltages which may be employed in the future, and suggestions have been made that such a limit might be about 1500 kV. As Fig. 5.3 indicates, the flashover voltage of E.H.V. system insulators at wavefronts of some hundreds of microseconds, can be appreciably less for positive waves than for negative waves (though for vertical strings the negative wet strength may be equal to or even less than the positive strength). For this reason the development of positive leader strokes in non-uniform asymmetrical fields has been studied in recent years, and it is clear that corona at the peak phase operating voltage should be avoided on transmission equipment, and to this end the characteristics of such devices as large corona shields at the end of an insulator string are being investigated (Aleksandrov et al., 1970).

The recent American literature on switching surges and the recommendations of an I.E.E.E. committee are given in their reports, Part III of which was published in 1970. The trend in system design is to limit surge voltage levels when switching overhead lines to maximum values of about 1·5 to 1·7 per unit of peak phase voltage, and limitation to 1·5 p.u. or less has even been proposed. Methods by which switching voltages may be reduced include multi-step resistors (see section 2.12 where single-step resistors are mentioned), time-varying resistors and more sophisticated synchronisation of the closing of all three poles (see section 2.12). The design of the insulation of an E.H.V. overhead line and the extent of the limitation of switching surges, involves the statistics of possible flashovers, e.g. a design basis could be 1 switching surge flashover per 1000 breaker operations or 1 flashover during a given period of time. A recent paper (Hileman et al., 1970) considers such criteria and permits determination of line performance under switching surges for given types of tower and switching surge distribution. The conductor arrangement will be dictated by radio interference effects. The more conductors there are in the bundle the greater the risk of mechanical oscillation troubles and the difficulty of spacer design.

In the long term the situation may perhaps be radically changed by other developments. It is possible that eventually d.c. cable links (see Chapter 6 and section 5.9 of Volume 1) and superconducting a.c. cables (see next section) may be developed sufficiently to replace at least some E.H.V. overhead lines. Perhaps in the even longer term, small local direct-energy conversion plants may take the place of the present single network interconnecting large generating stations.

9.7 Superconducting cables

A considerable research effort has been applied in recent years in a number of countries attempting to produce an economic and sound engineering design for superconducting cables. A main objective of this work is to reduce the cost of high-voltage or E.H.V. cables to the point, where the cost difference between an overhead line and an underground cable has been so narrowed, that for some areas and routes they can be used where at present they would not be economic. This would go at least some way to relieving the difficulties of obtaining way-leaves in densely populated areas, and of opposition on conservation and amenity grounds in many parts of the country-side.

In Great Britain, work on helium-cooled cables with niobium conductors backed with 99·999 % pure aluminium, produced several years ago some encouraging estimates of cost for 750-MVA, 33-kV. Types of cable examined include interleaved sets of tubular conductors, or an arrangement of four coaxial conductors with vacuum insulation (Norris and Swift, 1967, and Rogers and Edwards, 1967). Two recent American papers (Graneau, 1970 and Afshartous et al., 1970), have suggested that cables developed by one company are competitive with oil-paper insulated pipe-type cables (see Chapter 5, Volume 1), at ratings above 1000 MVA at voltages between 138 and 345 kV. The capital cost is claimed to be nearer to that of an overhead line than to that of conventional underground cables, and it was said to be possible that the first of such cables could be in service before 1975. This cable design uses tubular aluminium conductors cooled internally with liquid nitrogen, and supported by spacers in high-voltage vacuum insulation. In a very lively discussion on these papers, a number of questions were raised by several engineers, some from a competing company working on liquid hydrogen and liquid nitrogen cables. It was suggested that the cross-over point for economical use might be as high as 2500 MVA, and uncertainties were expressed on the use of vacuum insulation in cable installations. It now appears to be unlikely that superconducting cables can be an economic proposition to enter service in the immediately foreseeable future, unless perhaps through the development of materials with good superconducting properties at somewhat higher temperatures than at present, though it might take a long time for them to become available at 500 kV and above. In the case of power transformers, superconductivity is unlikely to be a feasible proposition until new low-loss superconductors are developed (Harrowell, 1970).

9.8 Fluctuating loads

As stated in section 1.5.2 of Volume 1 dealing with synchronous compensators, arc furnaces and steel rolling mills constitute loads which fluctuate widely, frequently and rapidly, and the effects worsen as the sizes of economic industrial units increase. In the case of arc furnaces, it is the reactive power swing when the electrodes are short-circuited by the unmelted scrap charge which is particularly troublesome, and this swing on an 80 MVA furnace may be of the order of 80 MVAr. A single melting shop may contain units totalling several hundred MVA, but they are generally programmed so that the large swings do not occur together. Arc furnaces already make one sixth of the steel in the U.K. and arc currents now reach 75 kA in 180 ton capacity furnaces and 400 ton furnaces with 100 kA have been projected in the U.S. When the fluctuating current drawn by such loads passes through the supply system source impedance, a corresponding fluctuation is set up in the supply voltage. This voltage (or a proportion of it) will be communicated to all the consumers fed from that network. The equipment most sensitive to voltage fluctuation is the tungsten-filament lamp which can produce a perceptible light flicker with a voltage fluctuation of between 0.1% and 0.2% r.m.s. In order that large fluctuating loads shall not cause nuisance to other consumers through light flicker, it is necessary either to reduce the fluctuation of the current drawn by the load or to connect the load to the supply system at a point sufficiently near to the source to restrict the voltage fluctuation at the point of connection to an acceptable value. In this country, it is generally found more economic to adopt the latter alternative, but there are cases, particularly abroad, where it is necessary to reduce the fluctuation of the current drawn from the supply system.

Although synchronous compensators have been successfully used for this purpose for many years, they have the disadvantages of high capital and maintenance costs and of supplying an infeed which adds to switchgear short-circuit duty. Designs of static compensator, incorporating saturable reactors, have recently been proposed to avoid these disadvantages, a typical circuit being shown in Fig. 9.4(a). In order to eliminate the predominant harmonics in the magnetising current of the saturable reactor X_s, this comprises two or three compensated frequency tripler reactors connected in series and

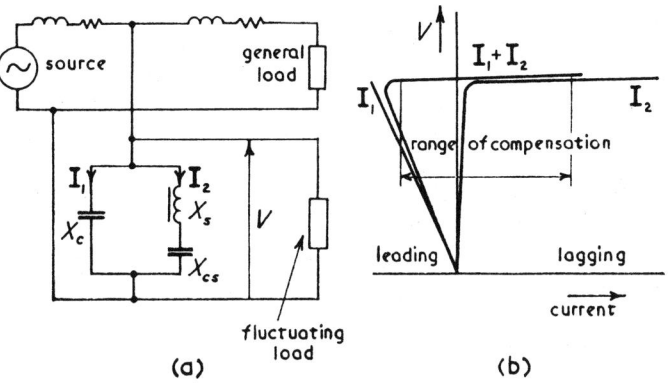

Fig. 9.4.   (a) Typical compensator circuit
(b) compensator characteristic.

combined with series and shunt capacitors X_{cs} and X_c to give the voltage/current characteristic shown in Fig. 9.4(b). It is seen that an increase in the load, reflected as a reduction in voltage at the terminals of the compensator, results in the compensator current becoming more capacitive which tends to correct the voltage. The response of the compensator to system voltage changes can be made within about one cycle of supply frequency, so that it is suited to deal with arc furnace loads whose predominantly troublesome effects lie between 2 and 10 Hz. Several commercial installations have recently been put into service and others will follow.

REFERENCES

Transient stability

BIRCH, F. H. Discussion on results of full scale stability tests on the British 132 kV grid system. *Proc. I.E.E.*, Part A, August 1958, p. 372.

HARLEY, R. G. 1970. Stability of synchronous machine with divided-rotor winding. *Proc. I.E.E.*, **117**, 933–947.

NAYLOR, J. H. Developments in transmission protection methods and equipment. *Electrical Times*, 21st August 1970, pp. 263–265.

RAMA MURTHI, M., WILLIAMS, D. & HOGG, B. W. 1970. Stability of synchronous machines with 2-axis excitation systems. *Proc. I.E.E.*, **117**, 1799–1807.

RAMARAO, N. & REITAN, D. K. 1970. Improvement of power system transient stability using optimal control (bang bang control). *I.E.E.E. Trans.*, **PAS-89**, 975–984.

SOPER, J. A. Divided winding rotor aids large set leading p.f. operation. *Electrical Times*, 17th April 1969, pp. 55–62.

SOPER, J. A. & FAGG, A. R. 1969. Divided-winding-rotor Synchronous Generator. *Proc. I.E.E.*, **116**, 113–126.

SOPER, J. A. *et al.* 1970. Dual axis excitation and control of synchronous turbo-generators. *International Conference on Large High-Tension Electric Systems*. Paper 11–01 (C.I.G.R.E.).

System conditions at high and low loads

BIRCH, F. H. & STALEWSKI, A. Costs of Supplying or Absorbing MVAr. *Electrical Times*, 10th October 1963, pp. 527–531.

HALL, J. H. & SHACKSHAFT, G. Developments in the Stability Characteristics of the Power System of England and Wales. Report 32–05, 1970, *International Conference on Large High-Tension Electric Systems* (C.I.G.R.E.)

SHACKSHAFT, G. Absorbing supergrid MVAr by transformer tap staggering. *Electrical Times*, 24th July 1969, p. 50.

Design criteria and equipment for transmission at 400 kV and higher voltages. *I.E.E. Conference Publication* No. 15, 1965.

Computer applications in power systems

BELCH, R., ALEXANDER, C. E. & WINCH, K. Computer control at Fawley. *Electronics and Power*. May 1969, pp. 152–155.

BREWER, C. *et al.* 1968. Performance of a predictive automatic load-despatching system. *Proc. I.E.E.*, **115**, 1577–1586.

FARMER, E. D., JAMES, K. W., MORAN, F. & PETTIT, P. 1969. Development of automatic digital control of a power system from the laboratory to a field installation. *Proc. I.E.E.*, **116**, 436–444.

PULSFORD, H. E. & GUNNING, P. F. 1967. Developments in power systems control. *Proc. I.E.E.*, **114**, 1139–1148 and 1717–1720.

PRAGNELL, K. N. & CORY, B. J. 1970. Security assessment in distribution systems. *Proc. I.E.E.*, **117**, 161–164.

SHEN, C. M. & LAUGHTON, M. A. 1970. Determination of optimum power system operating conditions under constraints. *Proc. I.E.E.*, **116**, 225–239.

SNELSON, J. K. & CARSON, M. J. 1970. Logical design of branched l.v. distributors. *Proc. I.E.E.*, **117**, 415–420.

Applying computers to power system operation. *Electrical Times*, 25th June 1970, p. 63.

Automatic operation of power stations. *Electrical Times*, 6th February 1964, p. 208.

Power system control—computers in command. *Electrical Times*, 10th September 1970, p. 89.

Load prediction

CUÉNOD, M., KNIGHT, U. G. *et al*. 1970. Present trends in automatic power system security assessment and control. Report 32–12, *International Conference on Large High-Tension Electric Systems* (C.I.G.R.E.).

FARMER, E. D. & POTTON, M. J. 1968. Development of on-line load-prediction techniques with results from trials in the south-west region of the C.E.G.B. *Proc. I.E.E.*, **115**, 1549–1558.

MATTHEWMAN, P. D. & NICHOLSON, H. 1968. Techniques for load prediction in the electricity supply industry. *Proc. I.E.E.*, **115**, 1451–1457.

TOYODA, J., CHEN, M. S. & INOUE, Y. 1970. An application of state estimation to short-term load forecasting, Part I: forecasting modelling. *I.E.E.E. Trans.* Power Apparatus and Systems PAS-89, 1678–1682. Part II: implementation. ibid. 1683–1686.

E.H.V. A.C. transmission

ALEKSANDROV, G. N., REDKOV, V. P., GORIN, B. N., STEKOLNIKOV, I. S. & SHKILYOV, A. V. 1970. Peculiarities of spark discharge in long air gaps of high-voltage structures. *Proceedings of the International Conference on Gas Discharges* (I.E.E., London, 1970). Conference Publication No. 70, 309–312.

HILEMAN, A. R., LEBLANC, P. R. & BROWN, G. W. 1970. Estimating the switching-surge performance of transmission lines. *Trans. I.E.E.E.*, Power Apparatus and Systems, PAS-89, 1455–1465.

NICHOLLS, W. J. The future of overhead lines. *Electrical Times*, pp. 2–6, 1st October 1970.

I.E.E.E. COMMITTEE REPORT: *Switching Surges*, Part I: Phase to ground voltages, A.I.E.E. Trans. PAS-80, 240–261, 1961. Part II: Selection of typical waves for insulation co-ordination, *Trans. I.E.E.E.*, PAS-85, 1091–1097, 1966. Part III: Field and analyzer results for transmission lines, past, present and future trends, *Trans. I.E.E.E.*, PAS-89, 173–189, 1970.

System reliability for the 1970's. *Electrical World*, 23rd June 1969.

Superconducting cables

AFSHARTOUS, S. B., GRANEAU, P. & JEANMONOD, J. 1970. Economic assessment of a liquid-nitrogen cooled cable. *I.E.E.E. Trans.*, Power Systems and Apparatus, PAS-89, 8–16.

GRANEAU, P. 1970. Economics of underground transmission with cryogenic cables. *I.E.E.E. Trans.*, Power Systems and Apparatus, PAS-89, 1–7.

HARROWELL, R. V. 1970. Feasibility of a power transformer with super-conducting windings. *Proc. I.E.E.*, **117**, 131–140.

NORRIS, W. T. & SWIFT, D. A. Developments augur design of superconducting cables. *Electrical World*, pp. 50–54, 24th July 1967.

ROGERS, E. C. & EDWARDS, D.R. Design for a 750 MVA superconducting power cable. *Electrical Review*, pp. 2–5, 8th September 1967.

Fluctuating loads

KENDALL, P.G. 1966. Light flicker in relation to power-system voltage fluctuation. *Proc. I.E.E.*, **113**, 471–479.

FRIEDLANDER, E. 1966. Static network stabilization. *G.E.C. Journal*, **33**, No. 2, 58–61.

General References

BROWN, S. 1968. The fascination of electrical power engineering. *Proc. I.E.E.*, **115,** 1–6.

E.H.V. Transmission Line Reference Book. Edison Electric Institute, New York, 1968.

FRANKEL, A. 1970. Large turbine generators: survey of progress. *Proc. I.E.E.*, **117,** 799–810.

HUMPAGE, W. D. & SAHA, T. N. 1967. Digital-computer methods in dynamic-response analysis of turbogenerator units. *Proc. I.E.E.*, **114,** 1115–30.

Progress in overhead lines and cables for 220 kV and above. I.E.E. Conference publication No. 44, 1968.

RAWCLIFFE, G. H. 1970. Research, education and the heavy electrical industry. *Proc. I.E.E.*, **117,** 961–967.

RIPPON, E. C. 1968. Power plant for the 1970s. *Proc. I.E.E.*, **115,** 103–112.

SAYERS, D. P. 1969. Twenty years of electricity distribution under public control. *Proc. I.E.E.*, **116,** 1527–1543.

Appendix A.1

MATRIX ALGEBRA

A.1.1 Introduction

This appendix is only intended to give a very brief statement of certain properties of matrices relevant to chapters of this book. For a more detailed treatment, reference may be made to such books as that of Lewis and Pryce (1966) or Tropper (1962).

A matrix is a systematic array of numbers (real or complex) which is useful in representing and manipulating the relationships between a number (possibly very large) of simultaneous linear equations. For example, equations such as [8.14] may be represented as

$$
\begin{bmatrix} I_1 \\ I_2 \\ \vdots \\ I_n \end{bmatrix}
=
\begin{bmatrix}
Y_{11} & Y_{12} & Y_{13} & \cdots & Y_{1n} \\
Y_{21} & Y_{22} & Y_{23} & \cdots & Y_{2n} \\
\vdots & \vdots & \vdots & & \vdots \\
Y_{n1} & Y_{n2} & Y_{n3} & \cdots & Y_{nn}
\end{bmatrix}
\begin{bmatrix} V_1 \\ V_2 \\ \vdots \\ V_n \end{bmatrix}
\qquad [A.1.1]
$$

or more compactly by $[I] = [Y][V]$. In this case the matrix of Y coefficients is square, i.e. it has n rows and n columns, but the number of rows and columns may differ in other examples of matrices. The matrices $[I]$ and $[V]$ having only one column are known as column matrices or column vectors. A matrix has no value but is an arrangement of numbers (real or complex) such as Y_{11}, Y_{12} etc. which are known as elements, and if this arrangement of elements is changed then a new matrix is formed. The elements along a horizontal line form a row and those along a vertical line form a column. The element Y_{32} may be identified by its double suffix as being in row 3 and column 2.

A.1.2 Addition or subtraction of matrices

If two matrices are of the same order, i.e. they have the same number of rows and the same number of columns, then it is possible to form their sum or their difference by adding or subtracting their corresponding elements. If

$$[A] = \begin{bmatrix} a & b \\ c & d \end{bmatrix} \quad \text{and} \quad [B] = \begin{bmatrix} e & f \\ g & h \end{bmatrix}$$

$$[A] \pm [B] = \begin{bmatrix} (a \pm e) & (b \pm f) \\ (c \pm g) & (d \pm h) \end{bmatrix} \quad [A.1.2]$$

It may sometimes (as for example in section 8.2.1) be necessary to note that there are missing rows and columns which should be regarded as having zeros in all their elements.

A.1.3 Multiplication of a matrix by a real or complex number

This is sometimes called multiplication by a scalar to distinguish it from multiplication by another matrix. If any matrix is multiplied by a scalar k, whether k is real or complex, then each element in the matrix is multiplied by k. Thus if $[A]$ is as defined above

$$k[A] = \begin{bmatrix} ka & kb \\ kc & kd \end{bmatrix} \quad [A.1.3]$$

A.1.4 Multiplication of two or more matrices

Two matrices can only be multiplied together if the number of columns of the first (left-hand one) is equal to the number of rows of the second. If the two matrices to be multiplied are $[A]$ which is $m \times n$ (i.e. has m rows and n columns), and $[B]$ which is $n \times p$, then the product $[A][B]$ is a matrix with m rows and p columns. The order of multiplication is important, i.e. $[A][B] \neq [B][A]$ in general.

The latter point may be illustrated by multiplying a row vector by a column vector, e.g.

$$[a_{11} \quad a_{12} \quad a_{13}] \begin{bmatrix} b_{11} \\ b_{21} \\ b_{31} \end{bmatrix} = (a_{11}b_{11} + a_{12}b_{21} + a_{13}b_{31}) \quad [A.1.4]$$

The result is a scalar (real or complex number) which is obtained by multiplying the first element of row 1 of the first matrix by the first element of column 1 of the second, and adding the corresponding products of the 2nd elements of the same rows and columns and adding the corresponding products of the 3rd elements and so on. On the other hand the result of multiplying a column vector by a row vector is

$$
\begin{bmatrix} b_{11} \\ b_{21} \\ b_{31} \end{bmatrix} \begin{bmatrix} a_{11} & a_{12} & a_{13} \end{bmatrix} = \begin{bmatrix} b_{11}a_{11} & b_{11}a_{12} & b_{11}a_{13} \\ b_{21}a_{11} & b_{21}a_{12} & b_{21}a_{13} \\ b_{31}a_{11} & b_{31}a_{12} & b_{31}a_{13} \end{bmatrix} \quad [A.1.5]
$$

and this is a square matrix with as many rows and columns as those in the original column and row vectors respectively (section 8.6 gives examples of these products). It can be seen that the position of any element in the product is determined by the number of the row in which the first term of the scalar product was located in the first of the matrices to be multiplied, together with the number of the column in the second matrix in which the second part of the scalar product was located. A second illustration of this is the following product of a 3×2 matrix with a 2×3 matrix

$$
[C] = \begin{bmatrix} a_{11} & a_{12} \\ a_{21} & a_{22} \\ a_{31} & a_{32} \end{bmatrix} \begin{bmatrix} b_{11} & b_{12} & b_{13} \\ b_{21} & b_{22} & b_{23} \end{bmatrix}
$$

$$
= \begin{bmatrix} (a_{11}b_{11}+a_{12}b_{21}) & (a_{11}b_{12}+a_{12}b_{22}) & (a_{11}b_{13}+a_{12}b_{23}) \\ (a_{21}b_{11}+a_{22}b_{21}) & (a_{21}b_{12}+a_{22}b_{22}) & (a_{21}b_{13}+a_{22}b_{23}) \\ (a_{31}b_{11}+a_{32}b_{21}) & (a_{31}b_{12}+a_{32}b_{22}) & (a_{31}b_{13}+a_{32}b_{23}) \end{bmatrix}
$$
$$[A.1.6]$$

In general any element c_{jk} of the product $[A][B]$ of two matrices is the scalar product of row j of $[A]$ and column k of $[B]$, so that the element c_{jk} is given by

$$
c_{jk} = a_{j1}b_{1k}+a_{j2}b_{2k}+a_{j3}b_{3k}+\ldots a_{jn}b_{nk}
$$

Thus the subscripts of any element in the product $[C]$ are the first and the last of the four subscripts in the element ab, e.g. $c_{32} = (a_{31}b_{12}+a_{32}b_{22})$.

Although the commutative law does not in general hold for

matrices so that pre-multiplication must be distinguished from post-multiplication, i.e.

$$[\mathbf{A}][\mathbf{B}] \neq [\mathbf{B}][\mathbf{A}] \qquad [A.1.7]$$

the associative law does hold, i.e.

$$([\mathbf{A}][\mathbf{B}])[\mathbf{C}] = [\mathbf{A}]([\mathbf{B}][\mathbf{C}]) \qquad [A.1.8]$$

and the distributive law also holds, i.e.

$$[\mathbf{C}]([\mathbf{A}]+[\mathbf{B}]) = [\mathbf{C}][\mathbf{A}]+[\mathbf{C}][\mathbf{B}] \qquad [A.1.9]$$

A.1.5 Transpose of a matrix

If the rows and the columns of a matrix are interchanged, i.e. row 1 becomes column 1, column 1 becomes row 1 and so on, then the new matrix is called the transpose of the first, and in general differs from it, e.g. if

$$[\mathbf{A}] = \begin{bmatrix} a_{11} & a_{12} \\ a_{21} & a_{22} \\ a_{31} & a_{32} \end{bmatrix}$$

then the transpose $[\mathbf{A}]_t$ is

$$[\mathbf{A}]_t = \begin{bmatrix} a_{11} & a_{21} & a_{31} \\ a_{12} & a_{22} & a_{32} \end{bmatrix} \qquad [A.1.10]$$

A.1.5.1 Transpose of the product of two matrices

The transpose of the product of two matrices is the product of the two transposed matrices but in the reverse order, i.e.

$$([\mathbf{A}][\mathbf{B}])_t = [\mathbf{B}]_t[\mathbf{A}]_t \qquad [A.1.11]$$

A.1.5.2 Transpose of partitioned matrices

If a matrix $[\mathbf{A}]$ is partitioned between rows as shown by the horizontal full line

$$[\mathbf{A}] = \begin{bmatrix} a_{11} & a_{12} & a_{13} & a_{14} \\ a_{21} & a_{22} & a_{23} & a_{24} \\ a_{31} & a_{32} & a_{33} & a_{34} \\ \hline a_{41} & a_{42} & a_{43} & a_{44} \\ a_{51} & a_{52} & a_{53} & a_{54} \end{bmatrix} = \begin{bmatrix} \mathbf{A}' \\ \hline \mathbf{A}'' \end{bmatrix} \qquad [A.1.12]$$

then

$$[A]_t = \begin{bmatrix} a_{11} & a_{21} & a_{31} & a_{41} & a_{51} \\ a_{12} & a_{22} & a_{32} & a_{42} & a_{52} \\ a_{13} & a_{23} & a_{33} & a_{43} & a_{53} \\ \hline a_{14} & a_{24} & a_{34} & a_{44} & a_{54} \end{bmatrix} = [A'_t \,|\, A''_t] \quad [A.1.13]$$

Similarly if it is partitioned between columns by the vertical dotted line in [A.1.12] so that

$$[A] = [A''' \vdots A''''] \qquad\qquad [A.1.14]$$

then

$$[A]_t = \begin{bmatrix} A'''_t \\ -- \\ A''''_t \end{bmatrix} \qquad\qquad [A.1.15]$$

If the matrix [A] is partitioned into four parts, e.g. if both full and dotted lines are used at once in [A.1.12] so that

$$[A] = \begin{bmatrix} B & C \\ D & E \end{bmatrix}$$

then

$$[A]_t = \begin{bmatrix} B_t & D_t \\ C_t & E_t \end{bmatrix} \qquad\qquad [A.1.16]$$

[A.1.13] illustrates the statements of [A.1.15] and [A.1.16] for the particular value of [A] given by [A.1.12].

A.1.6 Diagonal matrix

A diagonal matrix is a square matrix in which all elements except those in the main diagonal (top left corner to bottom right corner) are zero. Some, though not all (because it would then be a zero matrix) of the terms in the main diagonal may also be zero.

A.1.7 Unit matrix

The unit matrix is a special case of a diagonal matrix where all the main diagonal terms are unity and all the non-diagonal terms are zero. It is said to be the unit matrix of order n when it is an $n \times n$ matrix. The unit matrix of order 3 is

$$[U] = \begin{bmatrix} 1 & 0 & 0 \\ 0 & 1 & 0 \\ 0 & 0 & 1 \end{bmatrix} \qquad [A.1.17]$$

The unit matrix is denoted here by [U] rather than the more usual symbol to avoid possible confusion with a nodal current matrix [I]. For any square matrix [A]

$$[U][A] = [A][U] = [A] \qquad [A.1.18]$$

A.1.8 Inverse (or reciprocal) of a matrix

If from the nodal current equation ([8.20]), [I] = [Y][V], it is required to find the nodal voltages for given nodal currents, this cannot be done by dividing [I] by [Y] since division of matrices is not possible because a matrix has no value. If, however, both sides of [8.20] are pre-multiplied by a matrix such that its product with [Y] is the unit matrix, then [V] is obtained. This matrix is $[Y]^{-1}$ which is the inverse or reciprocal of [Y] so that

$$[Y]^{-1}[I] = [Y]^{-1}[Y][V] = [U][V] = [V] \qquad [A.1.19]$$

It may be noted that $[Y]^{-1}[Y] = [Y][Y]^{-1} = [U]$ for a square, non-singular matrix (that is one having a non-zero determinant) and if $[A] = [B]^{-1}$ then $[A]^{-1} = [B]$. (It may be noted that row and column matrices (vectors) are singular).

A.1.8.1 Inverse of the product of a number of matrices

If [A] and [B] are square non-singular matrices then

$$([A][B])([B]^{-1}[A]^{-1}) = [A][U][A]^{-1} = [U]$$

so that from the definition of an inverse [A][B] must be the inverse of

$[B]^{-1}[A]^{-1}$. This result can be extended to the product of any number of matrices and may be stated that the inverse of the product of any number of square non-singular matrices is the product of their separate inverses taken in the reverse order, e.g.

$$([A][B][C])^{-1} = [C]^{-1}[B]^{-1}[A]^{-1} \qquad [A.1.20]$$

A.1.9 Matrix inversion

For a square non-singular matrix $[Y]$, it is possible to find its inverse $[Y]^{-1}$. For a matrix to be non-singular its determinant $|Y|$ must not be zero. Thus if $[Y]$ is a non-singular matrix with the following elements

$$[Y] = \begin{bmatrix} a & b & c \\ d & e & f \\ g & h & i \end{bmatrix}$$

then its determinant $|Y|$ is given by

$$|Y| = (aei - afh + bfg - bdi + cdh - ceg) \neq 0 \qquad [A.1.21]$$

In order to find the inverse of a matrix by hand (as opposed to various methods which are available to minimise time and storage in a computer, see section 5 of Chapter 8), the following steps may be taken. These are illustrated for simplicity for the 3×3 matrix $[Y]$ given above.

(1) The matrix $[Y]$ is transposed (i.e. rows and columns are interchanged) to give

$$[Y]_t = \begin{bmatrix} a & d & g \\ b & e & h \\ c & f & i \end{bmatrix} \qquad [A.1.22]$$

(2) Each element in $[Y]_t$ is then replaced by its co-factor (or signed minor). The minor of any element is the determinant (of one order less than that of $|Y|$) which is obtained by removing the row and the column containing the element in question. The sign is given for element Y_{mn} (i.e. element in row m and column n) by $(-1)^{(m+n)}$. Thus in the particular case of $[A.1.22]$ the first

term of row 1 of the co-factor is positive: the first term of row 2 is negative etc. The co-factors of [Y], are

$$
\begin{array}{ccc}
+\begin{vmatrix} e & h \\ f & i \end{vmatrix} & -\begin{vmatrix} b & h \\ c & i \end{vmatrix} & +\begin{vmatrix} b & e \\ c & f \end{vmatrix} \\[2ex]
-\begin{vmatrix} d & g \\ f & i \end{vmatrix} & +\begin{vmatrix} a & g \\ c & i \end{vmatrix} & -\begin{vmatrix} a & d \\ c & f \end{vmatrix} \\[2ex]
+\begin{vmatrix} d & g \\ e & h \end{vmatrix} & -\begin{vmatrix} a & g \\ b & h \end{vmatrix} & +\begin{vmatrix} a & d \\ b & e \end{vmatrix}
\end{array}
$$

The new matrix formed by replacing each element of [Y], by its co-factor is called the adjoint of [Y].

(3) The inverse of [Y] is then given by dividing the adjoint matrix by the determinant $|Y|$.

$$
[\mathbf{Y}]^{-1} = \frac{1}{|Y|}\begin{bmatrix}
(ei-hf) & (hc-bi) & (bf-ec) \\
(gf-di) & (ai-gc) & (dc-af) \\
(dh-ge) & (gb-ah) & (ae-db)
\end{bmatrix} \quad [A.1.23]
$$

A.1.9.1 Orthogonal matrix

If the inverse of a matrix or the inverse of its conjugate is equal to its transpose then the matrix is said to be orthogonal. Thus [T] is an orthogonal matrix if

$$[\mathbf{T}]^{-1} = [\mathbf{T}]_t \quad \text{or} \quad [\mathbf{T}]^{-1} = [\mathbf{T}^*]_t$$

An example of an orthogonal transformation matrix is the symmetrical transformation matrix $1/\sqrt{3}[\mathbf{T}_s]$, as discussed in sections 7.5 and 7.6.

A.1.10 Eigenvalues of a matrix

The equations

$$
\begin{aligned}
a_{11}x_1 + a_{12}x_2 + a_{13}x_3 &= \lambda x_1 \\
a_{21}x_1 + a_{22}x_2 + a_{23}x_3 &= \lambda x_2 \\
a_{31}x_1 + a_{32}x_2 + a_{33}x_3 &= \lambda x_3
\end{aligned} \quad [A.1.24]
$$

where λ is a scalar (i.e. an ordinary number which may be real or complex) may be written in matrix form as

$$[A][X] = \lambda[X] \qquad [A.1.25]$$

If [X] is a vector in three dimensions the vector $\lambda[X]$ is a vector in the same direction as [X]. In general [A] is a square $n \times n$ matrix and [X] is a column vector of order n. Values of λ may be found which satisfy [A.1.25], by rewriting it as

$$[A - \lambda U][X] = 0 \qquad [A.1.26]$$

where [U] is the unit matrix of order n (Appendix A.1.7). If $[A - \lambda U] = [B]$ and if it were possible to have $|B| \neq 0$ then $[B]^{-1}$ would exist and [A.1.26] multiplied by $[B]^{-1}$ would become

$$[B]^{-1}[B][X] = 0 \qquad [A.1.27]$$

but since $[B]^{-1}[B] = [U]$, [A.1.27] only gives the trivial solution $[X] = 0$ if $|B| \neq 0$. On the other hand if $|B| = 0$ and $n = 1$, [A.1.26] becomes $b_{11}x_1 = 0$ with $|B| = b_{11} = 0$ and this is not a trivial solution.

For a non-trivial solution therefore

$$|A - \lambda U| = 0 \qquad [A.1.28]$$

Expanding [A.1.28] gives

$$\begin{vmatrix} (a_{11} - \lambda) & a_{12} & a_{13} \\ a_{21} & (a_{22} - \lambda) & a_{23} \\ a_{31} & a_{32} & (a_{33} - \lambda) \end{vmatrix} = 0 \qquad [A.1.29]$$

[A.1.29] is an equation of degree n (in this case a cubic) in λ, called the characteristic equation of [A], and its roots are called the eigenvalues (or characteristic or latent roots) of matrix [A]. For the present case where $n = 3$ and there are three eigenvalues, they may all be different or two or all three may be equal but if [A] is nonsingular none of them will be zero. The eigenvalues may be real or complex.

REFERENCES

Lewis, W. E. & Pryce, D. G. 1965. *The Application of Matrix Theory to Electrical Engineering*. Spon, London.

Tropper, A. M. 1962. *Matrix Theory for Electrical Engineering Students*, Harrap, London.

Appendix A.2

ECONOMICS OF ELECTRICAL POWER GENERATION

A.2.1 Introduction

This appendix will examine the economic operation of the well-established, national power system of the U.K. In this context economic means minimum total system costs per year and well-established means that the system has been operating for several decades so that ample statistical data exists to indicate trends in the errors between previous forward estimates and the actual event, in respect of maximum power demand, energy consumption and load factor. The load factor of a varying load is the ratio of the average power to the maximum power over a specified time interval, usually a year.

In England and Wales, generation and transmission are controlled by the Central Electricity Generating Board (C.E.G.B.), which supplies the twelve Area Boards which distribute to the consumers. These Boards, together with the equivalent Boards in Scotland, are responsible to the Electricity Council and hence to the Minister of Trade and Industry. Some of the data in this appendix is derived from the C.E.G.B. annual report to the Minister for the year ended 31st March 1970.

A.2.2 Forward estimates of future load growth

Since it takes about 6 years to construct a power station (some of this time being taken up by the legal procedure to obtain consent to build), about 4 years for a major transmission line and about 2 years for a distribution line, it is essential to forecast load growth several years ahead. Preliminary estimates are made 20 years ahead, corrected every 5 years, and final estimates are made about 7 years ahead and corrected each year. The estimates are a weighted mean based on information from government sources (in respect of national economic plans), from Area Boards (in respect of changes in consumers' loads), extrapolation of trends based on data covering many years past and the effect of competing forms of power, e.g. North-sea (natural) gas. Estimates are made of the maximum power demand, as this affects the installed capacity (rating) of plant to meet the demand (since electricity cannot be stored directly), and of

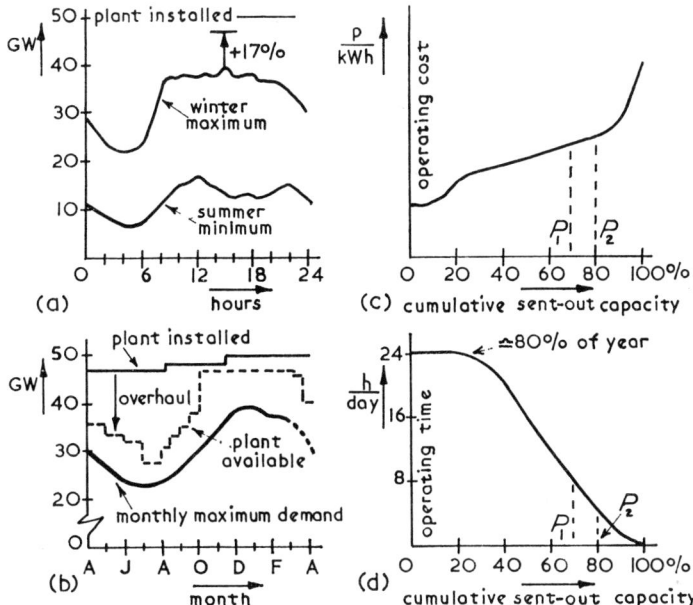

Fig. A.2.1. (a) and (b) C.E.G.B. load curves for 1969/70.
(In (b) the plant installed and available are approximate.)
(c) typical merit-order curve
(d) typical operating-time curve.

the consumption of energy, as this affects the annual requirements for fuel. Estimates are made not only as a national total but also in respect of each Area Board and particular types of consumers— industrial, commercial and domestic. Estimates are also made for typical days in the four seasons of the year. Fig. A.2.1(a), (b), shows typical demand curves. Since the maximum demand is dependent on the severity of the winter, the estimated maximum demand, for year by year comparisons, is based on an average cold spell condition.

Having estimated the average cold spell maximum demand, an estimate is made of the plant availability at the time of maximum demand. In addition to estimating plant breakdown, allowance is also made for the possibility of a severe winter and late commissioning of new plant, both mainly based on experience. The difference between the necessary installed capacity and the estimated maximum demand is referred to as the plant margin, and is (at present) usually 17% of the estimated maximum demand 6 years ahead. Of this figure, about 11% allows for plant breakdown while the remainder is a statistical allowance that demand is liable to exceed available capacity to the extent that for one winter in 4 some consumers will suffer a reduction in voltage and/or frequency and that for 3 winters in 100 some consumers will be disconnected. The chosen plant margin is an economic compromise between excessive capital expenditure on new plant and the risk of industrial chaos due to too frequent failure to meet the maximum demand.

If, as time passes, it appears that the estimates are in error by a significant amount, a few methods are available to minimise the error. For a probable shortage of plant, aircraft-engine type gas-turbine generators can be built in about 3 years: these have, at present, a maximum rating of 70 MW. These generator sets have a low capital cost but a high operating cost, so are mainly used at times of maximum demand: they are often referred to as peaking plant. For a probable surplus of plant, the obsolescence programme can be accelerated but, since a scrapped station may need to be replaced by reinforcement of the local transmission system, a 2-year time interval may be required. In the actual event, power may be obtainable from Scotland, or from France (and hence from the rest of Western Europe) via the 120/160 MW d.c. Channel cable providing they have surplus capacity.

The significance of the forward estimates can be judged from Fig. A.2.1(a), which shows the demand curve on the day of peak (maximum) demand during 1969/70. Although the peak demand lasts only for a short time, the plant installed (planned six years earlier) should be enough to give a plant availability greater than the peak demand. The demand for electricity, which during most of the 1960s was increasing at about 8% per annum, has over a few years been increasing at only about 4%, and has even temporarily declined somewhat. At one stage, due to an abnormally high breakdown of plant, a peak demand of 38·1 GW was met but only after a load reduction of 1·5 GW by reduction of some consumers' voltage. During emergency operation some consumers' voltage may be reduced to the legal minimum of 6 % below nominal thus reducing the power demand by 6–8 % if the reduction was applied to all consumers. System frequency may also be reduced to the legal minimum of 1 % below nominal, thus reducing the power demand by about 1 %.

The annual system load factor of 54·7 %, based on maximum demand actually met, is increasing slowly each year, indicating that energy consumption is rising faster than power demand, due in part to off-peak sales. If the peak demand can be met, then the annual consumption can be met providing the necessary fuel is available and time is available to overhaul the plant.

At modern stations with steam-driven boiler feed pumps, about 3·5 % of the power generated is used to feed the station auxiliaries. This percentage rises to between 5·3 and 7 at stations where all auxiliaries are electrically driven. The remainder, or station output, is referred to as being sent-out (s.o.) and quoted in GW s.o. Thus at 31st March 1970, the total installed capacity of C.E.G.B. stations was about 50·1 GW, the sent-out capacity was 46·9 GW s.o., and on the day of maximum demand the availability was 81·3 % of the sent-out capacity, i.e. 38·1 GW.

A.2.3 Economic principles

The annual total system expenditure of the C.E.G.B. system is made up of:

(a) capital charges (35) = interest (17) + depreciation (18),
(b) operating or generation works cost (59) = fuel (delivered) (44) + operation and maintenance (15),
(c) transmission (1),
(d) other (administration, education, research) (5).

The figures in brackets show the percentage parts where 100 % = 0·531 p/kWh sold for 1969/70.

The capital cost of new plant is paid for (at least in part) by borrowing from the Electricity Council and hence from the Government. Depreciation is paid over a number of years, referred to as the economic life of the plant, such that the total depreciation paid redeems the original loan. Norris discusses, and gives formulae for, various methods of allowing for depreciation. The simplest method is that of the sinking fund whereby interest and depreciation are combined as a fixed annual payment which is usually expressed as a percentage of the original loan. The fund accumulates, with interest, until at the end of the economic life the loan is redeemed. The economic life is less than the practical operating life by a safe margin. The capital charges must be paid regardless of the use made of the plant.

The revenue from the sale of electricity pays for all the costs listed above. Any surplus is used for self-financing, i.e. reducing the amount of a new loan for new plant. With the normal 2-part tariff, the first part represents each consumer's share of the annual capital and other fixed costs and is usually based, for industrial consumers, on his maximum demand, while the second part is his share of the operating cost and is based on his unit consumption. It is clearly unreasonable that consumers should be required to pay for the entire capital cost of a new station in the year it was commissioned: the method of annual depreciation, although somewhat arbitrary, spreads the capital repayments over a number of years which the accountants regard as safe, i.e. the station is not likely to become obsolete within its economic life due to technological progress.

Approximate (in general minimum) economic data is as follows:

	Coal	*Oil*	*AGT*	*Nuclear*	
		(4)	(3)	(1)	(2)
Capital cost, £/kW	55	50	35	100	70
Economic life, years	30	30	25	20	25
Operating cost, p/kWh					
Fuel	0·16	0·16	0·56	0·042	0·05
Capital (5)	0·083	0·075		0·19	0·14
Total (6)	0·25	0·24		0·24	0·22

(1) First generation (Mark I) magnox (1970),
(2) Second generation (Mark II) advanced gas-cooled reactor (AGR) (1970/75),

(3) Auxiliary gas turbine; fuel cost includes engine maintenance after each 2000 hours running; load factor usually very low,

(4) Including fuel-oil tax: without tax, fuel cost 0·108,

(5) Typical data; interest 8%; annual load factor 75%, (costs are very dependent on both),

(6) Discrepancy due to station operating costs.

A.2.4 Economic operation of generating plant

The economic efficiency of a set can be given as a graph relating the cost of the rate of energy input in the form of fuel (pence/hour) to a base of the rate of energy output or power output P. Assuming a quadratic form, the input cost rate equals $(a+bP+cP^2)$ where a is the input cost rate at no load. For any given value of P, the incremental fuel cost is the corresponding gradient of the graph and has a linear form $(b+2cP)$ pence/kWh. Thus if the output of a set is increased by a small amount from P to $P+\Delta P$, for a given time, the average incremental fuel cost is the ratio of the additional cost to the additional kWh output. Similarly, decremental costs refer to a reduction in output. Formulae are given by Taylor and Boal, p. 60.

If two sets having different incremental fuel costs are sharing a load, the total operating cost can be reduced by transferring load from the set having the higher incremental fuel cost to the other set. This decreases the incremental fuel cost of the former and increases that of the latter. The limit is reached when the two incremental fuel costs are equal, and the total operating cost is a minimum.

If the graph of fuel input cost rate to power output is assumed to be linear (Willan's line), then the incremental fuel cost is a constant, b, independent of load P, and referred to as its cost rate (pence/kWh). In this case economic operation is obtained when the base load—the load corresponding to the night minimum (Fig. A.2.1(a),—is taken by the sets having the lowest cost rate, and the dearer sets are loaded in the order of increasing cost rate as the load rises. This is referred to as merit-order loading. Modern sets have a maximum thermal efficiency (at best about 35%) at rated load and operate at rated load as base-load stations. Older sets have a maximum thermal efficiency at about 80% of rated load and operate at this load (in so far as this is possible, dependent on the total demand). The load range from 80 to 100% of full load could be regarded as an overload rating, and operation within this range should be costed at a higher fuel cost rate.

The operation of a grid-control area of the C.E.G.B. system to achieve minimum operating cost is on a merit-order basis. In practice, the operating cost includes allowances for the auxiliary power taken by the set and for the transmission losses. Since the annual capital cost of existing plant is a constant independent of operation, it does not affect the merit order. Fig. A.2.1(c) shows a smoothed and simplified histogram of the operating cost of each set in p/kWh to a base of cumulative sent-out rating, arranged in order of increasing cost. For a given total demand P_1, the sets to the left of P_1 will be (almost) fully loaded. The difference between the total sent-out rating of the sets operating and the actual total demand is called the spinning reserve. Each of the C.E.G.B. control areas must have a spinning reserve large enough to replace the sudden loss of the largest set in that area. The high-merit-order (low-cost) base load sets operate on a 3-shift basis, the intermediate-cost sets on a 2-shift basis and the low-merit-order sets on a 1-shift basis and are only loaded during times of maximum demand. Fig. A.2.1(d) shows typical average operating times to a base of cumulative MW s.o. rating. The merit-order of a set tends to fall with the passage of time: for example if a set is demoted from 3-shift to 2-shift operation, its operational costs increase due to the additional start-stop costs. Merit-order loading requires a 400 kV transmission system capable of carrying the necessary power transfers, and is therefore restricted by any deficiency or breakdown of the transmission system.

Referring again to Fig. A.2.1.(c) the average operating cost (p/kWh) equals

$$\frac{\text{total operating cost of all stations}}{\text{total kWh output}}$$

over a given time interval, e.g. a year.

If the demand is P_1 and is increasing, the incremental operating cost is the operating cost of the next set to be loaded in the merit order. Since the incremental cost is greater than the average cost, an increase in load increases the average generation cost.

The cost of new plant, expressed in £/kW s.o. capacity, falls as the capacity increases. Thus the current (1971) sets are rated at 500 and 660 MW and a 1300 MW, 3000 r.p.m., single-shaft set is envisaged in the next stage of development. The addition of high-capacity, low-operating-cost plant causes many smaller stations to be lower in the merit-order table, thus tending to lower their load factor and raise their operating cost. This is more than off-set by the low operating cost of the new station: the net operational savings can nearly offset the whole of the annual capital charges of an A.G.R. nuclear station and nearly half the cost of a thermal station (Berrie). Conversely, if new plant breaks down, especially if it is nuclear plant, the total operating cost can be seriously increased.

The discussion above relates to minimising the operating cost of a grid-control area. Some areas such as the Midlands and south Yorkshire have a low operating cost and export power to higher-cost areas. This inter-area transfer is set at such a level that the incremental cost of the exporting area equals the decremental cost of the importing area, due allowance being made for transmission losses and the rating of the transmission system. The international power transfers between England, Scotland and France are (ideally) governed by the same economic rule, and the resulting savings in operating costs are shared equally between each pair of countries. The potential savings by application of the method of incremental costs to every set in the C.E.G.B. system may be small relative to the cost of the computer programmes, and at the moment the method is restricted to grid areas and the power transfers between them. In large countries such as the U.S.A., the transmission losses warrant the calculation of the transmission loss coefficients (an introduction to work of Kirchmayer and others may be found in Stevenson, 1962).

A.2.5 Choice of fuel

The choice of fuel can be restricted by government policy; for example, that indigenous coal must be used when there is a surplus of coal or that imported oil consumption must be reduced as it adversely affects the national balance-of-payments position. Otherwise, the choice of fuel is based on economic considerations; these are usually expressed as the cost of heat delivered (p/kWh) or, including the efficiency of conversion, as the cost of electrical energy sent out (p/kWh s.o.).

Coal is the principal fuel used in the U.K. The coal output from mechanised mining is suitable for power stations but there is a growing shortage of the cheaper grades of coal: the dearest coal costs about 50% more than the cheapest. Most of the cheaper coal is mined in the Midlands and in south Yorkshire, so most of the larger power stations are sited in these areas. In 1969/70 coal accounted for about 77% of the total fuel requirements of the C.E.G.B. (weight for equal heat) and about 63% of that was delivered by rail. The ash content of coal is about 20% and although the ash is used increasingly for road making, concrete aggregate, and breeze blocks, its disposal into valleys and quarries during the operational life of the station is a problem. Coal is now the dearest fuel available, so a limited number of boilers are being converted to burn oil or gas while a few are suitable for dual or mixed firing.

Energy can be transported in the form of coal or electricity. The breakeven distance for which the two transport costs are equal depends on the data assumed: e.g. demand, consumption, load factor, voltage. For a station near a coal field, the costs should include coal transport over a few km, any new transmission lines linking the station to the grid, a share of the costs of any existing grid lines used and the cost of losses in these lines. For a modern 2000 MW power station, the breakeven distance is not likely to exceed about 30 km. Since the majority of transmission distances exceed 30 km, energy is normally transmitted in the form of electricity. This conclusion is valid for a base-load station but for new peak-load plant (e.g. gas turbines) it is more economic to site them near the load.

Residual *oil* is now being used in a few large stations and a limited number of boilers in existing stations have been converted from coal to residual oil burning. Oil is easier than coal to handle. Large residual-oil-burning stations are usually sited close to major refineries on the coast where heavy residual oil is separated from lighter fractions. Light (distillate) oil is also used for the gas-turbine sets installed in power stations, particularly in power importing areas, to generate during times of maximum demand, and to supply power for the modern station auxiliaries. Gas-turbine sets are compact and suitable for automatic remote control. They may be used to re-develop some old power station sites.

Gas, in the form of North-sea (natural) gas, is being used, like oil, in a few converted boilers of existing stations. Gas is easy to handle, but only limited quantities are as yet being made available for electricity generation.

Nuclear fuel is the cheapest fuel available, and fuel transport costs are also very low. Much of the total fuel cost is due to the fabrication of the fuel in the fuel cans and the loading of them into the reactor. On the credit side, is the recovery of reactive products from the irradiated fuel. First-generation or magnox reactors were commissioned during the later 1960s. Second-generation or advanced gas cooled (A.G.R.) reactors are now being commissioned. Third-generation reactors may be of the high-temperature gas-cooled type using inert helium and steam as the primary and secondary heat transfer media. It may even be possible to drive gas turbines directly and eliminate the steam cycle. A development to be completed in 1972 is the 250 MW liquid-metal cooled prototype fast-breeder reactor at Dounreay in North Scotland.

Most of the *hydro-electric* power is generated in Scotland and Wales, but it represents only a few per cent of the total electric power generated and most of the economic schemes have now been constructed. Hydro-electric sets, like gas-turbine sets, can be started and stopped quickly (relative to steam-turbine sets) and are suitable for automatic remote control. The energy which can be generated is dependent on a varying rainfall, and on the sizes of the catchment areas and reservoirs. The maximum generated power depends on the rating of the sets.

A *pumped-storage* scheme is a reversible hydro-electric scheme, and a method of storing electrical energy indirectly. In addition to the normal high-level reservoir a pumped-storage scheme requires a low-level reservoir, so that the water flow can be reversed and pumped back to the high-level reservoir. The civil-engineering costs can be reduced if one or both reservoirs already exist in a natural state. When load demand is low, the pumped-storage scheme pumps, and the cost per kWh used is the incremental cost of the next high-merit-order station to be loaded as a consequence of the additional demand, plus transmission costs. The scheme generates when demand is high. The overall cycle efficiency is about 75% so the regeneration cost equals input-cost $/0.75$, and if this cost is less than the decremental cost of the lowest-merit-order station operating, then regeneration is economic. The scheme is dependent on stored water being available for the appropriate duty. The amount of water in the upper reservoir determines the total kWh generated and that depends on the system demand being low enough for long enough to justify pumping the necessary water supply. Over a daily or weekly cycle, the overall costs should be minimised. Pumped-storage schemes are useful adjuncts to nuclear generation. It can often be shown to be economic to install more nuclear plant than is necessary to meet the base load, the surplus being used to pump water during the night. The water is then used to generate electricity during the peak load period the following day.

REFERENCES

AKHTAR, M. Y. 1969. New approach for economic load scheduling. *Proc. I.E.E.* **116,** 533–538.

BERRIE, T. W. 1967. The economics of system planning in bulk electricity supply. *Elec. Rev.,* 15th, 22nd, 29th, September.

CENTRAL ELECTRICITY GENERATING BOARD, Annual report. HMSO.

DALE, L. C. L. *et al.* 1968. Gas-turbine plant for peak-load generation and synchronous compensation. *Proc. I.E.E.* **115,** 969–979.

FANSHEL, S. & LYNES, E. S. 1964. Economic power generation using linear programming. *I.E.E.E.,* PAS-83, 347–356.

MITCHELL, J. M. *et al.* 1969. Steam plant for the 1970s. *Elec. Times,* **155,** 61–63.

NORRIS, T. E. 1970. Economic comparisons in planning for electricity supply. *Proc. I.E.E.* **117,** 593–605.

PIPE, E. J. 1969. Power station cost-trends underline nuclear claims. *Elec. Times,* **156,** 57–58.

STEVENSON, W. D. 1962. *Elements of Power System Analysis.* McGraw-Hill, New York.

TAYLOR, E. O. & BOAL, G. A. (Eds.) 1969. *Power System Economics,* Arnold.

WELLS, D. W. 1968. Method for economic secure loading of a power system. *Proc. I.E.E.* **115,** 1190–1194.

INDEX

Accelerated protection, 19
Acceleration factors, 282
Addition of matrices, 328
Advanced gas-cooled reactor costs, 340, 343
Air-blast circuit breakers, 47, 61, 63, 66, 79, 80
Air gaps, 185
Alpha, beta and zero components, 240
Angular momentum, 107, 108
Anode current, in rectifier, 200
Arc-back, in rectifier, 202
Arc furnaces, 322
Arcing earth, 174
Arc-suppression coil, 174
Attenuation coefficient, 146
Automatic machine control, 317
Automatic sectionaliser, 32
Automatic switching, 34
Automatic voltage regulator (A.V.R.), 125
Auto-reclosing circuit breaker, 30, 45, 121
    delayed slow speed, 34
    high speed, 33
A.V.R.: *see* automatic voltage regulator

Backfire, in rectifier, 202
Back-up protection, 5
Backward-travelling wave, 68, 145, 149, 176
Bang-bang' excitation, 305
Basic insulation level (B.I.L.), 180
Bewley lattice diagram, 163
B.I.L.: *see* basic insulation level
Bilateral network, 98, 245
Breakdown
    long gap, 181
    long wavefront, 181
    polarity effect, 179, 182
    spark, 48
    thermal reignition, 48, 79

Bridge connection of convertor, 194
Brushless excitation, 136
Bypass valve, in h.v.d.c. system, 203

Carrier protection, 23
Channel cable, 214, 338
Characteristic angle of distance relay, 13
Characteristic curve of relay, 11
Characteristic equation of control system, 118, 129
Characteristic impedance, 146, 149
Characteristic impedance ratio (C.I.R.), 22
Chopped-wave withstand voltage, 179
Circuit breakers (C.B.)
    air-blast, 47, 61, 63, 66, 79, 80
    arc extinction, 47
    arcing time, 45, 62
    arc reignition, 47, 48
    autoreclosing, 30, 31, 33, 45, 121
    breaking capacity, 46
    clearing time, 43
    closing an unloaded line, 72
    current chopping, 47, 63, 72
    dead time, 32
    first phase to clear, 54
    gas-blast, 47, 61, 63, 66, 79, 80
    interrupting time, 45
    interruption of line charging current, 64, 67
    kilometric fault, 74
    oil (O.C.B.), 47, 61, 63, 66, 80
    operating time, 304
    oscillatory (or natural) frequency, 52
    over-voltages, 62, 72, 181
    pressure immersed, 80
    protective time, 45
    rate of rise of restriking voltage (r.r.r.v.), 49, 61, 75, 78
    rating, 46
    reclosing time, 45
    recovery voltage, 49

Circuit breakers—*contd.*
  relay time, 45
  resistance switching, 60, 64, 78
  restriking voltage transient, 49
    single frequency, 49
    multi-frequency, 56
  series switching resistor, 74, 183
  short-line fault, 74
  spark breakdown, 48
  sulphur hexafluoride (SF_6), 81
  switching over-voltages, 62, 72, 181
  synthetic testing, 81
  thermal reignition, 48, 79
  transient recovery voltage, 49
    single frequency, 49
    multi-frequency, 56
  unit testing, 80
  vacuum, 81
Circuit reduction, 255
Clarke's components, 240
Close-up fault, 13
Coal-burning generating stations, 340, 343
Co-factor, of determinant, 97, 333
Commutation angle, 196
Commutation, in convertor, 194, 196, 204
Comparator relay, 3
Compensated distance relay, 7
Compounding, in rectifier and convertor, 208
Computers, 242, 255, 260, 262, 313
  off-line, 314
  on-line, 316
Connection matrix, 219, 223, 226, 245, 249, 252
Constant flux-linkages, theory of, 94
Control of system voltage, 309
Control system of
  turbine governor, 137
  voltage regulator, 125
Convergence of solutions, 282
Conversion: *see* rectification
Convertor: *see* rectification
Co-ordinate system transformations, 217
Co-ordinating air-gaps, 186
Co-ordination of insulation, 172
Cost rate, of generating station, 341
Creepage distance, 190
Critical fault-clearing angle, 105

Current chopping, 47, 63, 72
Current doubling, 35
Current transformers (C.T.), 34

Damping coefficient,
  for critical damping, 118
  small oscillations, 118
Damping ratio,
  control systems, 129
  small oscillations, 119
D.C.
  cables, 212
  links, advantages of, 211
  overhead lines, 212
Dead time, of auto-reclosing circuit breaker, 32
Depreciation of plant, 340
Determinants, 97, 333
Diagonal matrix, 331
Diakoptics, 284
Direct-axis,
  of salient-pole generator, 89
  transformation, 238
Direct current: *see* d.c.
Direct lightning stroke, 176
Direct solution of equations, 261
Discrimination, protective, 1, 4
Distance protection, 2
  accelerated, 19
  blocked, 20
  characteristic angle, 13
  characteristic impedance ratio (C.I.R.), 22
  compensated, 7
  interlocked, 20
  intertripped, 19
  line impedance angle, 11
  maximum torque angle, 13
  memory action, 13
  plain impedance, 5, 11
  reach, 3
  stage of, 4
  switched, 21
  system impedance ratio (S.I.R.), 22
  three-step, definite-distance, 5
  zone: *see* stage
Divided-winding rotor, 306
Double saturable reactor, 123
Double-star connection, of rectifier, 192
Dynamic stability, 88

Earth-fault compensated distance relay, 9
Earth-fault distance protection, 14
Earth-loop impedance, 6
Earth wire, 177
Economic life, 339
Effectively-earthed system, 174
Effective synchronous reactance, 89
E.H.V.-A.C. transmission, 319
Eigenvalues of a matrix, 155, 283, 334
Electronic machine control, 317
Equal-area criterion for transient stability, 101
  for change in load, 102
  fault, 105
  switching, 105
  proof of, 111
Equal transformation matrices, 227
Equivalent generator on the infinite busbar, 110
Excitation
  control, 125, 305
  systems, 125
  brushless, 136
  forced, 134, 305
  reactive winding, 307
  shaftmounted, 136
  torque winding, 307
Expulsion gap, 186
Externally generated over-voltages, 174

Factorized inverse matrix, 265
Fault-clearance time, 1, 45, 304
Fault location in transmission line, 166
Feeders, protection of, 1
Field forcing, 134, 305
First phase to clear in circuit breaker, 54
'Floating' busbar, 259, 260, 279
Fluctuating loads, 322
Footing resistance of tower, 177
Forced excitation, 134, 305
Forward estimates of load growth, 336
Forward-travelling wave, 68, 145, 149, 176
Forward voltage drop, 197
Four-terminal network analysis, 95
  transfer matrix, 96
Fuels for generation, 343

Full-wave impulse withstand voltage, 179

Gas-blast circuit breaker, 47, 61, 63, 66, 79, 80
Gas-burning generating station, 344
Gas-turbine generator, 338
Gate control, of rectifier, 196
Gaussian elimination, 264
Gauss–Seidel method, 276
Generating station costs, 340
'Generator' busbar, 260, 278
Governor, steam, 137
Grid control, of rectifier, 196
Grid delay angle, 196
'Grid' voltages, 308

H : see inertia constant
High-load system conditions, 308
High-voltage direct-current (h.v.d.c.), 124, 190, 208
Hybrid solution of equations, 261, 283
Hydro-electric generating station, 345

Impedance protection: see distance protection
Impulse
  flashover voltage, 176
  testing, 147, 173, 177
  voltage, 159, 177
  withstand strength oI voltage, 179
Incremental fuel costs, 341
Indirect lightning stroke, 176
Inertia constant (H), 109
Infinite busbar, 89
Input impedance, of network, 97
Insulation co-ordination, 172
Insulation for overhead lines, 184
Inter-area power transfer, 343
Inter-connector, 89
Interlocked distance system, 20
Internally generated over-voltages, 175
International power transfer, 343
Interphase reactor, of rectifier, 193
Interruption of line charging current, 64, 67
Intertripped distance system, 19
Invariance of power, 219, 220, 245, 252
Inverse matrix, 267, 332, 333
Inversion: see rectification

Inversion of a matrix, 262, 265, 284, 332, 333
Invertor, 203
Isoceraunic level, 177
Isolator, motorised, 21
Iterative solution of equations, 261, 276
Iterative solution of the swing curve, 112

Kilometric fault, 74

Lamp flicker, 322
Lattice (Bewley) diagram, 163
Leader stroke, of lightning, 175
Lightning, 175
    direct stroke, 176
    indirect stroke, 176
    over-voltages, 175
Light flicker, 322
Line closing over-voltages, 72
Line coupling equipment, of carrier protection, 28
Line impedance angle, 11
Line-outage matrix modification, 267, 269, 271
Load admittance matrix modification, 267, 270
Load angle, 91, 92, 101
Load angle/time curve, 112
'Load' busbar, 260
Load factor, 336, 339
Load flow analysis, 120, 242, 259
Load growth, 336
Load prediction, 314, 318, 336
Load reduction, 338
Long-gap breakdown, 181
Long-wavefront breakdown, 181
Low-load system conditions, 308

Matrix
    addition, 328
    diagonal, 331
    eigenvalues, 155, 283, 334
    inversion, 262, 265, 284, 332, 333
    modification, 267
    multiplication, 328
    orthogonal, 227, 334
    partitioning, 256, 265, 294
    subtraction, 328
        sformation, 154, 227

    transposition, 330
    unit, 332
Maximum torque angle, of relay, 13
Mean anode current, of rectifier, 201
Memory action, of relay, 13
Merit order loading, 317, 341
Mesh-current
    analysis using determinants, 96
    connection matrix, 222, 245
    equations, 220, 221, 243
    matrix, 243
Mesh driving-voltage matrix, 247
Mesh impedance matrix, 247
Mesh mutual impedance, 244
Mesh self impedance, 243
Modification matrix, 267
Momentum, angular, 107, 108
Multi-frequency transient recovery voltage, 56
Multiple earthing, 9
Multiplication of matrices, 328

Natural frequency of
    control systems, 129
    small oscillations, 118
    transient recovery voltage, 52
Negative-sequence impedance, 232
Neutral earthing, 174
New-load matrix modification, 267, 270
Newton–Raphson method, 279
Nodal
    admittance matrix, 222, 267
    current, 252
    impedance matrix, 262
    mutual-admittance, 252
    self-admittance, 252
    voltage equations, 221, 250
        advantages of, 260
Node elimination, 255
Non-effectively earthed system, 174
Non-unit protection, 5
Nuclear generating station, 340, 345

Off-line computers, 314
Off-set mho relay, 14
Oil-burning generating stations, 340, 344
Oil circuit breaker (O.C.B.), 47, 61, 63, 66, 80

One machine on the infinite busbar, reduction of a power system to, 110
On-line computers, 316
Orthogonal matrix, 227, 334
Orthogonal transformation, 227
Oscillatory (or natural) frequency of transient recovery voltage, 52
Overlap angle, of rectifier, 196
Overlap voltage drop, 200
Over-reach, of distance relay, 6
Over-voltages
    lightning, 175
    switching, 62, 72, 181
    system-frequency, 173
    transient, 173, 174
Over-voltage tests, 173
    impulse, 146, 173, 177
    power-frequency, 173, 180

Partitioned matrix, 256, 265, 294
Peak inverse voltage, of rectifier, 201
Persistent fault, 45
Peterson arc-suppression coil, 174
Phase-change coefficient, 147
Phase-comparison carrier-current protection, 23
Phase-fault compensated distance relay, 7
Plain impedance relay, 5, 11
Plant, margin, 338
Polarity effect in breakdown, 179, 182
Polarised-mho distance relay, 11
Pole-slipping, 104, 106
Positive-sequence impedance, 232
Potier reactance, 89
Power factor of
    invertor, 207
    rectifier, 200
Power formulae for
    four-terminal network, 96
    round-rotor generator, 91
    salient-pole generator, 93
    two-port network, 96
Power-frequency over-voltage tests, 173, 180
Power invariance, 219, 220, 245, 252
Pressure-immersed circuit breaker, 80
Primitive network, 218, 219, 246, 253
Propagation coefficient, 145, 156

Protection
    acceleration, 19
    back-up, 5
    carrier, 23
    characteristic angle of distance relay, 13
    characteristic curve of relay, 11
    characteristic impedance ratio, 22
    comparator relay, 3
    compensation of distance relays, 7
    discrimination, 1, 4
    distance, 2
    earth-fault compensated distance relay, 9
    earth-fault distance protection, 14
    fault-clearance time, 1, 45, 304
    feeders, 1
    impedance (see distance)
    interlocked distance system, 20
    intertripped distance system, 19
    memory action of relay, 13
    non-unit, 5
    off-set mho relay, 14
    over-reach of distance relay, 6
    phase-comparison carrier-current, 23
    phase-fault compensated distance relay, 7
    polarised-mho distance relay, 11
    plain impedance relay, 5, 11
    reach of distance relay, 6
    reactance relay, 16
    residual compensation, of distance relay, 9
    restraint of relay, 4
    setting of relay, 4
    sound-phase compensation of distance relay, 10
    starting (fault-detecting) relay, 20, 21, 27
    switched-distance system, 21
    system impedance ratio, 22
    three-step definite distance, 5
    time-grading, 1, 5
    under reach, of distance relay, 6
Pumped-storage station, 345

Quadrature-axis
    of salient-pole generator, 89
    transformation, 238

Rate of rise of restriking voltage (r.r.r.v.), 49, 61, 75, 78

Rating of circuit breaker, 46
Rating of rectifier, 201
Reach of distance relay, 6
Reactance relay, 16
Reactive power
    absorption of, 304, 310, 312
    compensation equipment, 121, 323
    of invertor, 208
Reactive-winding excitation, 307
Reciprocal mutual inductances, 225
Reciprocal of a matrix, 332
Recovery voltage, 49
Rectification (or inversion)
    arc-back, 202
    backfire, 202
    by-pass valve, 203
    commutation, 194, 196, 204
    commutation angle, 196
    compounding, 208
    double-star connection, 192
    forward voltage drop, 197
    gate control, 196
    grid control, 196
    grid delay angle, 196
    interphase reactor, 193
    mean anode current, 201
    natural commutation, 194
    overlap angle, 196
    overlap voltage drop, 200
    peak inverse voltage, 201
    power factor, 200, 207
    ratios of current and voltage, 197
    reactive power, 208
    R.M.S. anode current, 201
    three-phase bridge connection, 194
    utilisation factor, 201
    volt-ampère rating, 201
Reflection coefficient or factor, 160, 163
Reheater, steam flow, 137
Relays: see protection
Residual compensation of distance relay, 9
Resistance switching, 60, 64, 78
Resonant links, 121, 213
Restraint of relay, 4
Restriking voltage transient, 49
    single-frequency, 49
    multi-frequency, 56
Return stroke, of lightning, 175
R.M.S. anode current, 201

Routh–Hurwitz criterion, 130

Salient-pole generator
    direct-axis, 89
    power formulae, 91, 93
    quadrature-axis, 89
    transient analysis, 91
Saturable reactor, 123, 323
Saturation, of generator, 89
Self-financing, 340
Semiconductor rectifier, 197, 214
'Sent-out' power, 337
Sequence impedances, 232
Series switching resistor, 74, 183
Setting, of a protective relay, 4
Shaft-mounted excitation, 136
Short-circuit limiting coupler, 123
Short-circuit ratio, 304
Short-line fault, 74
Shunt reactors, 312
Sinking fund, 340
'Slack-take-up' busbar, 259, 260, 279
Small oscillations, of synchronous
    machines, 117
Solid earthing, 9, 174
Sound-phase compensation, of dis-
    tance relay, 10
Spark breakdown, 48
Spinning reserve, 314, 342
Stability
    dynamic, 88
    steady-state, 88, 92, 99, 119
    transient, 88, 101
Stability factor, 88
Stability limit, 88
    methods of improving, 120
Stage, of distance protection, 4
Staggering of transformer taps, 312
Static compensator, 208, 323
Steady-state stability limit, 88, 92, 99, 119
Steam-flow control, 137, 304
Step-by-step solution of the swing
    curve, 112
Stored-energy constant, 109
Sub-matrix, 255, 265
Sub-station insulation, 185
Substitute variables, 218
Subtraction of matrices, 328
Subtransient conditions in generator, 91

Sulphur hexafluoride (SF$_6$) circuit breaker, 81
Superconduction, 321
'Supergrid' voltages, 308
Surge diverter, 187, 191
Surge impedance, 146, 149
Surges in transmission lines: see travelling waves
'Swing' busbar, 259, 260, 279
Swing curve, 112
Switched-distance system, 21
Switching over-voltages, 62, 72, 181
Switching surge strength of insulation, 63, 181
Symmetrical component
    mesh impedance matrix, 230
    transformation matrix, 227, 229
Synchronising power coefficient, 93, 117, 120
Synchronous compensator, 208
Synchronous generator, 89
Synthetic testing of circuit breaker, 81
System frequency over-voltages, 173
System impedance ratio (S.I.R.), 22

Tap-changing matrix modification, 272
Tariff, 340
Tearing, 284, 291
Thermal reignition, 48, 79
Three-phase bridge connection, of converter, 194
Three-step, definite-distance protection, 5
Thyristor, 197, 214
Tie-line, 89
Time-graded protection, 1, 5
Torque-winding excitation, 307
Tower footing resistance, 177
Transfer function of
    excitation control system, 127
    turbine control system, 137
Transfer matrix, of power system, 96
Transfer reactance
    during a fault, 99
    short-circuit, 97
Transformation matrix, 154, 227
Transformer tap change matrix modification, 272
Transformer tap staggering, 312

Transient conditions in generator, 91
Transient fault, 45
Transient over-voltages, 173, 174
Transient recovery voltage, 49
Transient stability, 88, 101, 242, 304
    of generator, 91
Transit time, 159
Transmission coefficient, 164
Transmission line fault location, 166
Transmission line insulation, 184
Transmission-loss coefficients, 343
Transpose of
    matrix, 330
    partitioned matrices, 330
    product of matrices, 330
Travelling waves in transmission lines, 68, 144
    attenuation coefficient, 146
    backward-travelling wave, 68, 144, 149, 176
    Bewley lattice diagram, 163
    characteristic impedance, 146, 149
    fault location, 166
    forward-travelling wave, 68, 105, 149, 176
    impulse testing, 147
    impulse voltage, 159
    junctions of lines and cables, 166
    lattice (Bewley) diagram, 163
    phase-change coefficient, 147
    propagation coefficient, 145, 156
    reflection coefficient or factor, 160, 163
    surge diverter location, 189
    surge impedance, 146, 149
    terminations, 158
    transformation matrix, 154
    transit time, 159
    transmission coefficient, 164
    velocity of propagation, 147
Turbine governors, 120, 137
    control loops, 137
    electronic, 139
Two-machine system, reduction of a power system to, 110
Two-port network analysis, 95
    transfer matrix, 96

Unit matrix, 332
Under-reach, of distance relay, 6

Unit testing, of circuit breaker, 80

Vacuum circuit breaker, 81
Velocity of propagation, 147
Voltage control, 309
Voltage regulators, 125

Willan's line, 341
Wood-pole lines, 184

Zero-sequence impedance, 232
Zone of distance protection, *see* stage